Two-phase flows in chemical engineering

T0292200

Two-phase flows in chemical engineering

DAVID AZBEL

Professor of Chemical Engineering, University of Missouri, Rolla

WITH EDITORIAL ASSISTANCE BY
PHILIP KEMP-PRITCHARD

CAMBRIDGE UNIVERSITY PRESS

Cambridge

London New York New Rochelle

Melbourne Sydney

CAMBRIDGE UNIVERSITY PRESS
Cambridge, New York, Melbourne, Madrid, Cape Town, Singapore, São Paulo, Delhi

Cambridge University Press
The Edinburgh Building, Cambridge CB2 8RU, UK

Published in the United States of America by Cambridge University Press, New York

www.cambridge.org
Information on this title: www.cambridge.org/9780521104241

© Cambridge University Press 1981

This publication is in copyright. Subject to statutory exception
and to the provisions of relevant collective licensing agreements,
no reproduction of any part may take place without the written
permission of Cambridge University Press.

First published 1981
This digitally printed version 2009

A catalogue record for this publication is available from the British Library

Library of Congress Cataloguing in Publication data
Azbel, David.
Two-phase flows in chemical engineering.
Includes index.
1. Two-phase flow. 2. Chemical engineering.
I. Kemp-Pritchard, P. II. Title.
TP156.F6A98 660.2 80-20936

ISBN 978-0-521-23772-7 hardback
ISBN 978-0-521-10424-1 paperback

My wife Rachel and I
want this book to remind us always
of our good friends
Edward Fletcher and Herbert Isbin
who helped us in our first steps in America

CONTENTS

PREFACE

Problems in the hydrodynamics and mass- and heat-transfer processes of two-phase flows in the field of chemical engineering have received extensive treatment during the last thirty years. In writing this monograph I have not attempted to cover all the specific solutions available at the present time, and it is not intended to be exhaustive either in breadth or in depth of coverage. The problems under consideration are, to a certain extent, an account of my approach to the investigation of discrete flows of gas–liquid and solid–liquid systems, which I have developed in the course of time. Most of the analytical approaches have been selected for their generality and usefulness in deducing the parameters that are of engineering interest.

It is hoped that this monograph will be useful as a text in graduate courses, as well as in self-study by those practicing engineers who not only need to solve problems involving hydrodynamics and mass transfer in two-phase flows but also to gain a deeper understanding of the more advanced methods of analysis. It may also serve as a point of departure for further research.

The book is divided into three parts. Part I, Hydrodynamics of Two-Phase Flows, contains six chapters.

Chapter 1 reviews the phenomenon of two-phase flows and discusses the fundamental concepts important in setting the stage for subsequent treatment of more complex problems.

In Chapter 2, I consider the bubble-formation phenomenon for the simplified case of bubbling from a single orifice under a wide range of conditions, as well as the transition from a bubble-to-bubble regime to the continuous gas flow (jets) regime.

Chapter 3 is devoted to the hydrodynamic aspects of mass bubbling. On the basis of the theory of developed isotropic turbulence, an equation is derived for calculating the rise velocity of a bubble in a restricted flow. The bubble size generated as a result of dynamic interactions between liquid and gas, leading to the breakup and coalescence of gaseous bubbles, is evaluated.

The phase contact surface is examined, taking into account the polydispersity of the bubble system (considered as a statistical totality described by distribution functions).

Chapter 4 deals with the formation of dynamic two-phase flows. Since the hydrodynamic equations for one-phase flow are not applicable to calculations in two-phase flows (specifically, a gas-in-liquid dispersion), I develop a theory describing the physical mechanisms occurring in the dynamic two-phase layer and of the influence of individual parameters on the formation of the layer.

The principle of minimum total energy of a bubble layer is postulated; this energy is considered to be a function of the gas density distribution through the layer height. Starting from this principle, I derive equations suitable for calculating the hydrodynamic characteristics of the bubble layer.

The nonuniform velocity field in the bubble layer affects the intensity of the viscous forces and gives rise to convective accelerations and, consequently, inertial effects. Taking these effects into account, physical models of bubbling processes used in commercial practice are considered: "fast" bubbling, which corresponds to the ideal bubble layer having no viscosity; "slow" bubbling, which includes viscous, buoyancy, and surface forces; and "mixed" bubbling, which takes into account the combined influence of the viscous and inertia forces.

Chapter 5 analyzes two problems: cone stability and droplet entrainment. It is shown that the regime of open gaseous spray cones is unrealizable in practice in the range of gas flow rates acceptable in bubbling-flow processes, for a liquid of low viscosity.

Equations are proposed for determining the entrainment of liquid derived from the model of uniform isotropic turbulence, as is a criterion for determining the normal operating conditions of a bubble apparatus. I also examine a simplified model for the mechanism of entrainment for cellular foam conditions and present equations for calculating liquid entrainment in this case.

In Chapter 5, I also describe a statistical model for entrainment, based on the assumption that the distribution of droplet escape velocities from a bubbling layer is normal and independent of the size of the droplets.

In Chapter 6, I study the problem of steady and nonuniform motion of solid particles in liquid. I investigate the effect of system walls on particle velocities in a dilute suspension and obtain basic equations for the motion of micro- and macrosolid particles suspended in a turbulent flow.

Part II, Mass Transfer in Two-Phase Flows, contains Chapters 7 and 8.

In Chapter 7 are given the results of an investigation of the kinetic parameters for mass transfer in gas–liquid systems. Once again, using the concepts of developed isotropic turbulence, equations are presented for determining the velocity of a single bubble, and a group of bubbles, suspended in a turbulent liquid stream. Following these equations, diffusion flow to the surface and flow from a group of bubbles is calculated, including consideration of their distribution with respect to size.

In Chapter 8, I investigate the mass-transfer phenomenon in liquid–solid particle systems for a turbulent flow. Using the concepts of isotropic turbulence, equations are derived for calculating the velocities of solid particles whose size is considerably less, or more, than the internal scale of turbulence. On the basis of the diffusion boundary-layer theory, equations are obtained for calculating the coefficients of mass transfer to these particles, taking into account their distribution with respect to size.

Part III, Application to Chemical and Biochemical Processes, contains two chapters.

Chapter 9 is devoted to the design of a bubble-type chemical reactor. Liquid-phase oxidation reactions are referred to as chain-branched termination reactions with a square termination. This enables me to describe the complex chemical process of hydrocarbon oxidation in terms of a combination of elementary reactions, characterized by the numerical values of various constants.

I also investigate the influence of oxygen partial pressure on the oxidation rate and tar formation rate as a function of the dimensionless parameters of the process.

The macrokinetics of hydrocarbon oxidation in the liquid phase is studied in order to eliminate the diffusion limitation imposed on the process by the transfer of molecular oxygen.

By considering a material balance for the gaseous reagent I develop a method for designing a bubble-type reactor for hydrocarbon oxidation in the liquid phase, providing a scaling method for a commercial process in the kinetic regime.

In Chapter 10 these methods are used to investigate the design of a microbiological reactor (fermenter). I show that cultivation of microorganisms is associated with intensive mixing, which provides a uniform distribution of microorganisms in the bulk of a culture medium, air dispersion, and transfer of mass to the cell surface for utilization. It also provides for the removal of metabolism products from the reaction zone. By analyzing the intracellular processes, I show that the transfer of mass to the reaction surface is the limiting stage of the entire process.

An equation is proposed for determining the specific rate of growth of organisms as a function of the physical properties of the medium, the dimensions and density of the particles, the concentration of materials in the bulk of the solution, the liquid velocity, and the characteristic dimensions of the apparatus in which turbulence is produced. On the basis of these investigations, the design of a biochemical reactor is suggested.

The inclusion of practical aspects of the subject as an integral part of the monograph is intended to serve as a supplement to, rather than a substitute for, a strong foundation in chemical engineering fundamentals.

On the personal level, I would like to share the history of this book with its readers. This monograph was originally written in Soviet Russia after the author had been dismissed from his position for applying to leave the country and while waiting for the exit visa. The manuscript was then sent, page by page, by various routes, to Mr. Greville Janner, Q.C.M.P., in London, who passed it to Professor Kenneth Denbeigh, Member of the Royal Society and Principal of Queen Elizabeth College. While I was still in Moscow, Mr. Janner and Professor Denbeigh encouraged and helped me to find a publisher for this book. All thanks are due them, because without their help, this book would not have been possible.

Thanks are also due the large number of people who helped to transport the manuscript to England but who, unfortunately, must remain anonymous.

I would also like to thank my editor, Philip Kemp-Pritchard, for helping me to bring the monograph into its final form.

I owe a deep debt of gratitude to Professors Edward Fletcher and Herbert Isbin of the University of Minnesota for their help with the manuscript and valuable suggestions, and to the University of Minnesota and State University of New York at Stony Brook for supplying me with every possible practical help to facilitate my work.

I also wish to acknowledge with thanks the assistance given by the faculty and staff of the Chemical Engineering Department of the University of Missouri at Rolla and to thank Professors M. E. Findley, A. I. Liapis, G. K. Patterson, and D. J. Siehr for helpful comments regarding the manuscript. In particular, I am greatly indebted to Professor J. W. Johnson, chairman of the Chemical Engineering Department, who read the whole manuscript and who provided useful suggestions.

D. A.

LIST OF SYMBOLS

A	local cross-sectional area; interface contact area
A_b	cross-sectional area of bubble
A_m	cross-sectional area of membrane
A_l	cross-sectional area of laboratory reactor
Bo	Bond number
C_D	drag coefficient
C_W	constant that is numerically equal to maximum reaction rate
D	pipe diameter; diffusivity
D_m	mixer diameter
D_o	orifice diameter
D_{turb}	turbulent diffusivity
$\overline{D}$	longitudinal mixing coefficient
E	total kinetic energy of injected gas; total energy of mixture; spectral energy density/mass/wavenumber
E_b	energy of bubble
E_f	energy of liquid
Eu	Euler number
F_b	buoyancy force
F_D	drag force
F_σ	surface-tension force
F_0	form drag
Fo	Fourier diffusion number
Fr	Froude number
Ga	Galileo number
Ho	homochronity number
K	droplet free-ascent-velocity correction factor; respiratory coefficient
K_E	energy coefficient
K_f	friction factor
K_{tb}	recuperated catalyst
$K_{t_{ox}}$	oxidized catalyst
L_d	work done against drag
L_k	kinetic energy of bubble

L_σ	surface-tension energy of bubble
M	charge of *o*-xylene; mass flow rate/area of mixture; mass diffusion rate
M_b	liquid mass entrained/bubble
M_e	catalyst mass
M_f	droplet mass
M_g	gas mass flow rate
M_{ij}	mass flux tensor
N	valency of metal; number of bubbles/time; total number of bubbles; number of eddies/mass/wavenumber; number of cells
N_{max}	maximum number of cells
N_0	original number of cells
N_c	number of liquid cones
N_s	Stokes number
P	power input/volume; power consumption; steric factor
P_{O_2}	oxygen partial pressure
Pe	Peclet number
Q	gas flow rate; particle volumetric flow rate
Q_l	liquid volume; liquid flow rate
Q'	oxygen dissolution rate
R	equivalent radius of liquid cones
Re	Reynolds number
R	radical
S	plate surface area; total bubble surface area; total particle surface area; total cell surface area
Sc	Schmidt number
Sh	Sherwood number
T	tank diameter
U	characteristic velocity of flow; liquid flow rate; effective slip velocity
V	bubble volume; particle volume; working volume of apparatus
V_a	liquid volume equivalent to additional mass
V_{av}	average bubble volume
V_d	displaced bubble volume
V_E	final bubble volume
V_F	initial bubble volume
V_f	liquid volume
V_t	mixture volume
W	oxidation rate
W_e	enzyme reaction rate
W_i	initiation rate
W_{tf}	tar formation rate
W'	rate of oxygen diffusion; oxidation rate/system volume; catalytic oxidation rate
We	Weber number

Y	total mass of entrained drops/area/time
Y_0	total entrainment/area/time
Y_c	entrainment component independent of height/area/time
$\mathbf{Y}$	entrainment coefficient/area/time
Z	collision factor
a	specific surface area
a_e	eddy acceleration
a_f	fluid acceleration
a_p	particle acceleration
a_r	relative acceleration
b	weir length
c	local concentration; concentration of solute or hydrocarbon
c_0	bulk concentration
c_e	oxygen concentration in exhaust
c_{eq}	equilibrium concentration
c_i	initial hydrocarbon concentration; concentration at wall
c_{in}	concentration at interface
c_k	catalyst concentration
c_p	particle concentration; concentration of products
c_s	concentration of saturated solution; speed of sound in gas
c'	reaction product concentration
d	droplet diameter; diameter of pipe, apparatus, or reactor
d_{av}	average bubble diameter
d_b	bubble diameter; bubble diameter at release
d_m	impeller diameter
d_p	particle diameter
d_{pr}	propeller diameter
$\bar{d}$	dimensionless bubble diameter
dE	total energy of layer
dE_1	potential energy of layer
dE_2	kinetic energy of layer; dissipation in layer
dE_3	surface tension energy
e	eddy energy; total bubble energy density
e_b	bubble energy dissipation
f	frequency of particles crossing cross-sectional area; probability density
h	liquid column height
k	work of suspension; coefficient of apparent additional mass; mass-transfer coefficient
k_d	correction factor for equipment diameter
k_f	liquid mass-transfer coefficient
k_{f_s}	liquid surface mass-transfer coefficient
k_{f_v}	liquid volume mass-transfer coefficient

k_g	gas mass-transfer coefficient
k_h	correction factor for equipment height
k_i	reaction-rate constants; rate constant of desorption
k_s	mass-transfer coefficient/effective area
k_v	mass-transfer coefficient/effective volume
l	flow characteristic length
l_c	axis ratio
m	bubble instantaneous mass; diffused mass of bubble; diffused mass in cell model
m_i	diffusional flow of mass to particle
$\dot{m}$	mass transfer/time
m'	mass transfer/area
$\dot{m}'$	mass transfer/area/time
n	rotation rate; number of revolutions; turbulent frequency; number of bubbles/volume
p_a	ambient pressure
p_e	exhaust pressure
p_f	liquid pressure
p_g	gas pressure
p_o	gas pressure at orifice
$\bar{p}$	mean pressure
$\bar{p}_g$	time-averaged gas pressure
p'	fluctuating pressure
q	relative gas flow rate; average gas flow rate into bubble
r	eddy size
r_b	instantaneous bubble radius
r_c	cell radius; critical bubble size
r_d	eddy size of viscous effect; drop radius
r_e	bubble radius at detachment; energy-containing eddy size
r_0	maximum radius of entrained eddies
r_p	particle radius
r^*	radical size
s	distance of bubble center below surface; fractional rate of renewal of elements
s_{av}	average bubble surface area
s_b	bubble surface area
s_0	contaminated area of bubble
u	velocity parallel to surface
u_{f_i}	fluid velocity vector
u_i	velocity vector
u_o	velocity outside boundary layer
u_{p_i}	particle velocity vector
u_r	particle terminal velocity

u_i' fluctuating velocity vector
$\overline{u}_i$ mean velocity vector
v instantaneous local gas velocity; velocity perpendicular to surface; gas velocity in annular space
v_b bubble velocity
v_d droplet absolute velocity
$v_{d\infty}$ droplet absolute velocity in infinite medium
v_e small eddy characteristic velocity
v_f liquid velocity
v_g gas velocity
v_{gr} bubble surface velocity relative to liquid
v_i velocity vector
v_o gas mean velocity at orifice
$v_{o(\mathrm{crit})}$ critical mean velocity at orifice
v_p particle velocity
v_r bubble velocity relative to liquid; relative velocity of droplet or particle
v_s gas superficial velocity; solid superficial velocity; flow superficial velocity
v_{sf} liquid superficial velocity
v^i ideal fluid flow velocity
v_τ velocity defining turbulent motion
v^* shear velocity
$\overline{v}_p$ average particle velocity
x_1 height of two-phase mixture
y_e element thickness

α longitudinal turbulent intensity; acceleration coefficient for chemical reaction
β universal constant; measure of radical spent in tar formation
γ universal constant
δ membrane thickness; diffusion sublayer; boundary layer thickness
δ_b turbulent buffer layer thickness
δ_c critical membrane thickness
δ_{eff} effective film thickness
δ_0 viscous sublayer
ϵ energy dissipation/mass/time
ϵ_t dissipated turbulent energy
$\bar{\epsilon}$ mean viscous energy dissipation/mass/time
ζ dimensionless cell count
η Kolmogoroff length scale; dimensionless relative tar formation rate; order of reaction rate with respect to oxygen
θ relative output/reactor volume
θ_k active center "density"
κ wavenumber

κ_e wavenumber of energy-containing eddies
κ_d wavenumber of eddies with viscous effect
λ microscale of turbulent eddies
μ dimensionless cell growth rate; mixture dynamic viscosity
μ_f liquid dynamic viscosity
μ_g gas dynamic viscosity
μ_{max} maximum rate of microorganism growth
ν kinematic viscosity; dimensionless oxidation rate
ν_f liquid kinematic viscosity
ν_{tf} tar formation rate
ν' dimensionless catalytic oxidation rate
ρ dimensionless radical concentration; mixture density
ρ_f fluid density
ρ_g gas density
ρ_p particle density
$\overline{\rho}$ mean mixture density
ρ' fluctuating mixture density
σ dimensionless catalyst concentration; surface tension
σ_b bubble diameter standard deviation
σ_{b_s} bubble surface area standard deviation
σ_{b_v} bubble volume standard deviation
τ characteristic time; eddy duration at surface; eddy time scale
τ_f bubble formation time
τ_{ij} stress tensor
τ_0 time between bubbles
τ_s shear stress
ϕ gas void fraction; solid content of mixture; velocity potential
ϕ_{av} average gas void fraction
$\overline{\phi}$ mean void fraction
ϕ' fluctuating void fraction
ψ stream function; foam specific gravity
ω dimensionless oxygen concentration
ω_n gas motion natural frequency
Δ determinant
Δc concentration gradient
Δp pressure difference
ΔE activation energy
ΔU change in average velocity
Ω_n membrane natural frequency

PART I

Hydrodynamics of two-phase flows

1

The phenomenon of two-phase flows

In Part I we will be concerned exclusively with the fluid dynamics of two-phase flows. A *two-phase flow* is one in which we have dynamic (and sometimes chemical) reactions between two phases or components in a flowing system (e.g., liquid–liquid, liquid–gas, liquid–solid particles, gas–solid particles). Sometimes the two phases consist of the same chemical substance, as in distillation equipment, and sometimes the two phases are of unrelated chemical substances, as for dust particles in air. Flows with two distinct substances are often designated as "two-component flows," to distinguish flows of two phases of a single substance, but this distinction is not of great significance in many flow systems, where the reactions are only of a fluid-dynamical nature.

Theoretical and experimental studies of two-phase flows are becoming increasingly important because of their widespread applications in industry. This relevance is being given a great stimulus by the expanding needs of modern industrial societies for energy supplies from various sources. The application of two-phase-flow research to problems in the petrochemical industries is clear – one only has to consider such systems as boilers, evaporators, distillation towers, and turbines. Transportation of and extraction of the products of oil are other obvious applications. As we move toward such alternative energy sources as coal gasification, nuclear energy, and solar energy, new applications of two-phase-flow technology will become ever more important. Quite apart from industries concerned with the development of energy supplies, two-phase flows occur in such varied industries as food processing, paper manufacturing, and steel manufacture.

Finally, we can find many applications outside industry. Rain (and rain-making), snow, dust storms, and fog and cloud formation all involve interactions of two phases. Bioengineering, from the study of blood flow to the inhalation of air-suspended particulate matter, finds a need for understanding of two-phase-flow phenomena. It is thus abundantly clear that the subject of this book is not without practical importance.

We now define two of the more important quantities that can be used to characterize two-phase flows, or quantities that we may wish to be able to determine or predict, for any such given flow system.

The *void fraction,* ϕ, is defined as the ratio of the volume of gas to the total volume of gas and liquid in a flow (or section of it):

$$\phi = \frac{\text{volume of gas}}{\text{total volume}}$$

Note that the fractional volume of liquid is then $(1 - \phi)$.

The *superficial gas velocity,* v_s, is the ratio of the gas volumetric flow rate, at a given flow cross section, to that cross-sectional area:

$$v_s = Q/A$$

where Q is the gas flow rate and A the local cross-sectional area. In subsequent chapters, methods will be developed to enable these two quantities to be determined, as well as the relationships between them, for various flow conditions, such as gas and liquid densities, viscosities, surface tension and flow rates, and system geometries. The chapters in Part I are concerned primarily with flows in which we have a liquid-containing chamber through which a gas is being forced. However, the results of the various approaches adopted in the study of this kind of flow often have immediate application to the more general (and common) case of the flow of a liquid–gas mixture in a chamber or a duct. Further, the dispersed flow of solid particles in a fluid is considered in Part I, and this will also be found to have wide application.

Traditionally, in the study of two-phase-flow systems, because of the great complexity of such flows compared to single-phase flows, assumptions are made that break the research down into three broad areas of approach. In one approach the two-phase flow is decomposed into what is hoped to be a representative single-phase flow, in which the fictitious fluid's properties are defined in such a way as to maximize the effectiveness of the representation. This is the simplest approach to the problem and, in its simplicity, cannot be expected to furnish results on the detailed behavior of each phase. It is at best a crude approximation to the actual flow.

In "separated" flow models, fluid-dynamic equations are developed separately for each phase. These equations can then be combined to describe the total flow, or boundary conditions can be assumed between the two phases to couple the two sets of equations. The total number of equations will be 12; for each phase there will be three scalar momentum equations, a mass conservation equation, an energy equation, and an equation of state. This corresponds to the 12 unknowns: the gas void fraction, the velocity vectors, temperatures and densities of each phase, and the pressure. Obviously, without the use of many simplifications, this general problem is practically intractable, but with suitable simplifications, this approach has been more productive of results than has the simpler approach described above, and Part I tackles two-phase-flow problems by various techniques with this basic mode of description.

The third, and most phenomenological approach breaks the analysis of two-phase flows down to several flow regimes. Some regimes are almost continuous with others, whereas between other regimes there is a discrete, pronounced structural change.

We adopt a description given elsewhere,[1] which breaks the flow regimes down in terms of the actual structure existing between the two phases. This description is therefore distinctly different from corresponding regime characterizations in single-phase flows. In single-phase flows, theoretical and experimental investigations have led to an ability to characterize flows in terms of nondimensional factors such as the Reynolds number and the Prandtl number. In two-phase flows we cannot, at the present degree of knowledge, so characterize regimes. Instead, we can only define flow regimes in terms of actual resultant flows: there are too many operative factors to be able to predict, with accuracy, which of these regimes will occur in a given flow system. This description of regimes is therefore necessarily of a qualitative, subjective nature.

We describe first flow in a vertical direction in a pipe or duct. Perhaps the most obvious example of a two-phase flow is that of gas bubbles in a flowing liquid (see Figure 1.1). In a *bubbly regime*, the bubbles may be small and spherical, or larger, with their shapes nonspherical because of the flow of liquid around them. As the flow rates of the gas and liquid are increased, we develop *slug flow*. Here the gas elements are large enough to almost span the duct diameter, thus creating discrete "slugs" of liquid, connected by liquid flowing along the walls of the duct. Under certain conditions slug flow can make a transition into *churn flow*, where the flow is much more irregular and disturbed, primarily as a result of the breakdown of the gas bubbles. Some sources designate this flow as "slug-annular flow," because of oscillations between the slug-flow condition and annular-like flow, in which the central region of the flow is virtually liquid-free.

In *wispy annular flow*, the central region of the flow is gaseous except for "wisps" of liquid droplets bunched into discrete groupings. Along the walls of the duct we find bubble-impregnated liquid flow. Wisp-annular flow usually occurs at high mass-flow rates, and this explains why it has only recently been distinguished as a distinct flow regime.[2] Finally, in *annular flow*, the central region is gas-dominated, with sparse liquid droplet entrainment from the liquid lying on the walls of the duct. There is no agglomeration of droplets in the central region as in wispy-annular flow.

As earlier described, there is no satisfactory way of predicting which of these regimes will exist in a given system, primarily because so many factors are involved. For example, not only do the physical properties of the phases have an effect, but such things as the mode of gas injection and duct geometry can also have pronounced effects. (It is worth noting that these regimes, in the order described above, have been found to be generated along the direction of flow in a heated duct. The flow is initially wholly liquid, but as heat is absorbed, vapor bubbles appear which expand into slug flow and annular flow, finally becoming a single-phase gas flow.)

However, even with this qualification, attempts have been made to develop flow-pattern maps for specific situations when most parameters are kept constant, resultant regime changes then being a function of fewer parameters. An example is shown in Figure 1.2 for low-pressure air–water and high-pressure steam–water flows in small-diameter (1 to 3 cm, 0.4 to 1.2 in.) vertical tubes.[3] The parameters in which the regimes are mapped are the gas and liquid superficial momentum fluxes, $\rho_g v_s^2$ and $\rho_f v_{sf}^2$, respectively. Here ρ_g and ρ_f are the gas and liquid densities and v_s and v_{sf} are the corresponding local superficial velocities. It should be stressed that the boundaries between regimes are very approximate.

Figure 1.1 Flow patterns in vertical flow. (From J. G. Collier, *Convective Boiling and Condensation,* McGraw-Hill, New York, 1972.)

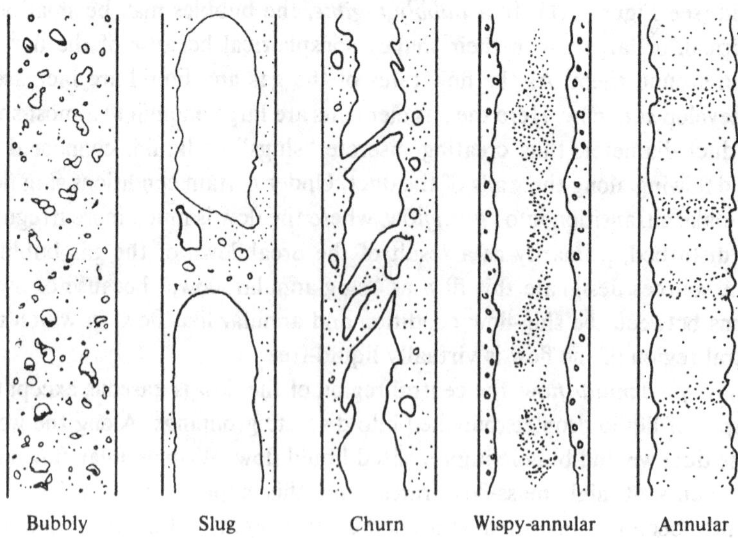

Although the transition from one regime to another is ill-defined, owing to the subjective visual classification, attempts have been made to determine criteria for predicting the transition points from one regime to another.

Bubbly flows begin to make a transition to slug flow as the likelihood of collision (and thus coalescence) of bubbles increases. Clearly the void fraction ϕ is a significant determinant here, and it has been found[4] that for a void fraction of less than about 0.1, the collision frequency of bubbles is low; hence there is not enough bubble coagulation to generate slug flow. As the

Figure 1.2 Flow pattern map for vertical flow. (From J. G. Collier, *Convective Boiling and Condensation,* McGraw-Hill, New York, 1972.)

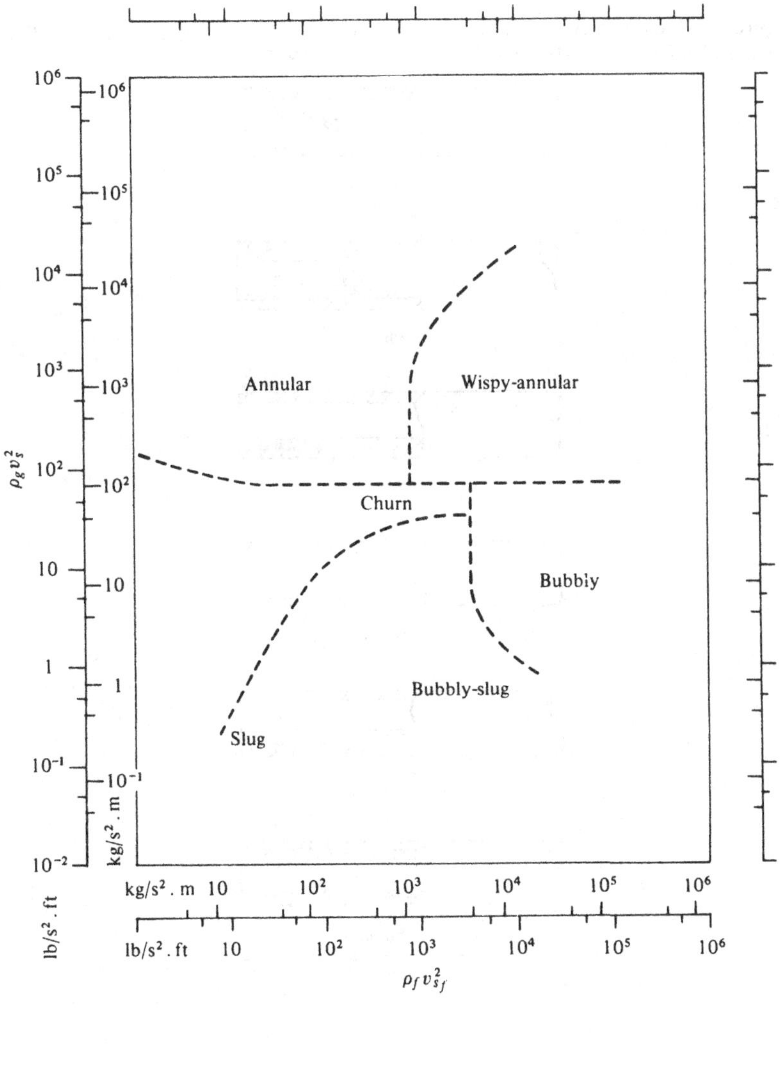

void fraction is increased, we find that for $\phi \simeq 0.3$, the likelihood is great that slug flow will be created.

As the void fraction is further increased, the slug flow begins to break down into *churn flow* at some value of ϕ less than about 0.8. Interactions between the rising gas bubbles and the falling liquid on the walls of the duct or pipe lead to instabilities, generating churn flow as the large bubbles begin to break up. A semiempirical theory has been developed[5] which yields the following approximate criterion for the establishment of churn flow:

$$v_s = 0.105\left[\frac{gD(\rho_f - \rho_g)}{\rho_g}\right]^{1/2}$$

Figure 1.3 Flow patterns in horizontal flow. (From J. G. Collier, *Convective Boiling and Condensation*, McGraw-Hill, New York, 1972.)

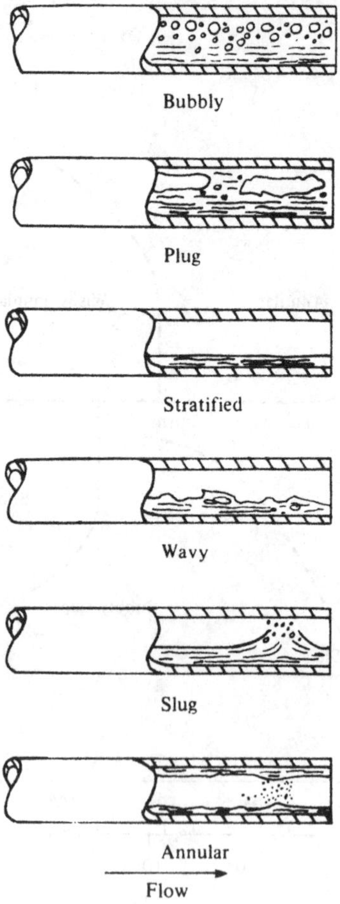

where D is the pipe diameter and ρ_f and ρ_g are the liquid and gas densities, respectively.

It is easiest to describe the transition from churn flow to annular flow by considering the effect of the flowing gas on the direction of flow of the liquid on the walls of the pipe.[1] There is a critical gas flow rate for which an increase ensures that the liquid flows upward with the gas and for which a decrease allows the liquid to fall under gravitational force. This *flow-reversal point* has been found to be determined by the following formula:

$$v_s = 0.9\left[\frac{gD(\rho_f - \rho_g)}{\rho_g}\right]^{1/2}$$

This can also be used as an approximate criterion for the churn flow–annular flow transition.[1]

We have concentrated our discussion on upward vertical flows. However, similar results have been found for flows in horizontal pipes or ducts. Given the very approximate nature of the mapping in Figure 1.2, it is known that this mapping also describes horizontal flow transitions. The flow regimes in horizontal flows are slightly different from those in the vertical flow, as can be seen in Figure 1.3.[6] The correspondences between the two flows are quite obvious.

In the following chapters we develop several different (and often very simple) approaches to the types of problems generated in an understanding of the flows we have described. Some of the techniques have been developed by other authors; the remainder constitute new and novel approaches toward obtaining simple, but realistic formulas and conclusions describing two-phase flows.

References

1. Collier, J. G., *Convective Boiling and Condensation,* McGraw-Hill, New York, 1972.
2. Bennett, A. W., Hewitt, G. F., Kearsey, H. A., Keeys, R. K. F., and Lacey, P. M. C., "Flow visualization studies of boiling at high pressure," Paper presented at *Symposium on Boiling Heat Transfer in Steam Generating Units and Heat Exchangers, Manchester, Sept. 15–16, 1965.*
3. Hewitt, G. F., and Roberts, D. N., *Studies of Two Phase Flow Patterns by Simultaneous X-ray and Flash Photography,* AERE-M2159, Her Majesty's Stationery Office, London, 1969.
4. Radovich, N. A., and Moissis, R., "The transition from two-phase bubble flow to slug flow," Rep. no. 7-7673-22, Department of Mechanical Engineering, Massachusetts Institute of Technology, Cambridge, Mass., 1962.
5. Porteous, A., *Br. Chem. Eng., 14*(9), 117 (1969).
6. Wallis, G. B., *One-Dimensional Two-Phase Flow,* McGraw-Hill, New York, 1969.

2

Single-bubble formation

2.1 Introduction

In the study of the dispersion of gas in a liquid, it is not usually necessary to consider the effect of molecular diffusion on the mixing process. This is because in most practical applications of the two-phase dispersion process, other dynamical effects are present that play a dominant role in mixing. Several such possible dispersion mechanisms are considered in this chapter.

A simple example of a mixing process is one in which gas is forced through an orifice (or orifices) submerged in a liquid (Figure 2.1). Depending on the gas flow rate, the gas will exit from the orifice as individual bubbles (low flow rate) or continuously as a jet (high flow rate). The jet subsequently breaks up into bubbles of various sizes. In this chapter we consider the low-flow-rate bubble-formation phenomenon for the simplified case of a single orifice. Also, we take the gas flow rate through the orifice into the liquid to be constant.

There have been numerous studies on bubble formation from a single orifice submerged in a liquid.[1–29] However, the results obtained in these studies, especially as far as the development of a general theory for bubble size for a two-phase system is concerned, have been inconclusive. Therefore, the need arises for a theoretical investigation of the phenomenon to gain some understanding of the experimental data that have been obtained.

2.2 Buoyancy and surface tension alone

We start with a consideration of the forces acting on each bubble as it forms at the orifice. Each bubble is acted upon by buoyancy, convection currents in the fluid, and by the surface-tension force acting on that section of the bubble that is still in contact with the orifice (Figure 2.2).

For the case of low gas flow, and when the viscosity of the liquid is small and convection is negligible, we can say that the buoyancy and surface-tension forces approximately balance, so that

$$\tfrac{1}{6}\pi d_b^3 g(\rho_f - \rho_g) = \pi D_o \sigma \tag{2.1}$$

where d_b is the diameter of the bubble at the instant of release; ρ_f and ρ_g the densities of the liquid and gas, respectively; D_o the diameter of the orifice; σ the surface tension; and g the acceleration due to gravity. Note that the factor $\cos\theta$ (where θ is the angle between the perpendicular and the bubble surface at the orifice), which is usually included on the right-hand side of equation (2.1), has been set equal to unity for simplicity. This gives us

$$d_b = \left[\frac{6\sigma D_o}{(\rho_f - \rho_g)g}\right]^{1/3}$$

The usefulness of this equation is confirmed by experiment for $\mathrm{Re}_o \leq 200$, where $\mathrm{Re}_o = \rho_f v_o D_o/\mu_f$ is the Reynolds number based on the orifice diameter. Here μ_f is the dynamic viscosity of the liquid and v_o the mean velocity of the gas at the orifice. For instance, if we take air ($\rho_g = 1.2$ kg/m^3, 7.49×10^{-2} lb/ft^3) bubbling into water ($\rho_f = 10^3$ kg/m^3, 62.4 lb/ft^3, $\sigma = 7.36 \times 10^{-2}$ N/m, 4.2×10^{-4} lbf/in.) through an orifice 1.3 mm (0.05 in.) in diameter, we find that the diameter of the bubbles will be, according

Figure 2.1 Example of a gas–liquid system.

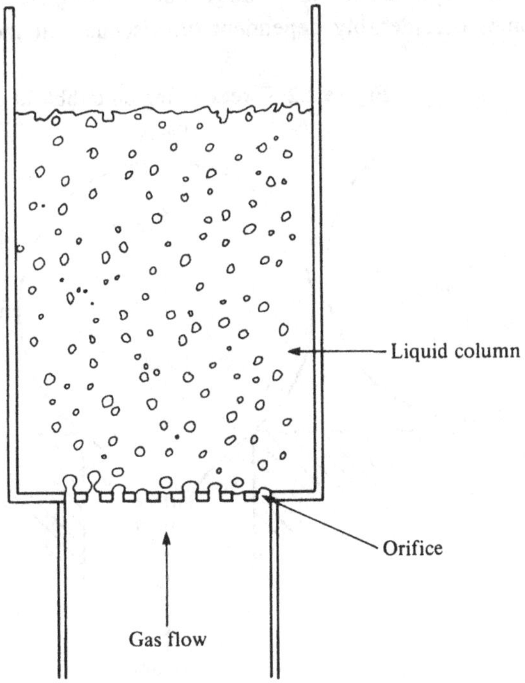

to the preceding equation, $d_b = 4.06$ mm (0.16 in.), and this result is supported by experiment.

The preceding equation can be rewritten in the form

$$\frac{d_b}{D_o} = c\left[\frac{4\sigma}{(\rho_f - \rho_g)gD_o^2}\right]^{1/3} \tag{2.2}$$

where $c = (1.5)^{1/3} = 1.15$.

In practice, the bubble is acted upon by forces in addition to buoyancy and surface tension, so that c will not have the value shown above. Experimental work[2] has shown that equation (2.2) is fairly good in determining the size of bubbles but that a better value for c in the equation would be approximately unity. This is shown in Figure 2.3 where, for different values of d_b/D_o and for different liquids,

$$c = d_b\left[\frac{g(\rho_f - \rho_g)}{4\sigma D_o}\right]^{1/3}$$

is plotted.

2.3 Buoyancy, drag, surface tension, and inertia

For the Reynolds number range $200 < \text{Re}_o < 1000$, the bubble size becomes considerably dependent on viscous and inertial forces. For this

Figure 2.2 Forces acting on bubble at orifice.

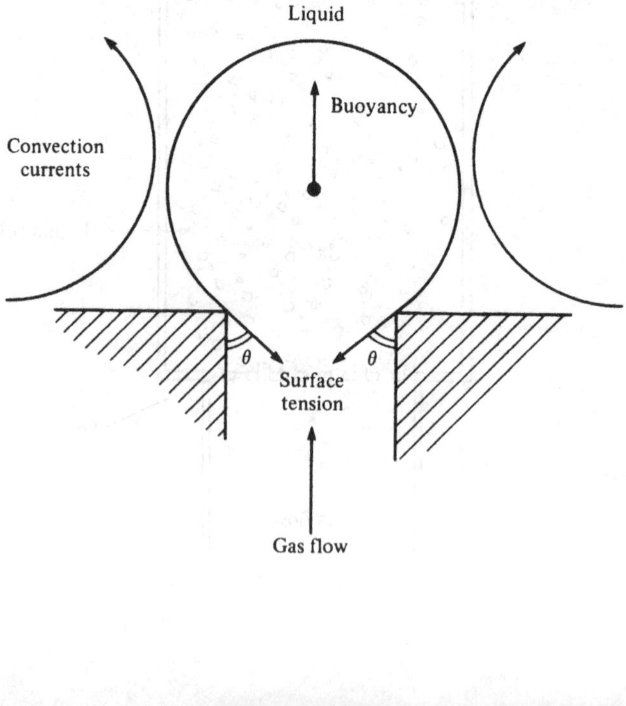

range a model has been proposed[30] in which bubble formation takes place in two stages, *expansion* and *detachment*. During the first stage the bubble remains attached to the orifice, and during the second stage the bubble moves away from the orifice, keeping in contact with the orifice through a "neck." These two stages are shown in Figure 2.4. The final volume of the bubble is the sum of the volumes generated in each of the two stages,

$$V_F = V_E + Q\tau_f \tag{2.3}$$

where V_F is the final bubble volume, V_E the volume generated in the first stage, Q the gas volumetric flow rate, and τ_f the detachment-time duration. For the first stage a force balance (allowing for buoyancy, viscous drag, surface tension, and inertia) leads to

$$V_E^{5/3} = \frac{11Q^2}{192\pi(3/4\pi)^{2/3}g} + \frac{3\mu_f Q V_E^{1/3}}{2(3/4\pi)^{1/3}g\rho_f} + \frac{\pi D_o \sigma V_E^{2/3}}{g\rho_f} \tag{2.4}$$

Equation (2.4) takes into account the "added mass" effect, by which the effective mass of the bubble is the sum of the actual bubble mass and the mass of liquid is equal to $\frac{11}{16}$ of the volume of the bubble.[4,31] The volume of the bubble after the expansion stage, V_E, can be found from equation (2.4) using an iterative procedure. Equation (2.4) is general in nature, and several simple cases can be obtained by neglecting certain terms on the right-hand side. When the second term is neglected, we have the case of an inviscid

Figure 2.3 Factor c versus d_b/D_o.

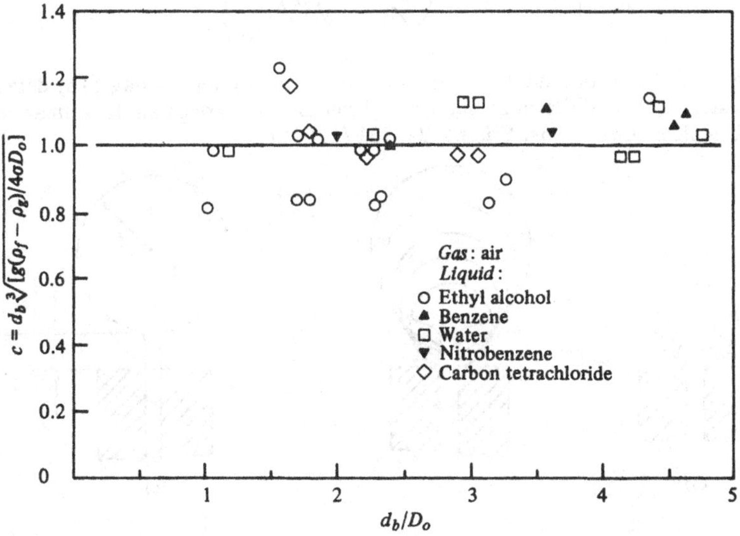

liquid with surface tension.[32] Neglecting the final term as well gives an equation describing behavior in an inviscid liquid without surface tension.[33,34] When the last term alone is eliminated, we have the case of a viscous liquid without surface tension. Finally, when the first two terms are neglected, equation (2.4) is reduced to equation (2.2), obtained by balancing buoyancy and surface-tension forces.

When the upward forces exceed the downward forces (the second stage) the bubble moves away from the orifice, being connected by the bubble neck. The length of this neck is taken to be equal in magnitude to r_E, the bubble radius at detachment (i.e., after the first stage), on the assumption that this condition avoids the possibility of the next bubble coalescing with the first. When the bubble base has traveled this distance, the bubble neck is broken and the bubble floats up.

For the second stage of bubble formation, the final bubble volume, V_F, is obtained using the second law of motion, taking into account viscosity, buoyancy, added mass, and surface tension. The resulting equation for V_F is

$$\begin{aligned} r_E = {} & \frac{B}{2Q(A+1)}(V_F^2 - V_E^2) - \left(\frac{C}{AQ}\right)(V_F - V_E) \\ & - \frac{3G}{2Q(A-\frac{1}{3})}(V_F^{2/3} - V_E^{2/3}) - \frac{3E}{Q(A-\frac{2}{3})}(V_F^{1/3} - V_E^{1/3}) \\ & - \frac{(V_F^{-A+1} - V_E^{-A+1})}{Q(-A+1)}\left[\left(\frac{B}{A+1}\right)V_E^{A+1} - \left(\frac{C}{A}\right)V_E^{A} \right. \\ & \left. - \left(\frac{G}{A-\frac{1}{3}}\right)V_E^{A-1/3} - \left(\frac{E}{A-\frac{2}{3}}\right)V_E^{A-2/3}\right] \end{aligned} \tag{2.5}$$

Figure 2.4 Stages of bubble formation at orifice: (a) expansion stage; (b) detachment stage; (c) condition of detachment. [From S. Ramakrishman, R. Kumar, and N. R. Kuloor, *Chem. Eng. Sci.*, no. 24, 731 (1969).]

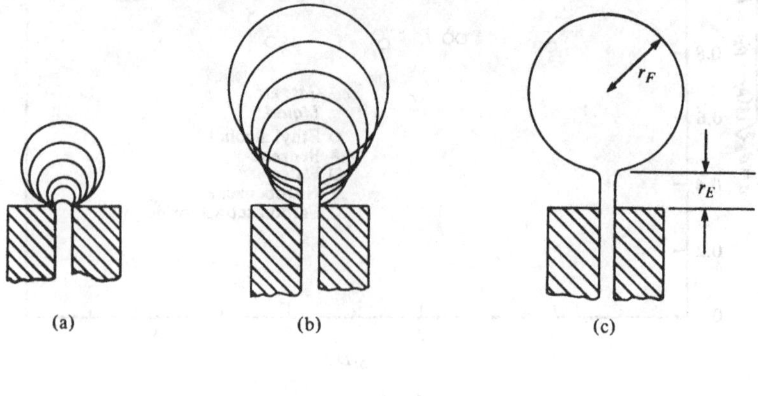

where

$$A = 1 + \frac{96\pi(1.25)r_E\mu_f}{11\rho_f Q}$$

$$B = \frac{16g}{11Q}$$

$$C = \frac{16\pi D_o\sigma\cos\theta}{11\rho_f Q}$$

$$E = \frac{Q}{12\pi(3/4\pi)^{2/3}}$$

$$G = \frac{24\mu_f}{11(3/4\pi)^{1/3}\rho_f}$$

It is found from calculations under various conditions of bubble formation that the contribution of the last two terms in equation (2.5) is negligible. Hence we can write

$$r_E = \frac{B}{2Q(A+1)}(V_F^2 - V_E^2) - \left(\frac{C}{AQ}\right)(V_F - V_E) - \frac{3G}{2Q(A-\frac{1}{3})}(V_F^{2/3} - V_E^{2/3}) \tag{2.6}$$

The final bubble radius, $r_F = (3V_F/4\pi)^{1/3}$, is obtained in the following manner. Equation (2.4) is used to obtain V_E and also $r_E = (3V_E/4\pi)^{1/3}$. The final bubble volume, and hence the final bubble radius, is then obtained by trial and error using equation (2.6).

Several important conclusions can be drawn from equations (2.4) and (2.6). From equation (2.4), for vanishingly small flow rates ($Q \approx 0$), the first two terms on the right-hand side of the equation become insignificant and the detachment bubble volume V_E depends only on buoyancy and surface-tension forces. Further, since Q is vanishingly small V_F, the final bubble volume, is approximately equal to the detachment volume V_E, because of equation (2.3). Hence we can conclude that for very small gas flow rates, bubble size is determined only by buoyancy and surface tension. Again, from equation (2.4), as the gas flow rate is increased, the first two terms increasingly dominate the last term on the right-hand side of the equation. This means that as Q increases, the contribution of surface-tension forces becomes smaller. This tendency is more pronounced for higher-viscosity liquids. As the flow rate increases, we reach a situation in which the surface tension makes no contribution to the bubble size. The minimum flow rate for this condition is reached earlier for a viscous, as opposed to an inviscid, liquid.

From equations (2.4) and (2.6) it is clear that as the viscosity increases, the bubble size also increases, and that the effect of viscosity is higher at high flow rates, low surface tension, and small orifice diameters.

Finally, using equation (2.4), we can determine the effects of liquid density. For low gas flow rate and low viscosity, only the last term on the right-hand side is effective, and the bubble size is decreased for increasing liquid density. As the flow rate increases, the effect of liquid density decreases. For the case of highly viscous liquids and orifices of small diameter, the bubble again decreases in size with increasing liquid density, because under these conditions the first and third terms on the right-hand side of equation (2.4) can be ignored.

Equations (2.4) and (2.6) have been compared with experimental data.[30] In Figure 2.5, note that not only do the formulas correctly indicate the general form of the dependence of the bubble volume on the volumetric flow rate, but they are also correct in their indication that the surface-tension effect is suppressed for higher flow rates. Figures 2.6 and 2.7 show that the formulas predict reasonably well the effect of orifice diameter and liquid viscosity on bubble size.

Figure 2.5 Effect of volumetric flow rate and surface tension on bubble size. [From S. Ramakrishman, R. Kumar, and N. R. Kuloor, *Chem. Eng. Sci.*, no. 24, 731 (1969).]

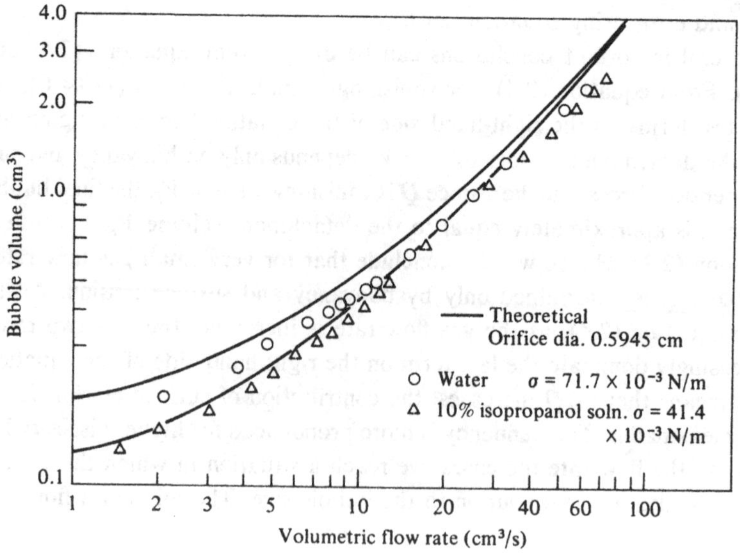

Equations (2.4) and (2.6) are clearly of considerable utility in determining the bubble sizes produced under various conditions of liquid viscosity, surface tension, gas flow rate, orifice diameter, and other factors.

2.4 Time-dependent gas flow rate

We now consider bubble formation under the additional complication that the gas flow rate and the pressure in the feeding chamber are both changing in time. Taking these factors into account, a model has been developed[35] in which it is assumed that each spherical bubble remains in the plane of the orifice exit during the bubble's formation, the upward velocity of the sphere's center of mass being[36]

$$v_C = \frac{q}{\pi d_b^2} \tag{2.7}$$

Figure 2.6 Effect of volumetric flow rate and orifice diameter on bubble size. [From S. Ramakrishman, R. Kumar, and N. R. Kuloor, *Chem. Eng. Sci.*, no. 24, 731 (1969).]

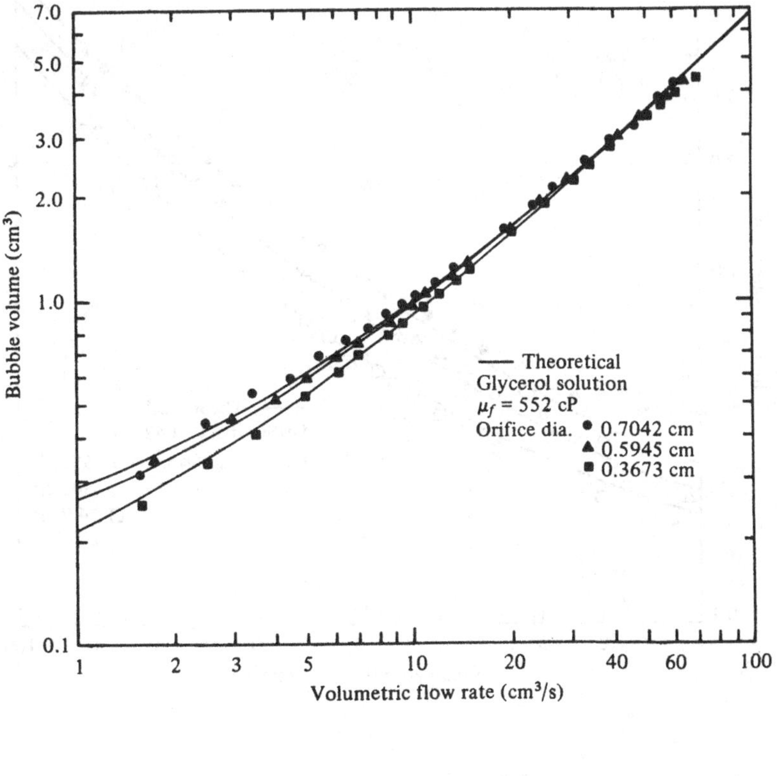

where q is now the average volumetric gas flow rate into the bubble and d_b the diameter of the bubble. If we consider each bubble to be acted on by buoyancy, excess pressure force, surface tension, and viscous drag, we can write Newton's second law of motion for the bubble:[37]

$$\frac{d}{dt}(mv_c) = v_o\frac{dm}{dt} + \frac{\pi D_o^2(p_f - p_o)}{4} - \pi D_o\sigma + gV(\rho_f - \rho_g) - 3\pi d_b\mu_f v_c\left(\frac{2\mu_f + 3\mu_g}{3\mu_f + 3\mu_g}\right) \tag{2.8}$$

where m is the instantaneous bubble mass, V the bubble volume, p_f the pressure in the liquid at the orifice, p_o the gas pressure at the orifice, and μ_g the gas dynamic viscosity. The last term on the right-hand side of the equation is the drag on a sphere of fluid moving in a second fluid.[38]

Figure 2.7 Effect of volumetric flow rate and liquid viscosity on bubble size. [From S. Ramakrishman, R. Kumar, and N. R. Kuloor, *Chem. Eng. Sci.*, no. 24, 731 (1969).]

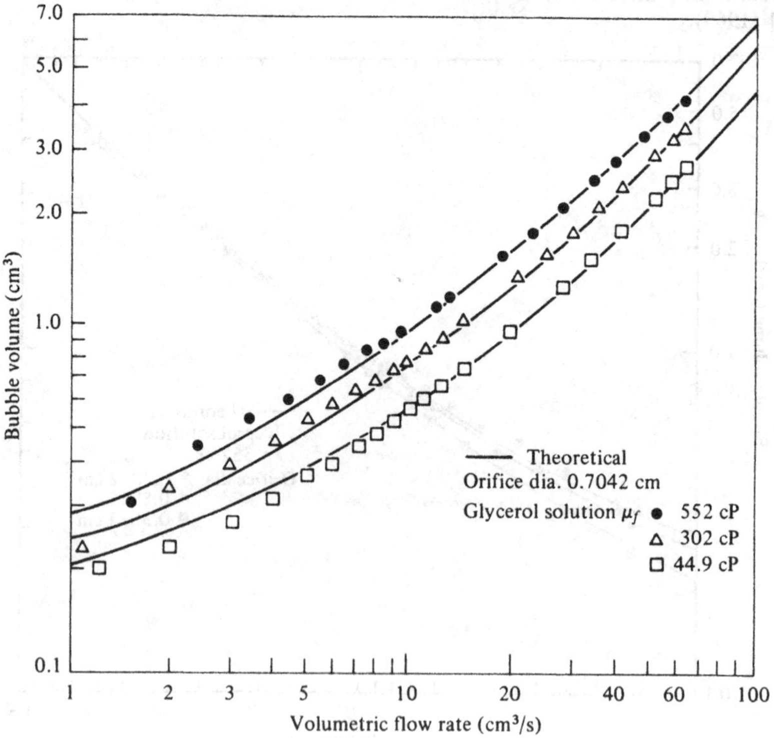

Now it is clear that

$$\frac{dm}{dt} = q\rho_g, \qquad m = V\rho_g \quad \text{and} \quad p_o - p_f = \frac{4\sigma}{d_b}$$

Using these in equation (2.8), we have

$$\frac{d}{dt}(\rho_g V v_c) = v_o q \rho_g + gV(\rho_f - \rho_g) + \frac{\pi D_o^2 \sigma}{d_b} - \pi D_o \sigma - 3\pi d_b \mu_f v_c \left(\frac{2\mu_f + 3\mu_g}{3\mu_f + 3\mu_g}\right) \tag{2.9}$$

Combining equations (2.7) and (2.9) and introducing a dimensionless bubble diameter $\bar{d} = d_b/D_o$, we obtain after simplifying, the following non-dimensional equation for predicting the bubble diameter:

$$\mathrm{Fr}\,\bar{d}^5 + \left(1 - \frac{1}{\mathrm{We}}\right)\bar{d}^2 + \left(\frac{1}{\mathrm{We}} - \frac{1}{\mathrm{Re}}\right)\bar{d} - \frac{1}{12} = 0 \tag{2.10}$$

where Fr = Froude number $= \dfrac{g\pi^2 D_o^5(\rho_f - \rho_g)}{24\rho_g q^2}$

We = Weber number $= \dfrac{4\rho_g q^2}{\pi^2 D_o^3 \sigma}$

Re = Reynolds number $= \dfrac{4\rho_g q(3\mu_f + 3\mu_g)}{3D_o\mu_f\pi(2\mu_f + 3\mu_g)}$

Equation (2.10) is a general expression for determining the diameter of the bubble, the Froude, Weber, and Reynolds numbers indicating the relative effects of inertia and gravity, surface tension, and viscosity, respectively. Consider the earlier example, where $D_o = 1.3$ mm (0.05 in.), $\rho_f = 10^3$ kg/m^3 (62.4 lb/ft^3), and $\rho_g = 1.2$ kg/m^3 (7.49×10^{-2} lb/ft^3); and taking $\mu_f = 1 \times 10^{-3}$ kg/m·s (6.72×10^{-4} lb/ft·s), $\mu_g = 1.81 \times 10^{-5}$ kg/m·s (1.22×10^{-5} lb/ft·s), with $q \simeq 6.75 \times 10^{-6}$ m^3/s (0.4 in.^{3} /s), equation (2.10) gives $\bar{d} \simeq 1.2$, or the diameter of the bubbles $d_b = 1.56$ mm (0.06 in.).

Note that, in the limit of very small gas flow rates, it can be shown that equation (2.10) reduces to equation (2.2).

2.5 Energy loss due to surface tension and drag forces

It is of some interest to ascertain the ways bubbles consume energy in their formation. This understanding is useful in indicating conditions under which bubbles can form and also gives insight into how the energy of the bubble will be released when it eventually bursts.

If we neglect inertia forces, on the assumption that the energy consumed in giving the bubble momentum is relatively small, the work of formation of the bubble is comprised of the work done against surface tension and the work done against drag forces.

In the case of a spherical-shaped interface, the work of formation against surface tension is[1]

$$dL_\sigma = 8\pi\sigma r_b dr_b \tag{2.11}$$

where r_b is the instantaneous bubble radius.

Further, the drag force on the bubble motion is

$$F_D = C_D \frac{\rho_f v_{gr}^2 \pi r_b^2}{2} \tag{2.12}$$

where C_D is the drag coefficient and v_{gr} the velocity of the bubble surface relative to the liquid. Before equation (2.12) can be used to obtain the work done due to drag, we need to know the behavior of the drag coefficient. For the Reynolds number range (based on v_{gr} and the diameter of the bubble) $10^{-4} < \text{Re} < 0.4$, we find that

$$C_D = \frac{24}{\text{Re}} \tag{2.13a}$$

For $2 < \text{Re} < 500$, we have

$$C_D = \frac{18.5}{\text{Re}^{0.6}} \tag{2.13b}$$

and in the range $500 < \text{Re} < 2 \times 10^5$,

$$C_D = 0.44 \tag{2.13c}$$

It will be noted that, as the Reynolds number increases, the effect of molecular viscosity diminishes.

Using equations (2.13) in equation (2.12) makes this equation applicable over a wide range of Reynolds numbers. We then obtain the work done against drag forces:

$$dL_d = C_D \frac{\rho_f v_{gr}^2 \pi r_b^2}{2} dr_b \tag{2.14}$$

Combining equations (2.11) and (2.14), we obtain the ratio of the elementary work done in the formation of the bubble surface to the elementary work done in overcoming drag:

$$\frac{dL_\sigma}{dL_d} = \frac{16\sigma}{C_D \rho_f v_{gr}^2 r_b} \tag{2.15}$$

Let us consider the case that the convective currents in the fluid are weak. When we do this, it is clear that the relative velocity, v_{gr}, is the same as the rate at which the bubble surface is growing in time, in other words,

$$v_{gr} = \frac{dr_b}{dt}$$

and hence that equation (2.15) becomes

$$\frac{dL_\sigma}{dL_d} = \frac{16\sigma}{C_D \rho_f r_b}\left(\frac{dt}{dr_b}\right)^2 \tag{2.16}$$

We need to know the value of (dt/dr_b), and this is obtained from the continuity equation for the gas. We have a balance between the mass of gas passing through the orifice and the mass of gas taken up in the expansion of the bubble:

$$v_o \frac{\rho_g \pi D_o^2 dt}{4} = \rho_g 4\pi r_b^2 dr_b$$

or

$$\frac{dt}{dr_b} = \frac{16 r_b^2}{D_o^2 v_o} \tag{2.17}$$

where v_o is the gas velocity at the orifice.

Using equation (2.17) in equation (2.16), we obtain finally

$$\frac{dL_\sigma}{dL_d} = \frac{4096 r_b^3 \sigma}{C_D \rho_f D_o^4 v_o^2} \tag{2.18}$$

It is possible, using equation (2.18), to obtain an estimate of the relative consumption of energy by surface tension and drag during the formation of a bubble.

Let us consider some typical examples illustrating the use of equation (2.18). We take first an example of air dispersion in water through an orifice (D_o = 1.3 mm, 0.05 in.) under various conditions.

Example 1. Take the velocity of a gas at the orifice as v_o = 0.2 m/s (7.87 in./s), the density of water, ρ_f = 10^3 kg/m^3 (62.4 lb/ft^3), the surface tension, σ = 7.36 × 10^{-2} N/m (4.2 × 10^{-4} lbf/in.), the radius of bubbles r_b = 1.95 mm (0.077 in.), and the dynamic viscosity of water, μ_f = 1 × 10^{-3} kg/m·s (6.72 × 10^{-4} lb/ft·s).

To use equation (2.18), we need to know the appropriate value of C_D. This is obtained from equations (2.13) once the Reynolds number is known. Now

$$\mathrm{Re} = \frac{v_{gr}\rho_f 2 r_b}{\mu_f}$$

The relative gas velocity used in this definition of the Reynolds number is obtained by the following reasoning. The number of bubbles produced per unit time is clearly obtained from

$$N = \frac{v_o(\pi/4)D_o^2}{\frac{4}{3}\pi r_b^3} = \frac{3v_oD_o^2}{16r_b^3}$$

and, on the assumption that the bubbles are produced with their centers one bubble diameter apart, we have

$$v_{gr} = 2r_bN = \frac{3v_oD_o^2}{8r_b^2}$$

Substituting this in the formula for the Reynolds number, we find that

$$\mathrm{Re} = \frac{\rho_f 2r_b}{\mu_f}\frac{3v_oD_o^2}{8r_b^2} = \frac{3v_oD_o^2\rho_f}{4\mu_f r_b}$$

and computing this value for this example, we obtain Re = 130. This means that equation (2.13b) should be used for the drag coefficient; in other words,

$$C_D = \frac{18.5}{\mathrm{Re}^{0.6}} \simeq 1$$

Equation (2.18) then yields the following result for the ratio of work done against surface-tension forces to work done against drag:

$$\frac{dL_\sigma}{dL_d} \simeq 2 \times 10^4$$

Example 2. Take $v_o = 1$ m/s (3.28 ft/s) and $r_b = 2.4$ mm (0.095 in.). As in Example 1, we obtain

$$\mathrm{Re} = \frac{\rho_f 2r_b v_{gr}}{\mu_f} = \frac{3v_oD_o^2\rho_f}{4\mu_f r_b} \simeq 530$$

Again, from equation (2.13c), $C_D = 0.44$ and

$$\frac{dL_\sigma}{dL_d} \simeq 3.3 \times 10^3$$

Example 3. Take $v_o = 15$ m/s (49 ft/s) and $r_b = 5$ mm (0.2 in.). Here

$$\mathrm{Re} = \frac{3v_oD_o^2\rho_f}{4\mu_f r_b} = 3.8 \times 10^3$$

and therefore $C_D = 0.44$ and

$$\frac{dL_\sigma}{dL_d} = 133$$

These examples illustrate that under a wide range of conditions, the work done in producing individual bubbles is used up mainly in overcoming surface-tension forces; the work done against drag is usually many orders of magnitude less than the work done against surface tension.

2.6 Transition to continuous gas flow (jets)[39]

When gas is discharged into a liquid, a phase contact surface forms that, depending on the ratio of gas to liquid flow rates, takes the form of either bubbles or spray cones. At high gas flow rates through the orifice we find spray cones forming, with bubbles breaking off from the tips of the spray cones at a rate depending on the liquid properties.[40,41] The study of the spray-cone mode of gas discharge is of considerable interest, because most bubbling equipment operates in this range of flow rates. It is important to determine the critical gas flow rate at which the bubble-by-bubble regime ends and the jet flow regime begins.

Let us, as in Section 2.5, consider the work of bubble formation. This is mainly consumed in overcoming surface tension, and if we assume that this is done at the expense of the kinetic energy of the gas, we find that[1]

$$L_\sigma = \int_0^{\tau_f} \frac{\pi D_o^2}{4} \frac{\rho_g v^2}{2} v \, dt \tag{2.19}$$

where v is the instantaneous local gas velocity and τ_f the time of bubble formation. The mean velocity of the gas during bubble formation at the orifice is

$$v_1 = \frac{1}{\tau_f} \int_0^{\tau_f} v \, dt \tag{2.20}$$

with

$$\tau_f = \frac{16 r_F^3}{3 D_o^2 v_1} \tag{2.21}$$

The mean velocity of the gas in the orifice, v_o (averaged over a time period much greater than τ_f), is related to the mean velocity during formation by

$$v_o = \epsilon v_1 \tag{2.22}$$

where

$$\epsilon = \frac{\tau_f}{\tau_f + \tau_0} \tag{2.23}$$

and τ_0 is the time between the release of one bubble and the formation of the next bubble. If we assume that

$$\frac{dr_b}{dt} = ct^n$$

where c and n are constants, then

$$r_b = \frac{ct^{n+1}}{n+1}$$

Therefore,

$$v = \frac{4\pi r_b^2(dr_b/dt)}{\frac{1}{4}\pi D_o^2} = \frac{16c^3t^{3n+2}}{D_o^2(n+1)^2}$$

and using this in equation (2.20) we obtain

$$v_1 = \frac{1}{\tau_f}\int_0^{\tau_f}\frac{16c^3}{D_o^2(n+1)^2}t^{3n+2}\,dt = \frac{16c^3\tau_f^{3n+2}}{3D_o^2(n+1)^3} = \frac{v}{3(n+1)}\left(\frac{\tau_f}{t}\right)^{3n+2}$$

or

$$v = 3v_1(n+1)\left(\frac{t}{\tau_f}\right)^{3n+2}$$

This result can now be used in equation (2.19), bearing in mind that $L_\sigma = 4\pi r_F^2\sigma$,

$$4\pi r_F^2\sigma = \frac{27v_1^3\rho_g\pi D_o^2\tau_f(n+1)^3}{8(9n+7)}$$

Introducing equations (2.21) and (2.22), we find that

$$4\pi r_F^2\sigma = \frac{18r_F^3v_o^2\rho_g\pi(n+1)^3}{\epsilon^2(9n+7)}$$

or

$$v_o = \epsilon\left[\frac{2(9n+7)\sigma}{9(n+1)^3\rho_g r_F}\right]^{1/2}$$

If we take the case of weak convective currents, we can use equation (2.2) to obtain r_F, and then

$$\frac{v_o(\rho_g)^{1/2}}{[\sigma(\rho_f-\rho_g)]^{1/4}} = \epsilon\left[\frac{2(9n+7)}{9(n+1)^3}\right]^{1/2}\left[\frac{16g^2\sigma}{9D_o^2(\rho_f-\rho_g)}\right]^{1/12}$$

$$\frac{v_o(\rho_g)^{1/2}}{[\sigma(\rho_f-\rho_g)]^{1/4}} = 0.495\epsilon\left[\frac{9n+7}{(n+1)^3}\right]^{1/2}\left[\frac{g^2\sigma}{D_o^2(\rho_f-\rho_g)}\right]^{1/12} \tag{2.24}$$

In the special case when the velocity of bubble growth is constant ($n = 0$), equation (2.24) becomes

$$\frac{v_o(\rho_g)^{1/2}}{[\sigma(\rho_f - \rho_g)]^{1/4}} = 1.3\epsilon\left[\frac{g^2\sigma}{D_o^2(\rho_f - \rho_g)}\right]^{1/12} \tag{2.25}$$

Now, we are looking for that value of the mean gas velocity (at the orifice) which is the critical velocity separating the bubble-by-bubble regime from the jet flow (continuous bubbles) regime. The latter regime is reached when the time between bubbles, τ_0, goes to zero. From equation (2.23), this means that $\epsilon = 1$, and equation (2.25) becomes

$$\frac{v_{o(\text{crit})}(\rho_g)^{1/2}}{[\sigma(\rho_f - \rho_g)]^{1/4}} = 1.3\left[\frac{g^2\sigma}{D_o^2(\rho_f - \rho_g)}\right]^{1/2} \tag{2.26}$$

where $v_{o(\text{crit})}$ is now the critical mean gas flow rate, separating the regime of bubbling from the regime of jets or spray cones. As an example, consider again air bubbling into water through an orifice 1.3 mm (0.05 in.) in diameter. Equation (2.26) gives a value for the critical mean gas velocity at the orifice, $v_{o(\text{crit})} \simeq 7$ m/s (23 ft/s). Equation (2.26) indicates that there is only a weak dependency of the critical velocity on the surface-tension coefficient, density differential, and orifice diameter, but a fairly strong dependency on the gas density.

References

1. Kutateladze, S. S., and Styrikovich, M. A., *Hydraulics of Gas–Liquid Systems*, Gosudazstvennoe Energeticheskoye Izdatelstvo, Moscow, 1958. Wright Field trans. F-TS-9814v.
2. Smirnov, N. I., and Poluta, S. E., *Zh. Prikl. Khim. (Moscow)*, *22*, 11 (1949).
3. Datta, R. L., Napier, D. H., and Newitt, D. U., *Trans. Inst. Chem. Eng.*, *28*, 14 (1950).
4. Davidson, J. F., and Schuler, B. O. C., *Trans. Inst. Chem. Eng.*, *38* 114 (1960).
5. Benzing, R. J., and Myers, J. E., *Ind. Eng. Chem. Eng. Des. Equip.* *47*, 2087 (1955).
6. Davidson, L., Thesis, Columbia University, 1951.
7. Eversole, W. G., Wagner, G. H., and Stackhouse, E., *Ind. Eng. Chem.*, *33*, 1495 (1941).
8. Leibson, I., Holcomb, E. G., Cacoso, A. G., and Jacmic, J. J., *AIChE J.*, *2*, 296 (1956).
9. Davidson, J. F., and Harrison, D., *Fluidized Particles*, Cambridge University Press, Cambridge, 1963.
10. Siemes, W., *Chem.-Ing.-Tech.*, *26*, 479 (1954).
11. Davidson, J. F., and Schuler, B. O. C., *Trans. Inst. Chem. Eng.*, *38*, 335 (1960).
12. Hayes, W. B., Hardy, B. W., and Holland, C. D., *AIChE J.*, *5*, 319 (1959).
13. Ho, G. E., Muller, R. L., and Prince, R. G. H., *Proc. Int. Symp. Distill. 1969, Inst. Chem. Eng.*, *2*, 10 (1969).
14. Spells, K. E., and Bakowski, S., *Trans. Inst. Chem. Eng.*. *28*, 38 (1950).
15. Spells, K. E., and Bakowski, S., *Trans. Inst. Chem. Eng.*, *30*, 189 (1952).
16. Spells, K. E., *Trans. Inst. Chem. Eng.*, *32*, 167 (1954).

17. Quigley, C. J., Johnson, A. K., and Harris, B. L., *Chem. Eng. Prog. Symp. Ser., 51,* 31 (1955).
18. Davidson, L., and Amick, E. H., *AIChE J., 2,* 337 (1956).
19. Brown, R. S., Univ. Calif. Rep. UCRL 8558, 1958.
20. McCann, D. J., and Prince, R. G. H., *Chem. Eng. Sci.,* no. 47, 241 (1969).
21. McCann, D. J., and Prince, R. G. H., *Chem. Eng. Sci.,* no. 26, 1505 (1971).
22. Kupferberg, A., and Jameson, G. J., *Trans. Inst. Chem. Eng., 47,* 241 (1969).
23. L'Ecuyer, M. R., and Murthy, S. N. B., NASA Tech. Note no. D-2547, 1965.
24. Muller, R. L., 1970. Ph.D thesis, University of Queensland.
25. Nielsen, R. D., 1965. Ph.D. thesis, University of Michigan.
26. Porter, K. E., and Wong, P. F., *Proc. Int. Symp. Distill. 1969, Inst. Chem. Eng., 2,* 22 (1969).
27. Burgess, R. G., and Robinson, K., *Proc. Int. Symp. Distill. 1969, Inst. Chem. Eng., 2,* 34 (1969).
28. Muller, R. L., and Prince, R. G. H., *Chem. Eng. Sci. 27,* 1583 (1972).
29. Letan, R., *Chem. Eng. Sci., 29*(2), 621 (1974).
30. Ramakrishman, S., Kumar, R., and Kuloor, N. R., *Chem. Eng. Sci., 24,* 731 (1969).
31. Milne Thomson, L. M., *Theoretical Hydrodynamics,* 3rd ed., Macmillan, New York, 1955.
32. Kumar, R., and Kuloor, N. R., *Chem. Tech. 19,* 78 (1967).
33. Kumar, R., and Kuloor, N. R., *Chem. Tech. 19,* 657 (1967).
34. Kumar, R., and Kuloor, N. R., *Chem. Tech. 19,* 733 (1967).
35. Swope, R. D., *Can. J. Chem. Eng., 49,* 2, 169 (1971).
36. Sullivan, S. L., Jr., Hardy, B. W., and Holland, C. D., *AIChE J., 10* (6), 848 (1964).
37. McCuskey, S. W., *An Introduction to Advanced Dynamics,* Addison-Wesley, Reading, Mass., 1959, p. 144.
38. Wallis, G. B., *One-Dimensional Two-Phase Flow,* McGraw-Hill, New York, 1969, p. 245.
39. Azbel, D. S., *Theor. Found. Chem. Eng. (USSR), 5*(5), 645 (1971).
40. Azbel, D. S., *Khim. Mashinostr. (Moscow),* no. 6, 14 (1960).
41. Marfenina, I. V., Candidate's dissertation, Moscow Higher Technical School, 1948. (In Russian.)

3

Mass bubbling

3.1 Introduction

When a gas is mixed with a liquid, at sufficiently high flow rates of the gas and liquid a mass motion of gas bubbles is produced, which, in turn, gives rise to an intensive mixing of the liquid. Depending on the flow rates of the gas and liquid in the mixture, different hydrodynamic regimes arise and, consequently, changes in the structure of the mixture occur. This structure can be characterized by its hydraulic resistance, which for the particular case of gas being forced into a liquid through an array of orifices depends on the height of the liquid column in the chamber and also on the surface contact area, this being defined by bubble size, gas content, and column height.

At low flow rates, the gas passes through the liquid mixture at regular intervals in the form of bubbles generated from the orifices,[1–6] as we saw in Chapter 2. Under this condition a layer of cellular foam, consisting of greatly enlarged bubbles that are close-packed and deformed, is formed on the free surface of the liquid. As the superficial gas velocity is increased, this layer of cellular foam thickens and finally, at a certain ratio of gas and liquid flow rates, the entire bubbling mixture is transformed into cellular foam.

We will see in Chapter 5 that the mode of gas injection into the liquid is also changed when we increase the gas flow rate. Instead of bubbles, the orifices produce gas jets, which are broken up into bubbles as a result of the dynamic influence of the liquid.[1,7–9] Increasing the gas flow rate retards the circulatory downflow of liquid, and at a certain critical gas velocity the amount of liquid descending to the orifices is insufficient to generate new bubbles,[1] at which point the height of the cellular foam layer reaches its maximum and the foam starts to burst.[1,10] On further increase in superficial gas velocity (to about 1.0 m/s, 3.28 ft/s) there is an abrupt change in the structure of the mixture, from the cellular foam regime to a regime of developed turbulence. In this regime the bubbling process is accompanied by liquid breaking up into drops. As we increase the superficial gas velocity once again (to about 3.0 m/s, 9.84 ft/s) the bubbling mixture is totally destroyed and liquid is completely transformed into drops entrained by the gas flow.

The regimes arising after destruction of the cellular foam condition have great practical importance.[10] The lower gas flow limit of these regimes is defined by the Froude number being greater than unity ($Fr = v_s^2/gh > 1$, where v_s is the superficial gas velocity, h the height of the liquid column, and g acceleration due to the gravity) and transition from one regime to the other is, as qualitatively described above, defined by the gas and liquid flow rates and geometric properties of the flow system.

In this chapter we confine ourselves to the regime of intensive (mass) bubbling. We particularize our study to the flow of gas, through an array of orifices, into quiescent liquid, but, as will become clear, the results of this analysis have applicability for a flowing dynamic gas–liquid mixture. We begin with some considerations of the turbulent properties of mass bubbling.

3.2 Turbulence in mass bubbling

We shall examine mass bubbling on the basis of the theory of homogeneous and isotropic turbulence, since the motion of a large swarm of bubbles through a liquid produces a practically complete mixing; in other words, a regime of developed turbulence is created, and we will further assume that the energy of the gas passing into the liquid is transmitted to the liquid and dissipated by its turbulent flow. In this section we describe several of the quantities usually used to define the turbulent flow.

It is well known that there are always turbulent eddies in flows with large Reynolds numbers, such eddies being usually classified by their size. Much of the energy of the flow is contained in large-scale eddies, this scale having an order of magnitude comparable to the largest dimension of the flow system (e.g., the diameter of a pipe). Also, the velocity of these large-scale eddies is comparable to the mean velocity occurring in the system, so that the Reynolds number of these motions is therefore of the same order of magnitude as the Reynolds number of the mean or time-averaged flow.

The kinetic energy of the turbulent motion is continuously being transferred from large eddies to eddies of smaller sizes until it is dissipated in the smallest eddies, where the liquid motion has a viscous character.

According to the theory of homogeneous isotropic turbulence,[11] all eddies, the scales of which are significantly less than the large-scale eddies, are statistically independent of the larger ones, and the properties of these smaller eddies are defined solely by the local energy dissipation. This homogeneous isotropic turbulence exists only in a small region of the flow, much smaller than the volume of the enclosing system, whereas the large-scale turbulent motion is obviously far from being isotropic, being influenced by wall effects. Liquid viscosity is significant only for small-scale motion, and, conversely, quantities describing large-scale motion are independent of liquid viscosity.

The dissipation energy is defined by properties of the large-scale motion, and the relationship between the dissipation and these properties can be expressed[12,13] by the following equation:

$$\epsilon \simeq U^3/l \tag{3.1}$$

where ϵ is the energy dissipation per unit mass and time, U the characteristic (or typical) velocity of the large eddies, l the characteristic size of the large eddies, and ρ_f the liquid density.

The characterizing pressure difference in turbulent motions is usually defined[12] as

$$\Delta p \simeq \rho_f U^2 \tag{3.2}$$

The velocity v_τ, defining the motion of the turbulent liquid itself, can be expressed as the change of velocity of an element of liquid in a (small) time, τ; is dependent on local properties of the turbulence and the time interval; and can be written as

$$v_\tau \simeq (\epsilon\tau)^{1/2} \tag{3.3}$$

For the practical use of equations (3.1), (3.2), and (3.3), it is first necessary to know the characteristic (or typical) velocity and size of the large eddies. The size of large eddies is, as described above, comparable with the largest dimension of the flow system, either the height of the gas–liquid mixture or the diameter of the liquid chamber.[13] Let us obtain the characteristic velocity of the large eddies by considering the work done by the gas during the bubbling process. This loss of energy consists of work done against gravity force and work done against surface forces. The work against surface forces during intensive bubbling is so small that it can be ignored. The total kinetic energy, E, of the gas injected into the liquid is converted into potential energy of the liquid, which, in turn, is transformed into kinetic energy of the descending flow of liquid. It can be shown that, for this process, the work done by the gas during its passage through the liquid can be written as[14]

$$E = \rho_f Q_l g h \left(\frac{\phi}{1-\phi}\right)$$

where Q_l is the volume of liquid and ϕ the gas void fraction (defined as the ratio of gas volume to total volume), and this work when divided by the volume $Q_l[\phi/(1-\phi)]$ of gas in the chamber is numerically equal to the liquid static pressure in the chamber, Δp.

Taking equation (3.2) into consideration, the velocity of large-scale eddies is then given by

$$U = \text{const.}\left(\frac{\Delta p}{\rho_f}\right)^{1/2} = \text{const.}(gh)^{1/2} \tag{3.4}$$

Now that we have defined various quantities characterizing turbulent flows, we are in a position to study the effect of liquid turbulence on bubble motion.

3.3 Maximum velocity of a bubble in the turbulent flow

Large-scale eddies may entrain a bubble suspended in the turbulent flow, but because the density of a bubble differs significantly from the density of the liquid, the inertia forces are of different magnitudes, and this entrainment cannot be complete. At the other extreme the small-scale eddies cannot entrain a bubble, and in relationship to these eddies the bubble may be considered to be an immobile body. Liquid taking part in this small-scale motion will flow around the surface of the gas bubble. Hence, we need to determine the intermediate scale that is capable of entraining bubbles.

The equation of motion determining the scale of eddies that can entrain a rising bubble can be written[15]

$$(\rho_g + k\rho_f)V\frac{dv_r}{dt} = (\rho_f - \rho_g)V\frac{dv_\tau}{dt} - F_D \tag{3.5}$$

where ρ_f is the liquid density, ρ_g the gas density, V the volume of a bubble, v_r the velocity of the bubbles relative to the liquid, F_D the drag force, v_τ the liquid velocity, and k the coefficient of "apparent additional mass" (for spherical bubbles $k = \frac{1}{2}$).

Here the left-hand side of the equation is the inertia force of the bubble, the first term on the right-hand side of the equation expresses the inertial effect of the surrounding turbulent liquid, and the last term is the drag force. Usually, this force F_D is taken to be equal to the drag of a free-rising bubble of a moderate size (Re $\simeq$ 800); in other words,

$$F_D = 12\pi\mu_f r_b v_r \tag{3.6}$$

where μ_f is the dynamic viscosity of the liquid and r_b the radius of the bubble.

However, in mass-bubbling operations, where the turbulent motion is generated by the bubbly flow, each bubble is influenced by other bubbles, both close to it and farther removed from it, and it is obvious that in such a case the drag of a bubble must differ from the drag of a single free-rising bubble.

The estimation of the drag exerted on an individual bubble in a swarm of bubbles is very difficult[16–21] and cannot be accomplished without the use of some simplifying assumptions. Let us assume a uniform bubble distribution in the volume of liquid.[22–24] In this case, we shall suppose that every bubble is located in the center of a *spherical compartment* formed by adjacent bubbles, so using this model we can use spherical coordinates and consider the motion to be equivalent to that of liquid between two concentric spheres, the inner surface being the bubble and the outer constituting a *free surface* boundary condition.

For this flow system, the radial and tangent velocity components of the enclosed liquid can be written as follows:[16,24]

$$v_R = \frac{v_r}{1-\phi}\left(1 - \frac{r_b^3}{r^3}\right)\cos\theta \tag{3.7a}$$

$$v_\theta = \frac{v_r}{1-\phi}\left(1 + \frac{r_b^3}{2r^3}\right)\sin\theta \tag{3.7b}$$

where ϕ is the gas void fraction, r the radial coordinate, θ the polar angle, and r_b the bubble radius. Let us estimate the drag from knowledge of the energy dissipation in the flowing liquid. This energy, per unit time, will be

$$\frac{dE_f}{dt} = \mu_f \int (\text{curl } v_i)^2\, dV + \mu_f \int \frac{dv_i^2}{dr}\, ds + 2\mu_f \int (v_i \text{ curl } v_i)\, ds \tag{3.8}$$

where v_i denotes the velocity vector whose components are given by equations (3.7) and the first integral is a volume integral, the last two integrals being surface integrals. Combining equations (3.7) and (3.8) results in

$$\frac{dE_f}{dt} = 9\mu_f\left(\frac{v_r}{1-\phi}\right)^2 r_b^6 \int_0^\pi \int_{r_b}^{r_b/\phi^{(1/3)}} \frac{1 + 2\cos^2\theta}{r^8}\, 2\pi r^2 \sin\theta\, d\theta\, dr$$

Note that as an upper boundary condition on the radial coordinate r, we have $r = r_b/\phi^{1/3}$, so that the volume of integration and the volume of the bubble are consistent with the flow system void fraction. After integrating, we find that

$$\frac{dE_f}{dt} = 12\pi\mu_f v_r^2 r_b \frac{1 - \phi^{5/3}}{(1-\phi)^2}$$

and therefore the drag force is simply

$$F_D = 12\pi\mu_f v_r r_b \frac{1 - \phi^{5/3}}{(1-\phi)^2} \tag{3.9}$$

Thus the expression for the drag on a bubble in a swarm of bubbles, given by equation (3.9), differs from the drag on a single free-rising bubble, given by equation (3.6), by a factor that takes into account the flow system void fraction.

Now, using the value of the drag force, F_D, from equation (3.9) and v_r from equation (3.3) in equation (3.5), we obtain

$$\frac{4}{3}\pi r_b^3(\rho_g + k\rho_f)\frac{v_r}{\tau} = \frac{4}{3}\pi r_b^3(\rho_f - \rho_g)\left(\frac{\epsilon}{\tau}\right)^{1/2} - 12\pi\mu_f v_r\left[\frac{1-\phi^{5/3}}{(1-\phi)^2}\right]$$

and solving for v_r, we find

$$v_r = \frac{\rho_f - \rho_g}{\rho_f}(\epsilon\tau)^{1/2}\left\{k + \frac{9\nu_f\tau}{r_b^2}\left[\frac{1-\phi^{5/3}}{(1-\phi)^2}\right]\right\}^{-1} \tag{3.10}$$

where ν_f is the liquid kinematic viscosity. Here we have said that $\rho_g/\rho_f \ll 1$.

The characteristic period of the liquid motion, τ, can be expressed by the scale of entraining eddies, λ, and the energy dissipation of the turbulent flow ϵ in the form[33]

$$\tau = \left(\frac{\lambda^2}{\epsilon}\right)^{1/3} \tag{3.11}$$

Now, the velocity of the bubbles is a maximum when we have eddies of size λ obtained by the following condition:

$$\frac{dv_r}{d\lambda} = 0$$

or its equivalent by equation (3.11),

$$\frac{dv_r}{d\tau} = 0$$

Performing the differentiation on equation (3.10) and using the foregoing condition, with the assumption that $\rho_g \ll \rho_f$ gives the time duration for maximum bubble velocity,

$$\tau = \frac{kr_b^2(1-\phi)^2}{9\nu_f(1-\phi^{5/3})} \tag{3.12}$$

The maximum velocity of the bubbles is obtained by substituting the values of ϵ and τ from equations (3.1) and (3.12) into equation (3.10), and we find

$$v_{r_{\max}} = \frac{U^{3/2}r_b(1-\phi)}{6(kl\nu_f)^{1/2}(1-\phi^{5/3})^{1/2}} \tag{3.13}$$

The velocity given by equation (3.13) is the characteristic velocity of a bubble in restricted motion. In the case of a spherical bubble, this equation can be written

$$v_{r_{\max}} = \text{const.} \frac{U^{3/2} r_b}{(l\nu_f)^{1/2}} \frac{1 - \phi}{(1 - \phi^{5/3})^{1/2}} \tag{3.14}$$

3.4 Size of bubbles in mass bubbling

In intensive mass-bubbling operations, bubbles of a fairly uniform size are formed in a gas–liquid mixture, this being a result of the breaking up and coalescing of bubbles, which in turn is due to the dynamic interaction between the liquid and gas. It has been shown experimentally that the size of a bubble depends on surface tension[25–30] and the viscosity of the liquid,[31–35] but that the physical properties of the gas[25] have no significant influence on the size of the bubble. At low gas flow rates the bubble radius increases with an increase in viscosity,[31] as we saw in Section 2.3. For example, at a superficial velocity of gas of 4 mm/s (0.16 in./s), the bubble radius is doubled with an increase in viscosity from 5×10^{-2} to 1 g/cm·s (3.5×10^{-3} to 6.9×10^{-2} lb/ft·s).

In mass bubbling the bubble radius is practically independent of the geometric characteristics of the flow system (e.g., the diameter of the orifice in a gas-distributing device[36]), so, for example, increasing the diameter of the orifice 100 times increases the bubble diameter only twice for an air–water system.[37] On the other hand, the *wall effect* has some influence on bubble size when the diameter of an apparatus is less than 200 mm (8 in.).

In a mass-bubbling process, bubbles have an oblate ellipsoid shape whose short axis is parallel with the direction of motion,[30,31] and the rise velocity of such bubbles has been estimated by equating drag to buoyancy force.[15]

$$v_c \simeq \left(\frac{\sigma^2 g}{3\pi \mu_f \rho_f} \right)^{1/5} \tag{3.15}$$

where v_c is the velocity of the center of the bubble. Note that the surface tension, σ, figures in this equation insofar as it affects the shape of the bubble and hence the drag force.[15] This equation can be used to obtain a rough estimate of the bubble velocity.

We are interested in obtaining information about bubble size in mass bubbling, and equation (3.15) is of use in this quest, as it defines a dynamic property of the bubbles, the velocity, in terms of system properties. Now, it is known[38,39] that viscous liquid moving under the action of constant forces has an unambiguous potential, and in this case the velocity distribution is such that it gives a minimum of dissipative energy.

With this fact in mind, let us use the equation of relative motion of a gas bubble in a gas–liquid mixture to estimate the diameter of an average bubble:

$$g(\rho_f - \rho_g)V - C_D \frac{\rho_f v_c^2}{2} A_b - (\rho_g + k\rho_f)V \frac{dv_c}{dt} = 0 \qquad (3.16)$$

where ρ_f is the density of liquid, ρ_g the density of gas, V the volume of a bubble, C_D the drag coefficient, A_b the cross-sectional area of a bubble, and k the "apparent additional mass" coefficient (for a spherical bubble $k = 0.5$).

It has been found,[40–42] for gas bubbling into a liquid-containing chamber through an orifice (or orifices), that in the bulk of the bubbling mixture, the void fraction ϕ (defined as the ratio of gas volume to the total volume) and so also the velocity of bubble rise, v_c, are only slowly varying functions of the position in the mixture, $\phi \simeq$ const. and $v_c \simeq$ const. This is illustrated in Figure 3.1 for a steam–water system. Ignoring ρ_g, which is usually three orders of magnitude less than ρ_f, and taking into account that under steady-state conditions $dv_c/dt = 0$, we can rewrite equation (3.16) in the form

$$g\rho_f V - C_D \frac{\rho_f v_c^2}{2} A_b = 0$$

Figure 3.1 Void and velocity as a function of height. System: steam–water (p = 17 atm). The dashed lines show the average height of the liquid, h, measured with a planimeter.

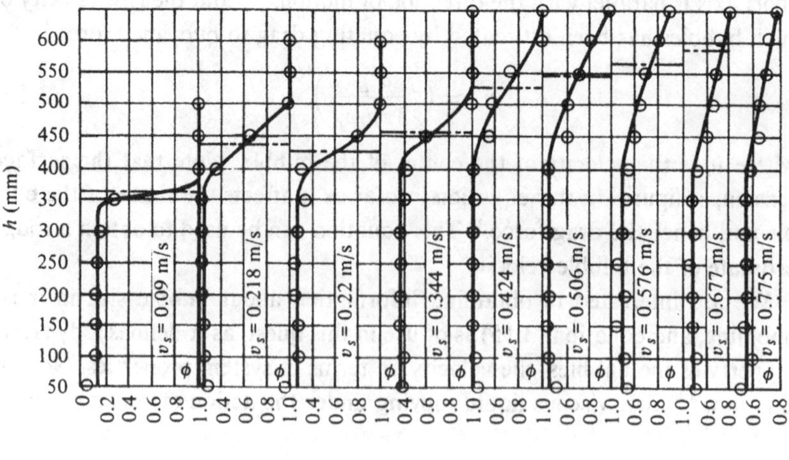

To convert this equation into an energy equation, we multiply it by dz (where z is the perpendicular coordinate) and obtain

$$g\rho_f V\,dz - C_D \frac{\rho_f v_c^2}{2} A_b\,dz = 0$$

Integrating between limits from any value z to h, we find an expression for the bubble energy dissipation:

$$e_b = g\rho_f V(h - z) + C_D \frac{\rho_f v_c^2}{2} A_b z = \text{const.} \tag{3.17}$$

where h is the total travel distance of the bubble (equal to the depth of the liquid). We have here used the condition that the total energy dissipation is $e_b = C_D(\rho_f v_c^2/2)A_b h$, this being an expression of the work done by the bubble against the force $[C_D(\rho_f v_c^2/2)A_b]$ over the distance h.

Note that the energy spent on formation of the surface of a bubble is ignored in equation (3.17) because it is minimal compared to the energy $C_D(\rho_f v_c^2/2)A_b z$ spent on overcoming resistance of the fluid, or, in other words, the energy spent on formation of the contact surface is

$$4\pi r_b^2 \sigma \ll \frac{\rho_f v_c^2}{2} A_b h$$

where the right-hand side of the equation is a measure of the total work done against drag forces over the distance h, r_b the average bubble radius, and σ the surface-tension coefficient. The formula for the average bubble radius may now be obtained by varying equation (3.17) with respect to the radius and by setting the variation at zero, which will give the radius of bubbles having minimum energy dissipation, the most likely state of the bubbles:

$$4\pi\rho_f z r_b^2 g - C_D \pi \rho_f v_c^2 z r_b = 0$$

or

$$r_b \simeq \frac{C_D v_c^2}{4g} \tag{3.18}$$

Note that we have ignored the term $g\rho_f Vh$ in equation (3.17) to eliminate the dependency of r_b on z in equation (3.18). This is, of course, an approximation equivalent to saying that $z/(h - z) \simeq 1$. It follows from equations (3.15) and (3.18) that the radius of a bubble increases with increase of surface tension and decrease of liquid viscosity.

To determine the bubble radius, we need to know the drag coefficient C_D, and this can be obtained from experimental data,[43,44] for instance that shown in Figure 3.2. Comparison of experimental data and data calculated using equations (3.15) and (3.18) show[45] that this formula is very good for estimating the average bubble size in mass bubbling.

Now, instead of determining the average bubble size in mass bubbling by using the momentum equation, we can again use the same concept of minimum energy dissipation, but in a different context. When experimental data are available for bubble velocity as a function of bubble size, it is possible to calculate the average bubble size in the following manner.

When a bubble is rising under the action of buoyancy and drag forces and the terminal velocity is reached, the buoyancy and drag forces balance. The dissipation of energy per unit time of the bubble is given by

$$\frac{de_b}{dt} = F_D v_c$$

where F_D is the drag force and, as before, e_b is the bubble energy dissipation. From what has just been stated, we know that $F_D = F_b$, where F_b is the buoyancy force. This means that $de_b/dt = F_b v_c$, and because F_b is a constant, $(de_b/dt) \propto v_c$, and from the principle of minimum energy dissipation it then follows that min (de_b/dt) takes place at min (v_c).

Thus, if we know the velocity v_c as a function of bubble radius, the most probable bubble radius is obviously that for which v_c is a minimum. This is illustrated in Figure 3.3 for an air–water system. Here we find $v_{c(\min)}$ at a bubble radius of approximately 2 mm (0.079 in.).

Figure 3.2 Drag coefficient for equation (3.18).

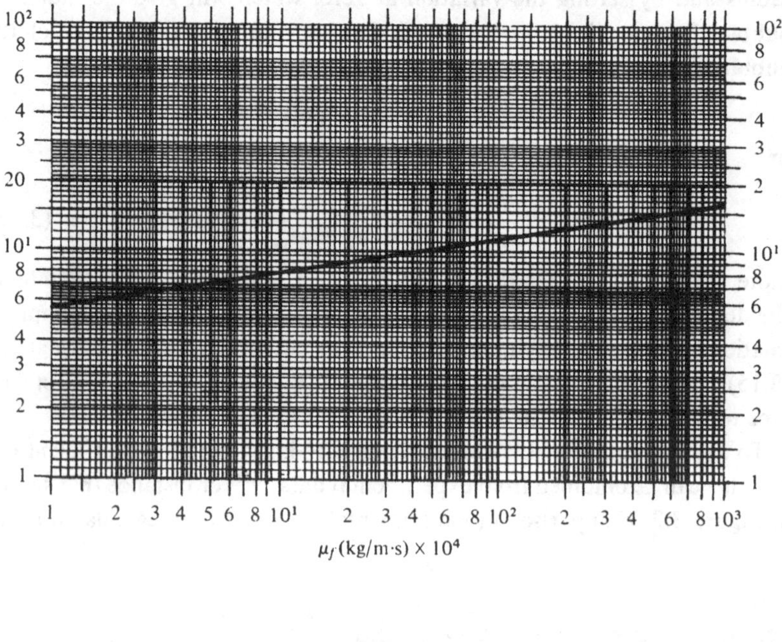

The average bubble size may be obtained from yet another approach. During the motion of sufficiently large single bubbles, they are not only deformed but also break up, and the condition for breakup of bubbles[15] may be written as

$$\frac{\rho_g v_r^2}{2} \geqslant \frac{3\sigma^3}{r_b^3 \rho_f^2 v_r^4}$$

The left-hand side of the foregoing equation represents the kinetic energy of the flow per unit volume, and the right-hand side expresses the surface-tension energy of the bubble (the surface tension σ is to the third power because of its effect on the bubble shape, the details of which have been described elsewhere[15]). The equation states that if the kinetic energy of the flow is greater than the energy of surface tension, the bubble is likely to break up. This equation can be rewritten as

$$r_c = \frac{6^{1/3}\sigma}{v_r^2(\rho_f^2\rho_g)^{1/3}} \tag{3.19}$$

where r_c is now the critical bubble size for breakup. Now, the velocity of the liquid changes from place to place in turbulent flow, and thus different dynamic heads are present in the flow, leading, by acting on the bubble

Figure 3.3 Bubble velocity versus bubble diameter.

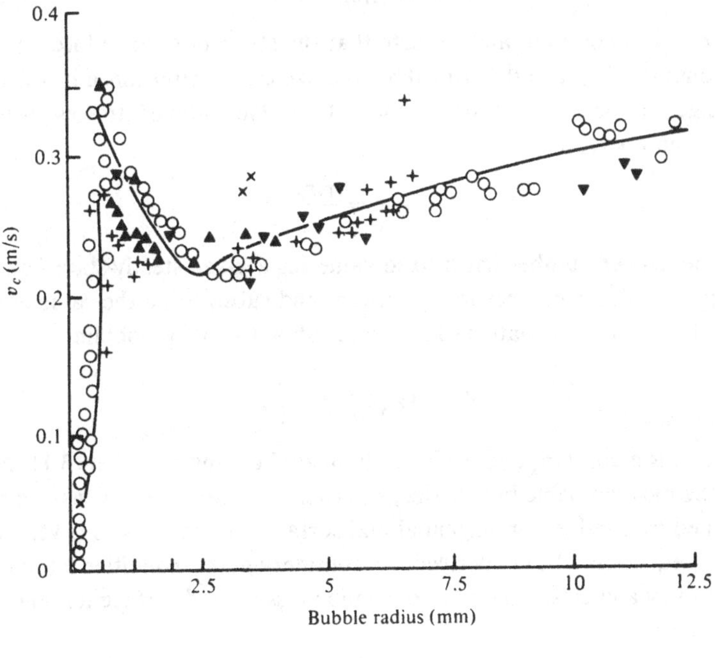

surface in successive instants, to the deformation and rupture of the bubble, as represented by equation (3.19).

Large-scale eddies hardly change their velocity from one end of a bubble to the other, and thus have no influence on the bubble, so that deformation and breakup of a bubble is caused by comparatively small turbulent eddies[15] whose characteristic velocity may be written as

$$v_e \simeq (\epsilon\lambda)^{1/3} \tag{3.20}$$

where ϵ is the energy dissipation and λ the scale of eddies capable of breaking up the bubbles. We can say that this scale will be approximately that of the critical bubble diameter, so that $\lambda \simeq 2r_c$. Further, we can say that the relative velocity of the bubble is approximately the same as the eddy velocity, $v_r \simeq v_e$, which means that we are assuming that the absolute velocity of the bubble is small. Using this in equation (3.20) and taking into account equations (3.1) and (3.19), we find that

$$r_c = c\left(\frac{\sigma}{\rho_f}\right)^{3/5} \frac{l^{2/5}}{U^{6/5}} \left(\frac{\rho_f}{\rho_g}\right)^{1/5} \tag{3.21}$$

where l and U are the scale and velocity of large-scale motions, as described in Section 3.2, and c is a dimensionless constant. Equation (3.21) is an expression for the average bubble size in terms of known quantities.

Now, if we assume that the tangential stress as a result of turbulence is

$$\tau_s = c_1\rho_f(\epsilon r_c)^{2/3} \tag{3.22}$$

where c_1 is a constant, and we note that the stress of surface forces is proportional to σ/r_c, then the critical bubble size is also determined by the ratio of tangential stress to the surface force stress. This ratio of stresses gives the Weber number,

$$\mathrm{We} = \frac{\tau_s r_c}{\sigma} \tag{3.23}$$

and the Weber number has a fixed value for a given steady-state flow, so, taking the Weber number to be constant and substituting the value of tangential stress from equation (3.22) in equation (3.23), we obtain

$$r_c = c_2\left(\frac{\sigma}{\rho_f}\right)^{3/5} \frac{l^{2/5}}{U^{6/5}} \tag{3.24}$$

where c_2 is a constant and we have eliminated ϵ using equation (3.1). Note that the most probable bubble size, r_c, given in equation (3.24) that we have obtained by considering tangential and surface tension stresses is very similar to equation (3.21), derived by considering the condition of bubble breakup by a critical balance of dynamic pressure and surface tension. The velocity, U, and length, l, are taken from the overall flow system; for instance, U may be taken to be the gas superficial velocity, v_s, and l might be the disturbed liquid level in the chamber, $h/(1 - \phi)$. Assuming that the velocity of the large-scale fluctuations is given by equation (3.4), we can rewrite equation (3.24) in the form

$$r_c = c_3 \left[\frac{(\sigma/\rho_f g)^3}{h(1 - \phi)^2} \right]^{1/5} \tag{3.25}$$

where ϕ is the gas void fraction. In actual flows, because of the "wall effect," coalescence, and viscous effects, the average bubble radius may differ from its value determined by equation (3.24), which was derived from breakup conditions. Analysis of experimental data on bubbles of differing sizes in systems with differing physical properties has shown that equation (3.25) does not correlate the experimental data correctly within a constant. Therefore, we introduce into the equation factors ϕ^n and $(\mu_g/\mu_f)^n$, which take into account the influence of coalescence and viscosity.

Experimental data[44] has been obtained, by chemical and gamma-ray methods,[18,45] of the specific surface area (see Section 3.5) for flows under various conditions, and the average bubble diameter was then estimated from the average void fraction and the specific phase contact area. This made it possible to obtain experimental data on bubble size not only in the region near the wall but also in the bulk of the gas–liquid flow. From such an analysis the following equation was derived for calculating the average radius of a bubble in mass-bubbling operations:

$$r_c = 2.56 \left(\frac{\sigma}{\rho_f g} \right)^{0.6} h^{-0.2} \left(\frac{\mu_g}{\mu_f} \right)^{0.25} \frac{\phi^{0.65}}{(1 - \phi)^{0.4}} \tag{3.26}$$

where σ is the surface tension, ρ_f the liquid density, g the acceleration due to gravity, h the height of the liquid column, and μ_g and μ_f are the dynamic viscosity of the gas and liquid, respectively. As we have noted, equations (3.15) and (3.18) give an alternative approach for determination of the probable bubble size in mass bubbling, by considering the conditions for minimum energy dissipation.

3.5 Specific phase contact area

In mass-bubbling operations, bubble size or droplet size is not uniform but varies over a certain range. In this section we shall examine the diameters of discrete elements (bubbles or droplets) as a continuous variable. This means that no matter how close the diameters of two discrete elements may be, it is always possible to find a bubble diameter or droplet diameter intermediate between those two diameters if a sufficiently large number of measurements are taken.

Bubble size (or droplet size) is a random variable, so that when we study the spectrum of bubbles or droplets, we are dealing with an assemblage of random values lying in a certain interval and, as is known from the laws of probability, the universal characteristic of any random value is its distribution function.

The diameter spectrum of bubbles in the case of gas forced into a liquid through, for instance, transverse porous plates,[30,31,46] or distributor grids with holes all of the same diameter,[47] has been found to fit quite closely to the Gaussian curve described by the equation[48]

$$f(y) = \frac{1}{\sigma_b(2\pi)^{1/2}} \exp\left[\frac{-(y - d_{av})^2}{2\sigma_b^2}\right] \tag{3.27}$$

where $f(y)$ is the fraction of bubbles with diameters between y and $(y + dy)$ (the probability density), and σ_b is the standard deviation from the average bubble diameter d_{av}. Note in equation (3.27) that y can take all values from minus infinity to plus infinity. This is physically meaningless but is a mathematical convenience that does not affect the appropriateness of the results obtained by its use because $f(y)$ will be essentially zero for y being negative. With this normal distribution law for bubble (or droplet) diameters, let us determine[48] the specific surface area, a, defined as the ratio of total bubble surface area to total mixture volume. This quantity is obviously of some interest for studies of heat transfer and chemically active processes.

Denoting the number of bubbles whose diameters lie in the interval between y and $(y + dy)$ by

$$dN = N f(y)\, dy$$

where N is the total number of bubbles, we find that the volume of the bubbles enclosed in this diameter range is

$$dV = \tfrac{1}{6}\pi y^3 N f(y)\, dy$$

and their total surface area is

$$dS = \pi y^2 N f(y)\, dy$$

The total volume of all the bubbles enclosed in the two-phase flow and their total surface area are then, respectively,

$$V_b = \int_{-\infty}^{+\infty} \frac{1}{6} \pi y^3 N f(y)\, dy \tag{3.28}$$

and

$$S = \int_{-\infty}^{+\infty} \pi y^2 N f(y)\, dy \tag{3.29}$$

Using equation (3.27) in equation (3.28), we obtain

$$V_b = \frac{\pi}{6(2\pi)^{1/2}\sigma_b} N \int_{-\infty}^{+\infty} y^3 \exp\left[\frac{-(y - d_{av})^2}{2\sigma_b^2}\right] dy$$

or

$$\begin{aligned} V_b = \frac{\pi}{6(2\pi)^{1/2}\sigma_b} N \Bigg[& \int_{-\infty}^{+\infty} t^3 \exp\left(-\frac{t^2}{2\sigma_b^2}\right) dt \\ & + 3d_{av} \int_{-\infty}^{+\infty} t^2 \exp\left(-\frac{t^2}{2\sigma_b^2}\right) dt \\ & + 3d_{av}^2 \int_{-\infty}^{+\infty} t \exp\left(-\frac{t^2}{2\sigma_b^2}\right) dt \\ & + d_{av}^3 \int_{-\infty}^{+\infty} \exp\left(-\frac{t^2}{2\sigma_b^2}\right) dt \Bigg] \end{aligned} \tag{3.30}$$

where we have used the substitution

$$y - d_{av} = t; \qquad dy = dt$$

The first and the third integrals in equation (3.30) are equal to zero because their integrand is odd and the limits of integration are symmetrical about the origin. Thus

$$\begin{aligned} V_b &= \frac{\pi}{6(2\pi)^{1/2}\sigma_b} N \Bigg[3d_{av} \int_{-\infty}^{+\infty} t^2 \exp\left(-\frac{t^2}{2\sigma_b^2}\right) dt \\ &\qquad + d_{av}^3 \int_{-\infty}^{+\infty} \exp\left(-\frac{t^2}{2\sigma_b^2}\right) dt \Bigg] \\ &= \frac{\pi}{6(2\pi)^{1/2}\sigma_b} N\, [3d_{av}(2\pi)^{1/2}\sigma_b^3 + d_{av}^3\sigma_b(2\pi)^{1/2}] \\ &= \frac{\pi d_{av}}{6} N(3\sigma_b^2 + d_{av}^2) \end{aligned} \tag{3.31}$$

Similarly, substituting equation (3.27) in equation (3.29) and integrating, we find

$$\begin{aligned} S = \frac{\pi}{(2\pi)^{1/2}\sigma_b} N \Bigg[& \int_{-\infty}^{+\infty} t^2 \exp\left(-\frac{t^2}{2\sigma_b^2}\right) dt \\ & + 2d_{av} \int_{-\infty}^{+\infty} t \exp\left(-\frac{t^2}{2\sigma_b^2}\right) dt \\ & + d_{av}^2 \int_{-\infty}^{+\infty} \exp\left(-\frac{t^2}{2\sigma_b^2}\right) dt \Bigg] \\ = \pi N(\sigma_b^2 + d_{av}^2) & \end{aligned} \tag{3.32}$$

The specific phase contact surface area is then given by

$$a = \frac{S}{V_t} \tag{3.33}$$

where

$$V_t = \frac{V_b}{\phi_{av}} \tag{3.34}$$

is the volume of the two-phase mixture and ϕ_{av} the average void fraction.

Substituting the values of S from equation (3.32) and V_b from equation (3.31) into the foregoing equations, we can determine the specific surface area as

$$a = \frac{6\phi_{av}}{d_{av}^2} \frac{\sigma_b^2 + d_{av}^2}{3\sigma_b^2 + d_{av}^2} \tag{3.35}$$

Equation (3.35) differs from the generally accepted equation for calculating the specific surface area of contact by a multiplicative factor that accounts for polydispersity of the system and is equal to the ratio

$$\frac{\sigma_b^2 + d_{av}^2}{3\sigma_b^2 + d_{av}^2}$$

In other words, the specific surface area is affected by the standard deviation of bubble diameters, σ_b, and by the average bubble diameter, d_{av}.

The standard deviation, σ_b, from the average bubble diameter is found to be independent of the equipment diameter and independent of the properties of the system[30,31,46] over a wide range of gas flow rates.[30] Various experiments[47] have established a relationship between the standard deviation σ_b and the average bubble diameter d_{av} which shows that σ_b is usually much

Figure 3.4 Bubble distribution function. Column, 120 × 120 mm; tubular plate, 2 to 12 mm; superficial gas velocity, v_s = 0.73 m/s; liquid flow rate, 9.0 m³/h.

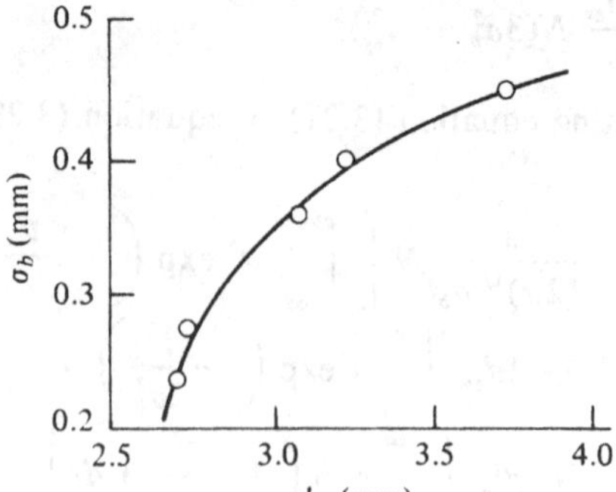

less than d_{av}, as shown in Figure 3.4, so that the factor can in fact be ignored, and then the specific (phase contact) surface area is to be found from the average bubble diameter and the average gas content:

$$a = \frac{6\phi_{av}}{d_{av}} \qquad (3.36)$$

This formula for the specific phase contact surface area has been obtained, in a somewhat roundabout way, by assuming a normal distribution for the droplet diameters, the assumption being validated by experimental data. However, several experiments have been performed to obtain the surface distribution and volume distribution of bubbles, and they have also been found[46,49] to satisfy the normal distribution law. The volume distribution function of the bubbles can be described in terms of the diameter probability distribution

$$P(y_1 < d < y_2) = P\left[y_1 < \left(\frac{6V}{\pi}\right)^{1/3} < y_2\right]$$

where y_1 and y_2 are upper and lower limits for diameter, d, and V is the volume of a bubble. This equation states the essential equality of the two distribution functions. This equality is illustrated in Figure 3.5, where curve 1 shows the bubble diameter density distribution, curve 2 shows the bubble area density distribution, and curve 3 shows the bubble volume den-

Figure 3.5 Bubble diameter distribution. Column, 120 × 120 mm; tubular plate, 2 to 12 mm; liquid flow rate, 6.3 m^3/h. 1, Bubble diameters; 2, bubble surface; 3, bubble volumes. (From A. M. Kashnikov, Candidate's dissertation, D. I. Mendeleev Moscow Institute of Chemical Technology, 1965.)

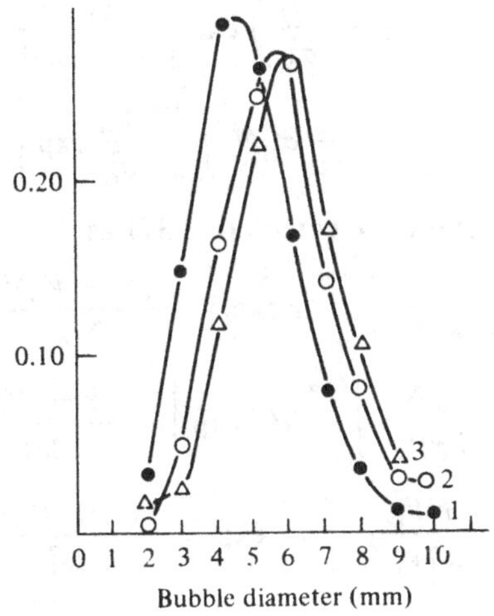

sity distribution for a particular two-phase flow. Now we can write the fraction of bubbles with volumes between the values of V_1 and V_2 as

$$P\left(V_1 < \frac{\pi y^3}{6} < V_2\right) = F\left(\frac{\pi y^3}{6}\right)$$

where $F(\pi y^3/6)$ denotes some function of $\pi y^3/6$.

The fraction (or probability density) of bubbles with diameters lying between y and $y + dy$ is therefore

$$f(y) = \frac{d}{dy} F\left(\frac{\pi y^3}{6}\right) = \frac{\pi y^2}{2} \frac{1}{\sigma_{b_v}(2\pi)^{1/2}} \exp\left\{-\frac{[(\pi y^3/6) - V_{av}]^2}{2\sigma_{b_v}^2}\right\}$$

where V_{av} is the average bubble volume and σ_{b_v} the standard deviation of bubble volume from the average value.

We then find that the volume of all the bubbles enclosed in the two-phase mixture is

$$V_b = \int_{-\infty}^{+\infty} \frac{\pi y^3}{6} N \frac{\pi y^2}{2} \frac{1}{\sigma_{b_v}(2\pi)^{1/2}} \exp\left\{-\frac{[(\pi y^3/6) - V_{av}]^2}{2\sigma_{b_v}^2}\right\} dy$$

Introducing the notation $\pi y^3/6 = z$, we can rewrite this equation in the form

$$V_b = \int_{-\infty}^{+\infty} \frac{Nz}{\sigma_{b_v}(2\pi)^{1/2}} \exp\left[-\frac{(z - V_{av})^2}{2\sigma_{b_v}^2}\right] dz \qquad (3.37)$$

Now consider the following:

$$-\sigma_{b_v}^2 N \int_{-\infty}^{+\infty} d\left\{\exp\left[-\frac{(z - V_{av})^2}{2\sigma_{b_v}}\right]\right\}$$

$$= -\sigma_{b_v}^2 N \int_{-\infty}^{+\infty} \left(-\frac{z}{\sigma_{b_v}^2}\right) \exp\left[-\frac{(z - V_{av})^2}{2\sigma_{b_v}^2}\right] dz$$

$$- \sigma_{b_v}^2 N \int_{-\infty}^{+\infty} \frac{V_{av}}{\sigma_{b_v}^2} \exp\left[-\frac{(z - V_{av})^2}{2\sigma_{b_v}^2}\right] dz$$

With this in mind we can write equation (3.37) as

$$V_b = \frac{N}{\sigma_b(2\pi)^{1/2}} \int_{-\infty}^{+\infty} z \exp\left[-\frac{(z - V_{av})^2}{2\sigma_{b_v}^2}\right] dz$$

$$= -\frac{\sigma_{b_v} N}{(2\pi)^{1/2}} \int_{-\infty}^{+\infty} d\left\{\exp\left[-\frac{(z - V_{av})^2}{2\sigma_{b_v}^2}\right]\right\}$$

$$+ \frac{NV_{av}}{\sigma_b(2\pi)^{1/2}} \int_{-\infty}^{+\infty} \exp\left[-\frac{(z - V_{av})^2}{2\sigma_{b_v}^2}\right] dz$$

or

$$V_b = NV_{av} \tag{3.38}$$

By a similar analysis, the total surface area of all the bubbles enclosed within the two-phase mixture is

$$S = \int_{-\infty}^{+\infty} \pi y^2 N 2\pi y \frac{1}{\sigma_{b_s}(2\pi)^{1/2}} \exp\left[-\frac{(\pi y^2 - s_{av})^2}{2\sigma_{b_s}^2}\right] dy$$

where σ_{b_s} is the standard deviation of the bubble surface areas from the average surface area, s_{av}. Introducing the notation $\pi y^2 = x$, this equation can be rewritten

$$S = \int_{-\infty}^{+\infty} \frac{Nx}{\sigma_{b_s}(2\pi)^{1/2}} \exp\left[-\frac{(x - s_{av})^2}{2\sigma_{b_s}^2}\right] dx$$

Now

$$-\sigma_{b_s}^2 N \int_{-\infty}^{+\infty} d\left\{\exp\left[-\frac{(x - s_{av})^2}{2\sigma_{b_s}^2}\right]\right\}$$
$$= -\sigma_{b_s}^2 N \int_{-\infty}^{+\infty} \left(-\frac{x}{\sigma_{b_s}^2}\right) \exp\left[-\frac{(x - s_{av})^2}{2\sigma_{b_s}^2}\right] dx$$
$$- \sigma_{b_s}^2 N \int_{-\infty}^{+\infty} \frac{s_{av}}{\sigma_{b_s}^2} \exp\left[-\frac{(x - s_{av})^2}{2\sigma_{b_s}^2}\right] dx$$

and it then follows that

$$S = \frac{N}{\sigma_{b_s}(2\pi)^{1/2}} \int_{-\infty}^{+\infty} x \exp\left[-\frac{(x - s_{av})^2}{2\sigma_{b_s}^2}\right] dx$$
$$= -\frac{\sigma_{b_s} N}{(2\pi)^{1/2}} \int_{-\infty}^{+\infty} d\left\{\exp\left[-\frac{(x - s_{av})^2}{2\sigma_{b_s}^2}\right]\right\}$$
$$+ \frac{N s_{av}}{\sigma_{b_s}(2\pi)^{1/2}} \int_{-\infty}^{+\infty} \exp\left[-\frac{(x - s_{av})^2}{2\sigma_{b_s}^2}\right] dx$$

or

$$S = N s_{av} \tag{3.39}$$

It follows from equations (3.33) and (3.34) that the specific phase contact surface area is given by the formula

$$a = \frac{S\phi_{av}}{V_b} \tag{3.40}$$

Substituting the values of V_b and S from equations (3.38) and (3.39) into equation (3.40), we obtain

$$a = \frac{Ns_{av}\phi_{av}}{NV_{av}} = \frac{6\pi d_{av}^2\phi_{av}}{\pi d_{av}^3} = \frac{6\phi_{av}}{d_{av}}$$

Hence in this case also, the standard deviations of volume, σ_{b_v}, and surface area, σ_{b_s}, have no effect on the specific phase contact surface area.

Additional investigations of dispersion in the two-phase mixture in rapid bubbling[43,44,46,50,51] have shown that the bubble diameter density distribution obeys the logarithmic-normal distribution law[48,52] described by the equation

$$f(y) = \frac{1}{y \ln \sigma_b (2\pi)^{1/2}} \exp\left[-\frac{(\ln y - \ln d_{av})^2}{2(\ln \sigma_b)^2}\right]$$

This distribution law is very common and is observed in, for example, solvent extraction equipment (packed columns, rotary disk columns, stirred reactors[51-54]), in equipment for aerosol particulate distribution,[55] in emulsification,[56] and in the distribution of powdered materials[57] in spray towers.[11]

In the case of the logarithmic-normal distribution, the entire volume of all the bubbles enclosed in the two-phase mixture will be

$$V_b = \tfrac{1}{6}\pi N \int_0^{+\infty} 2y^3 f(y)\, dy$$

$$= \frac{2\pi N}{6(2\pi)^{1/2}\sigma_1} \int_0^{+\infty} y^2 \exp\left[\frac{(\ln y - b)^2}{2\sigma_1^2}\right] dy \qquad (3.41)$$

where $\sigma_1 = \ln \sigma_b$ and $b = \ln d_{av}$. By substitution of variables, $x = \ln y$, $dx = dy/y$, and introduction of the notation $k = (N/3\sigma_1)(\pi/2)^{1/2}$, we reduce equation (3.41) to the form

$$V_b = k \int_{-\infty}^{+\infty} \exp\left\{-\frac{[x - (b + 3\sigma_1^2)]^2 - (6b\sigma_1^2 + 9\sigma_1^4)}{2\sigma_1^2}\right\}$$

$$= k \exp\left(\frac{6b\sigma_1^2 + 9\sigma_1^4}{2\sigma_1^2}\right) \int_{-\infty}^{+\infty} \exp\left\{-\frac{[x - (3\sigma_1^2 + b)]^2}{2\sigma_1^2}\right\} dx$$

Integrating this equation, we obtain

$$V_b = k \exp(3b + \tfrac{9}{2}\sigma_1^2)(2\pi)^{1/2}\sigma_1 = \frac{\pi N}{3} \exp(3b + \tfrac{9}{2}\sigma_1^2)$$

or

$$V_b = \frac{\pi N}{3} \exp(3 \ln d_{av}) \exp(\tfrac{9}{2} \ln^2 \sigma_b) = \frac{\pi N}{3} d_{av}^3 \sigma_b^{(9/2)\ln \sigma_b} \qquad (3.42)$$

Now the total surface area of all the bubbles enclosed in the two-phase mixture amounts to

$$S = 2\pi N \int_0^{\infty} y^2 f(y)\, dy = \frac{2\pi N}{\sigma_1(2\pi)^{1/2}} \int_0^{\infty} y \exp\left[-\frac{(\ln y - b)^2}{2\sigma_1^2}\right] dy$$

Repeating a substitution of variables, as before, we transform this equation to the form

$$S = \frac{2\pi N}{\sigma_1(2\pi)^{1/2}} \int_{-\infty}^{+\infty} \exp\left\{-\frac{[x - (b + 2\sigma_1^2)^2]^2 - (4b\sigma_1^2 + 4\sigma_1^4)}{2\sigma_1^2}\right\} dx$$

$$= \frac{2\pi N}{\sigma_1(2\pi)^{1/2}} \exp\left(\frac{4b\sigma_1^2 + 4\sigma_1^4}{2\sigma_1^2}\right) \int_{-\infty}^{+\infty} \exp\left\{-\frac{[(x - (b + 2\sigma_1^2)]^2}{2\sigma_1^2}\right\} dx$$

$$= \frac{2\pi N}{\sigma_1(2\pi)^{1/2}} \exp(2b + 2\sigma_1^2)(2\pi)^{1/2}\sigma_1$$

or

$$S = 2\pi N \exp(2 \ln d_{av}) \exp(2 \ln^2 \sigma_b) = 2\pi N d_{av}^2 \sigma_b^{2 \ln \sigma_b} \tag{3.43}$$

The specific phase contact surface area is then obtained from equations (3.42) and (3.43),

$$a = \frac{S}{V_t} = \frac{S\phi_{av}}{V_b} = \frac{6\phi_{av}}{d_{av}} \sigma_b^{-(5/2)\ln \sigma_b}$$

Since, in most practical applications, $\sigma_b \simeq 1.0$ mm (0.04 in.), we have $\sigma_b^{-(5/2)\ln \sigma_b} \simeq 1$, and the specific phase contact surface area for the case of logarithmic-normal bubble diameter distribution can be determined to an accuracy sufficient for most applications by using equation (3.36).

In addition to the normal distribution law and the logarithmic-normal distribution law, it has been demonstrated[58] that the density distribution

$$f(y, \alpha) = 4\left(\frac{\alpha^3}{\pi}\right)^{1/2} y^2 \exp(-\alpha y^2)$$

where

$$\alpha = \left[\frac{16(\pi N)^{1/2}}{3\phi}\right]^{2/3} > 0$$

also agrees closely with experiments performed on vapor–liquid and liquid–liquid systems,[59,60] and the specific phase contact surface area can be calculated in this case by using equation (3.36), as in the case of normal and logarithmic-normal distributions. Thus it is clear that equation (3.36) for the specific phase contact surface area is useful over a wide range of systems and conditions.

Figure 3.6 Bubble diameter distributions for different gas superficial velocities. (From A. M. Kashnikov, Candidate's dissertation, D. I. Mendeleev Moscow Institute of Chemical Technology, 1965.)

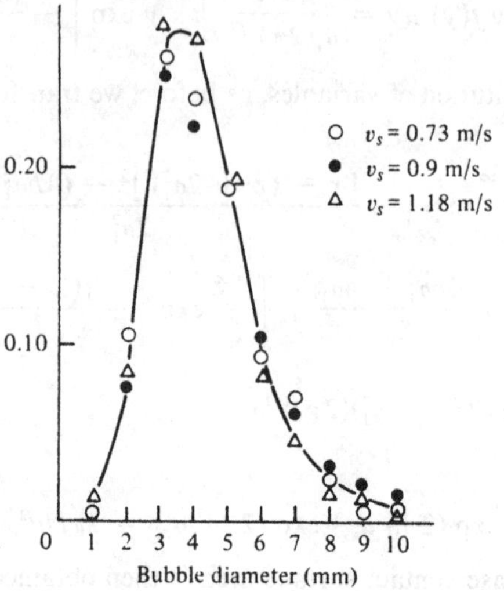

Figure 3.7 Bubble diameter distributions for different flow rates: 1, 3m^3/h; 2, 6.3 m^3/h; 3, 9 m^3/h. (1) Column, 40 × 60 mm; grid plate, 8 to 16 mm; free area, F_a, 18.8 percent. (2) Column, 120 × 120 mm; grid plate, 2 to 12 mm; F_a = 12.0 percent. (3) Column, 120 × 120 mm; grid plate, 2 to 12 mm; F_a = 12.0 percent. (From A. M. Kashnikov, Candidate's dissertation, D. I. Mendeleev Moscow Institute of Chemical Technology, 1965.)

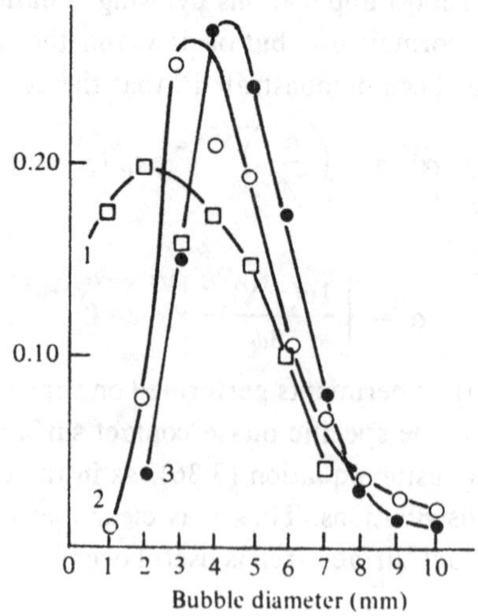

Before this formula can be used, however, the void fraction, ϕ_{av}, and average bubble size, d_{av}, must be obtained. A value for ϕ_{av} can often be obtained from flow conditions (gas and liquid flow rates), and the appropriate value for the average bubble size has been experimentally determined for various flows. Figure 3.6 shows data obtained for an air–water flow system for various gas superficial velocities. Figure 3.7 shows data for the same flow system but for differing liquid flow rates. Note that as the flow rate increases, the distribution moves toward higher values of the bubble diameter. If data of the same form as that in Figures 3.6 and 3.7 are available for a given flow, equation (3.36) becomes useful for computing the specific phase contact area. Equation (3.26) can also be used in equation (3.36) to obtain

$$a = 1.17\left(\frac{\rho_f g}{\sigma}\right)^{0.6} h^{0.2}\left(\frac{\mu_f}{\mu_g}\right)^{0.25}(1-\phi)^{0.4}\phi^{0.95} \tag{3.44}$$

This equation is a semiempirical formula for the specific phase contact area in terms of known flow parameters.

References

1. Akselrod, L. S., and Dilman, V. V., *Zh. Prikl. Khim. (Moscow)*, *27*, 5 (1954).
2. Akselrod, L. S., and Dilman, V. V., *Zh. Prikl. Khim. (Moscow)*, *29*(12), 1803 (1956).
3. Davidson, L., and Amick, E. H., *AIChE J.*, *2*(2), 337 (1956).
4. Jackson, I. R. W., *Ind. Chem.*, *28*(3), 68 (1952).
5. Siemes, W., and Kauffmann, J. F., *Chem. Eng. Sci.*, *5*, 127 (1956).
6. Chang, R. C., Schoen, H. M., and Grove, C. S., *Ind. Eng. Chem.*, *48*(11), 2035 (1956).
7. Spells, K. B., *Trans. Inst. Chem. Eng.*, *32*, 167 (1954).
8. Stabnikov, V. N., *Khim. Mashinostr. (Moscow)*, no. 1, 17 (1938).
9. Shabalin, K. N., *Gas Friction against Liquid in Absorption Processes*, Metallurgizdat, Moscow, 1943.
10. Ramm, V. M., *Absorption of Gases*, Chimia, Moscow, 1966. (In Russian.)
11. Kolmogoroff, A. N., *Dokl. Acad. Nauk SSSR*, *31*(2), 99 (1941).
12. Landau, L. D., and Lifshitz, E. M., *Fluid Mechanics*, Pergamon Press, London, 1959.
13. Hinze, I. O., *Turbulence*, McGraw-Hill, New York, 1959.
14. Pavlov, V. P., Candidate's dissertation, Moscow Institute of Chemical Engineers, 1963. (In Russian.)
15. Levich, V. G., *Physicochemical Hydrodynamics*, Prentice-Hall, Englewood Cliffs, N.J., 1962.
16. Lamb, G., *Hydrodynamics*, Dover Publications, New York, 1945.
17. Golovin, A. M., Levich, V. G., and Tolmachev, *Prikl. Mat. Teor. Fiz.*, *2*, 63 (1966).
18. Golovin, A. M., *Prikl. Mat. Teor. Fiz.*, *6* (1967).
19. Miasnikov, V. P., and Levich, V. G., *Khim. Prom. (Moscow)*, no. 6 (1966).
20. Gupalo, Yu. P., *Inzh.-Fiz. Zh.*, *1*, 16 (1962).
21. Leva, M., *Fluidization*, McGraw-Hill, New York, 1959.
22. Happle, J., *AIChE J.*, *4*, 197 (1958).
23. Uchida, S., *Ind. Eng. Chem.*, *46*, 1194 (1958).
24. Marrucci, G., *Ind. Eng. Chem. Fund.*, *2*, 224 (1965).

25. Halberstadt, S., and Praussnitz, P. H., *Angew. Chem.*, *2*(43), 970 (1940).
26. Praussnitz, P. H., *Kolloid-Z.*, *50*, 183 (1930).
27. Praussnitz, P. H., *Kolloid-Z.*, *76*, 227 (1936).
28. Praussnitz, P. H., *Kolloid-Z.*, *104*, 246 (1943).
29. Rudolph, H., *Kolloid-Z.*, *60*, 308 (1932).
30. Siemes, W., and Borchers, E., *Chem. Eng. Sci.*, *12*(2), 77 (1960).
31. Kolber, H., Borchers, E., and Langemann, H., *Chem.-Ing.-Tech.*, *33*(10), 668 (1961).
32. Siemes, W., and Borchers, E., *Chem.-Ing.-Tech.*, *28*, 783 (1956).
33. Houghton, G., McLean, P., and Ritchie, D., *Chem. Eng. Sci.*, *7*, 40 (1957).
34. Schnurmann, R., *Z. Phys. Chem.*, *143*, 456 (1929).
35. Schnurmann, R., *Kolloid-Z.*, *80*, 148 (1937).
36. Aizenbud, M. B., Candidate's dissertation, Moscow Institute of Chemical Engineers, 1961. (In Russian.)
37. Smirnov, N. I., and Poluta, S. E., *Zh. Fiz. Khim.*, *22*(11), 1208 (1949).
38. Helmholtz, H. L. F. von, *Verh. Naturhist. Med. Vereins.* (Oct. 30, 1868).
39. Rayleigh, Lord, *Phil. Mag.*, *26*, 776 (1913).
40. Sterman, L. S., *Zh. Tekh. Fiz.*, *26*(7), 1512 (1956).
41. Sterman, L. S., and Surnov, A. B., *Teploenergetika*, *8*, 39 (1955).
42. Vinokur, Ya. G., and Dil'man, V. V., *Khim. Prom. (Moscow)*, no. 7, 619 (1959).
43. Radikovskii, V. M., Candidate's dissertation, D. I. Mendeleev Moscow Institute of Chemical Technology, 1965. (In Russian.)
44. Kashnikov, A. M., Candidate's dissertation, D. I. Mendeleev Moscow Institute of Chemical Technology, 1965. (In Russian.)
45. Calderbank, P. H., and F. Rennie, *Trans. Inst. Chem. Eng.*, *40*, 1, 3 (1962).
46. Siemes, W., *Chem.-Ing.-Tech.*, *26*(11), 639 (1954).
47. Viviorovski, M. M., Dil'man, V. V., and Aizenbud, M. B., *Khim. Prom. (Moscow)*, no. 3, 204 (1965).
48. Gnedenko, V. V., *Treatise of Probability Theory*, Fizmatgis, Moscow, 1961. (In Russian.)
49. Siemes, W., and Ganther, K., *Chem.-Ing.-Tech.*, *28*(6), 389 (1956).
50. Koto, T., *Kagaku Kogaku* [*Chem. Eng. (Jap.)*], *26*, no. 11, 114 (1962). Abstracted in *Ref. Zh. Khim.*, no. 16, 110 (1963).
51. Rodionov, A. I., Kashnikov, A. M., and Radikovskii, V. M., *Khim. Prom. (Moscow)*, no. 10, 17 (1964).
52. Radionov, D. A., *Distribution Functions of the Content of Elements and Minerals in Igneous Rocks*, Nauka, Moscow, 1964. (In Russian.)
53. Thornton, J. D., *Ind. Chem.*, *39*(12), 632 (1963).
54. Kagan, S. Z., Doctoral dissertation, D. I. Mendeleev Moscow Institute of Chemical Technology, 1965. (In Russian.)
55. Yamaguchi, I., Kabuta, S., and Nagata, S., *Chem. Eng. (Jap.)*, *27*(8), 576 (1963).
56. Babanov, B. M., Candidate's dissertation, D. I. Mendeleev Moscow Institute of Chemical Technology, 1960. (In Russian.)
57. Bezemer, C., and Schwarz, N., *Kolloid-Z.*, *146*(1–3), 145 (1956).
58. Avdeev, N. Ya., *Analytical Method for Sedimentometric Dispersion Analysis Calculations*, Rostov University, Rostov-on-Don, 1964. (In Russian.)
59. Gel'perin, N. I., Sklokin, L. I., and Assmus, M. G., *Teor. Osn. Khim. Tekhnol.*, *1*(4), 463 (1967).
60. Gal-Or, B., and Hoelscher, H. E., *AIChE J.*, *12*(3), 499 (1966).

4

The dynamic two-phase flow

4.1 Introduction

In earlier chapters we considered two-phase flows in which we have bubble-by-bubble flow, and flows in which we have mass bubbling. We now consider the general behavior of two-phase flows in an attempt to obtain conclusions that will have applicability over a wide range of gas and liquid flow rates.

To simplify analysis of the hydrodynamics of the bubbling mixture, and to indicate at which flow rates the viscous and inertia effects appear, we first consider two limiting cases of bubbling regimes: rapid bubbling and slow bubbling. The regime of ***rapid bubbling*** is applicable, for example, in shallow liquid pools (about 3 to 15 cm, 1.2 to 5.9 in., deep) and for relatively large superficial gas velocities (more than 0.5 m/s, 20 in./s), in other words, for values of Froude number, Fr, greater than unity. The Froude number is defined as $\mathrm{Fr} = v_s^2/gh$, where v_s is the superficial gas velocity, h the undisturbed liquid height, and g the acceleration due to gravity. This kind of liquid pool is found in the use of sieve plates, bubble plates, and so on, in distillation and absorption processes. On the other hand, the regime of ***slow bubbling*** is characterized by a deep pool, which is often termed a ***bubble column*** (with a depth more than 15 cm, 5.9 in.) and by superficial gas velocity less than 0.2 m/s (7.9 in./s), in other words, for values of Froude number much less than unity. This regime occurs extensively in gas–liquid reactors, for example in reactors for oxidation, hydrogenation, fermentation processes, and so on.

Now, in the dynamic two-phase mixture, the flow consists of numerous discrete elements (bubbles or droplets) enclosed in a continuous fluid (liquid or gas, depending on the gas void fraction).

It is known[1–5] that formulation of the fundamental equations of hydrodynamics is possible only for single discrete elements of the two-phase flow, and therefore in a theoretical analysis of the total two-phase flow, we adopt the calculus of variations as one of the few methods of obtaining information on the flow characteristics. By taking account of the geometric characteristics of the system in which the bubbling process takes place, the physical properties of the gas and liquid, the gas and liquid flow rates, and also the boundary conditions, we can determine the fundamental characteristics of the dynamic two-phase mixture: for example, the distribution of the gas and liquid in the flow, the void fraction, and so on.

As an example of how the physical properties of the gas and liquid may affect the dynamics of the flow, consider a gas–liquid flow in which we have surface-active substances present; for example, some kind of contaminant present at the interface of the bubbles and the liquid. The presence of this substance can result in a change in the capillary force as well as in the appearance of additional surface forces, leading to a significant change in the hydrodynamical regime of the bubble process. Now, when surface-active substances are present in the gas–liquid mixture, the bubble surfaces are covered by a monolayer of surface-active molecules, and the energy dissipation in this layer is known to be small compared to the dissipation of energy in the two-phase mixture.[6]

The form drag, F_0, of a bubble, considered as a sphere for simplicity, will be

$$F_0 \simeq C_D \frac{\pi r_b^2 \rho_f v_r^2}{2}$$

where C_D is the drag coefficient (at large Reynolds numbers, $C_D \simeq 0.5 =$ const.), ρ_f the liquid density, r_b the bubble radius, and v_r the velocity of the bubble relative to the liquid.

As a result of the flow of the liquid around the bubble, a saturated monolayer of the surface-active substance is accumulated at the rear of the bubble and this area, s_0 (less than πr_b^2), of the bubble remains covered by a nondeforming monolayer of the surface-active substance while the bubble rises. During the entire period of motion this monolayer on the rear of the bubble will cause the liquid velocity on the bubble surface to become zero, just as on the surface of a solid body. Separation of liquid from the bubble surface then occurs, accompanied by a new factor in the form drag.[7]

With this in mind, the ratio of viscous drag F_D to form drag F_0 is

$$\frac{F_D}{F_0} \simeq \frac{12\pi\mu_f v_r r_b}{C_D s_0 \rho_f v_r^2} = \frac{12\pi r_b^2}{C_D s_0 \mathrm{Re}}$$

where μ_f is the liquid dynamic viscosity and $\mathrm{Re} = \rho_f v_r r_b/\mu_f$ is the Reynolds number. This ratio will be less than unity when

$$s_0 > \frac{12\pi r_b^2}{C_D \mathrm{Re}} = \frac{3 s_b}{C_D \mathrm{Re}}$$

where s_b is the bubble surface area. Hence

$$\frac{s_0}{s_b} > \frac{3}{C_D \mathrm{Re}}$$

This condition is fulfilled in the case where the bubble surface s_0 covered by a monolayer of surface-active substances is more than a fraction $6/\mathrm{Re}$ of the total bubble surface area, when $\mathrm{Re} \gg 6$, and in this case the problem of the motion of a gas bubble, with surface contaminants, in a continuous liquid may be reduced to the problem of the motion of a solid sphere in an ideal infinite liquid,[6,8] considerably simplifying the analysis.

4.2 Rapid bubbling with ideal liquid

In this section we obtain a very important characteristic of any two-phase flow, the gas void fraction, ϕ, defined as the ratio of gas volume to total volume. We will obtain the void fraction as a function of system properties for the case of rapid bubbling of gas in liquid when the viscosity of the liquid is negligible.

Now, for moderate and large values of the Reynolds number, which is typical of bubbling processes utilized in commercial distillation and absorption equipment, the viscous forces, being small compared to the inertia forces, have little influence on the hydrodynamics of the gas–liquid mixture and therefore may be ignored in a study of such flows. Theoretical[5,6] and experimental studies[9–12] concerning the influence of viscosity on the behavior of the two-phase flow in a bubbling process confirm this conclusion, giving validity to our assumption of ideal liquid.

There have been many detailed experimental studies[13–37] on mass-bubbling operations, for instance, in shallow pools, where three groups of parameters influencing the gas–liquid flow are apparent:

1. Geometrical characteristics of the flow system.
2. Physical properties of the gas and liquid.
3. Dynamic factors.

To the first group, for example, belong the height of the liquid in a flow system where the liquid is static, the diameter of the equipment, and the geometry of the gas-distributing device (e.g., the orifices). To the second group belong the viscosity and surface tension of the gas and liquid, and to the third group belong the gas and liquid flow rates.

Let us now examine a two-phase flow that is sufficiently distant from the walls of the equipment and from the gas-distribution devices that any effects they may have on the flow are eliminated.

We can then take the two-phase flow to be one-dimensional, and at a distance x from the gas inlet we consider a differential layer located perpendicular to the direction of motion of the gas stream, with thickness dx. The fraction of the volume of this differential layer occupied by the gas is

$$\phi = \tfrac{4}{3}\pi r_b^3 n \tag{4.1}$$

where r_b is the mean radius of the bubbles, n denotes the number of bubbles per unit volume, and ϕ is, of course, the gas void fraction.

Let us make an energy balance of a unit cross section of the differential layer during the bubbling process:

$$dE = dE_1 + dE_2 + dE_3 \tag{4.2}$$

where dE is the total energy of the layer, dE_1 the potential energy of the liquid, dE_2 the kinetic energy of the layer, and dE_3 the energy of surface tension of the bubbles in the layer. Taking into account the fact that buoyancy forces are balanced by the resistance forces, we will assume that the potential energy of the gas does not change during the motion of a bubble. The potential energy of the liquid per unit cross-sectional area is

$$dE_1 = (1 - \phi)\rho_f g x \, dx \tag{4.3}$$

where ρ_f is the liquid density and x the direction of the flow (vertical).

The kinetic energy of the layer is comprised of the kinetic energy of the bubbles of gas [equal to $(\rho_g v^2/2)\tfrac{4}{3}\pi r_b^3 n \, dx$, where v is the velocity of the gas in the differential layer and ρ_g the density of the gas] and the kinetic energy of the liquid carried along by the bubbles. In order to take both components into account, it is simplest to consider each bubble of gas as a rigid sphere, the mass of which consists of the mass of the gas enclosed in it, plus the mass of a volume of liquid equal to half the volume of the bubble (the "additional mass").[6,8,38] The total kinetic energy in the layer is then

$$dE_2 = \left(\rho_g + \frac{\rho_f}{2}\right)\phi \frac{v^2}{2}\, dx$$

where we have used equation (4.1). From mass conservation of the gas, we can write

$$\phi v = v_s \tag{4.4}$$

where v_s is the gas velocity in the region of the flow free of liquid; in other words, it is the gas superficial velocity.

Then we find that

$$dE_2 = \left(\rho_g + \frac{\rho_f}{2} \right) \frac{v_s^2}{2\phi} \, dx \tag{4.5}$$

Finally, the energy of surface tension can be calculated using the formula

$$dE_3 = 4\pi r_b^2 \sigma n \, dx = \frac{3\phi}{r_b} \sigma \, dx \tag{4.6}$$

where σ is the surface tension coefficient and we have used equation (4.1).

If we substitute into equation (4.2) the expressions for dE_1, dE_2, and dE_3 from equations (4.3), (4.5), and (4.6), respectively, we obtain

$$dE = \left[(1 - \phi)\rho_f g x + \left(\rho_g + \frac{\rho_f}{2} \right) \frac{v_s^2}{2\phi} + \frac{3\sigma}{r_b} \sigma \right] dx \tag{4.7}$$

The total energy of a two-phase mixture of height x_1 is then

$$E = \int_0^{x_1} \left[(1 - \phi)\rho_f g x + \left(\rho_g + \frac{\rho_f}{2} \right) \frac{v_s^2}{2\phi} + \frac{3\sigma}{r_b} \phi \right] dx \tag{4.8}$$

It is a fundamental axiom of physical theory that for any system the steady state of the system is that for which the available energy of the system is at a minimum. Hence the steady-state distribution of the gas void fraction ϕ occurs when this energy is at a minimum. Thus to find ϕ it is necessary to determine the minimum of the integral in equation (4.8). This problem[39] can be reduced to that of finding the minimum of the integral E under the condition of invariability of the amount of liquid in the system, in other words with

$$h = \int (1 - \phi) \, dx = \text{const.} \tag{4.9}$$

where h is the static liquid height. Equation (4.9) is a boundary condition on the variation of equation (4.8).

Now consider the method of the calculus of variations, beginning with the following function of a function:

$$E = \int f[\phi(x), \phi'(x), x] \, dx \tag{4.10}$$

where f is some known function of its argument, $\phi(x)$ an unknown function, and $\phi'(x) \equiv d\phi/dx$. The calculus of variations is a method to obtain a function $\phi(x)$ such that the variation of E is zero,

$$\delta E = 0 \tag{4.11}$$

which also means that the function E is at a minimum (or possibly a maximum) for that function $\phi(x)$. This is done subject to the constraint that

$$\psi(\phi, x) = 0 \tag{4.12}$$

where ψ is some known function. Using the method of variations on equations (4.10), (4.11), and (4.12), it can be shown that we obtain the Euler equation:

$$\frac{\partial f}{\partial \phi} - \frac{d}{dx}\left(\frac{\partial f}{\partial \phi'}\right) + \lambda \frac{\partial \psi}{\partial \phi} - \lambda \frac{d}{dx}\left(\frac{\partial \psi}{\partial \phi'}\right) = 0 \tag{4.13}$$

where λ is the Lagrange multiplier and is determined by the constraint equation (4.12). Equation (4.13), which is usually differential but is sometimes algebraic, is then used to obtain that function $\phi(x)$ which gives a minimum of E.

When we apply this analysis to equation (4.8), with the constraint equation (4.9), we find that

$$f(\phi, \phi', x) = (1 - \phi)\rho_f g x + \left(\rho_g + \frac{\rho_f}{2}\right)\frac{v_s^2}{2\phi} + \frac{3\sigma}{r_b}\phi$$

$$\psi(\phi, x) = 1 - \phi$$

and for the Euler equation, we obtain

$$-\rho_f g x - \left(\rho_g + \frac{\rho_f}{2}\right)\frac{v_s^2}{2\phi^2} + \frac{3\sigma}{r_b} - \lambda = 0 \tag{4.14}$$

We have ignored the variation due to the unknown upper limit, x_1, in the calculus-of-variation analysis given above. This upper limit is the point where the two-phase nature of the flow breaks down, in other words when $\phi = 1$. We now show that it is acceptable to ignore the variation due to x_1.

The problem of the floating limit of an integral is called the *problem with natural boundary conditions.*[39] If we take

$$E = \int_0^{x_1} f(\phi, \phi', x)\, dx$$

and we need to find the $\phi(x)$ that gives an extremum of E when $\phi(0) = \phi_0$ is set (the lower limit is fixed) and $\phi(x_1)$ is not set (no condition is made on the upper limit), then the value of δE will be given not only by the variation of f under the integral but also the variation at the upper limit. The most general expression of the variation δE can be written as

$$\delta E = [f\phi'\, \delta\phi]_0^{x_1} + \int_0^{x_1} \left[f\phi - \frac{d}{dx}(f\phi')\right] \delta y\, dx$$

where we are here not taking into account system constraints. In the case we are now considering, $f = f(\phi, x)$; in other words, we do not have a term $\phi' = d\phi/dx$, $f\phi' \equiv 0$, and the term integrated in brackets is equal to zero,

$[f\phi'\,\delta\phi]_0^{x_1} = 0$. Therefore, the floating limit of x has no effect and Euler's equation is as shown in equation (4.13).

Now, since the derivative ϕ' does not enter into equation (4.13), equation (4.14) is not differential, but algebraic, and we find

$$\phi(x) = \left[\frac{v_s^2(\rho_g + \rho_f/2)}{2[(3\sigma/r_b) - \rho_f g x - \lambda]}\right]^{1/2} \tag{4.15}$$

Further, since the derivative $\phi'(x)$ does not appear in equation (4.14), we cannot impose any additional conditions on the values of $\phi(x)$ at the limits of the interval 0 to x_1, and therefore this function may become discontinuous at the ends of the interval. However, from physical considerations it must be the case that the function $\phi(x)$ is continuous at all values of x between the limits, and therefore can be determined from equation (4.15). The value of x_1 can now be determined from the equality

$$1 = \left[\frac{v_s^2(\rho_g + \rho_f/2)}{2[(3\sigma/r_b) - \rho_f g x_1 - \lambda]}\right]^{1/2} \tag{4.16}$$

We still have to eliminate the undetermined Lagrange multiplier λ, and to do this we use ϕ from equation (4.15) in the constraint equation (4.9).

$$h = \int_0^{x_1} (1 - \phi)\,dx = \int_0^{x_1} \left\{1 - \left[\frac{v_s^2(\rho_g + \rho_f/2)}{2[(3\sigma/r_b) - \rho_f g x - \lambda]}\right]^{1/2}\right\} dx$$

If we integrate this expression and eliminate λ using equation (4.16), we get

$$h = x_1 + \frac{\rho_g + \rho_f/2}{\rho_f}\frac{v_s^2}{g} - \frac{2}{\rho_f g}\left\{\left[\frac{(\rho_g + \rho_f/2)v_s^2}{2}\right]\left[\rho_f g x_1 + \frac{(\rho_g + \rho_f/2)v_s^2}{2}\right]\right\}^{1/2} \tag{4.17}$$

Let us define the term

$$A = 1 + \frac{2\rho_f g x_1}{(\rho_g + \rho_f/2)v_s^2}$$

which means that

$$x_1 = \frac{1}{\rho_f g}\left(\rho_g + \frac{\rho_f}{2}\right) v_s^2(A - 1)$$

If we substitute this value of x_1 into equation (4.17), we obtain, after simplifying,

$$\tfrac{1}{2}A - \tfrac{1}{2} + 1 - A^{1/2} = \frac{\rho_f g h}{(\rho_g + \rho_f/2)v_s^2} \tag{4.18}$$

We now define

$$B = \frac{2\rho_f gh}{(\rho_g + \rho_f/2)v_s^2}$$

and equation (4.18) takes the form

$$A - 2A^{1/2} + 1 - B = 0$$

and, solving with respect to $A^{1/2}$,

$$A^{1/2} = 1 \pm B^{1/2}$$

We now replace A and B, and, after some manipulation, we find that

$$\left[\rho_f g x_1 + \frac{(\rho_g + \rho_f/2)v_s^2}{2}\right]^{1/2} = \left[\frac{(\rho_g + \rho_f/2)v_s^2}{2}\right]^{1/2} \pm (\rho_f gh)^{1/2}$$

In the equation above we select the + sign, since when $v_s = 0$, the condition $x_1 = h$ must be fulfilled.

The equation for the height of the dynamic two-phase mixture in terms of the static liquid level is then

$$x_1 = 2\left[\frac{h(\rho_f + \rho_f/2)v_s^2}{2\rho_f g}\right]^{1/2} + h$$

Assuming that $\rho_g \ll \rho_f$, we can write this in the form

$$x_1 = \left(\frac{v_s^2 h}{g}\right)^{1/2} + h = h(\mathrm{Fr}^{1/2} + 1) \tag{4.19}$$

where Fr is the Froude number. It follows from equation (4.19) that the height of the dynamic two-phase mixture varies linearly with the superficial velocity of the gas, v_s.

Let us now determine the value of $\phi(x)$. From equation (4.16) we can write

$$\lambda = \frac{3\sigma}{r_b} - \rho_f g\left(x_1 + \frac{v_s^2}{4g}\right)$$

where we have again used the inequality $\rho_g \ll \rho_f$. If we substitute the value of x_1 from equation (4.19), we get

$$\lambda = \frac{3\sigma}{r_b} - \rho_f g\left(h\mathrm{Fr}^{1/2} + h + \frac{v_s^2}{4g}\right)$$

Taking into consideration that

$$\frac{hv_s^2}{4gh} = \frac{h}{4}\mathrm{Fr}$$

we can write

$$\frac{3\sigma}{r_b} - \lambda = \rho_f g h \left(\mathrm{Fr}^{1/2} + \frac{\mathrm{Fr}}{4} + 1 \right)$$

and using this in equation (4.15),

$$\phi(x) = \left[\frac{\rho_f v_s^2}{4[\rho_f g h(\mathrm{Fr}^{1/2} + \mathrm{Fr}/4 + 1) - \rho_f g x]} \right]^{1/2}$$

or

$$\phi(x) = \left[\frac{\mathrm{Fr}}{4[\mathrm{Fr}^{1/2} + (\mathrm{Fr}/4) + 1 - (x/h)]} \right]^{1/2} \tag{4.20}$$

If we designate Fr/4 by F and substitute F in equation (4.20), we finally have

$$\phi(x) = \left[\frac{F}{(1 + F^{1/2})^2 - x/h} \right]^{1/2} \tag{4.21}$$

Thus we have obtained the local gas void fraction $\phi(x)$ solely in terms of the Froude number, Fr, and the location, x, for the case of an ideal liquid. Equation (4.21) is illustrated[40] for a particular flow configuration in Figure 4.1. Note the close correspondence to experimental data. All the results we will obtain in this section, where we are assuming ideal-fluid conditions, will

Figure 4.1 Void fraction against height.

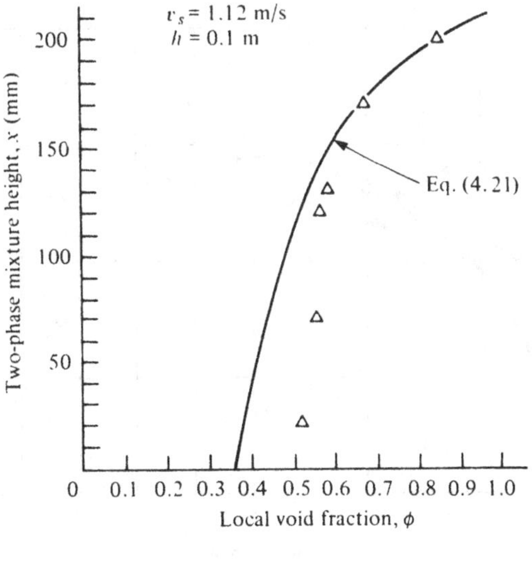

be shown to give such results for tests in equipment that has a typical dimension less than 20 cm (7.9 in.). It follows that the gas content (or void fraction) does not depend on the form and size of the bubbles of gas, since the bubble radius does not enter into the equation. Consequently, this equation can be used for calculations over a broad range of velocities of the gas.

Let us now determine the average gas content (or void fraction) of the two-phase mixture:

$$\phi_{av} = \frac{\int_0^{x_1} \phi(x)\,dx}{\int_0^{x_1} dx} = \frac{\int_0^{x_1} \{F^{1/2}/[(1+F^{1/2})^2 - x/h]^{1/2}\}\,dx}{x_1}$$

$$= -\frac{2h}{x_1} F^{1/2} \left\{ \left[(1+F^{1/2})^2 - \frac{x_1}{h} \right]^{1/2} - (1+F^{1/2}) \right\}$$

After substitution into this equation of the expression for x_1 from equation (4.19), we obtain

$$\phi_{av} = -2\frac{F^{1/2}}{1+2F^{1/2}}(F^{1/2} - 1 - F^{1/2})$$

$$= 2\frac{F^{1/2}}{1+2F^{1/2}} = \frac{1}{1+(1/2F^{1/2})}$$

Figure 4.2 Mean void fraction against Froude number for air–water mixture.

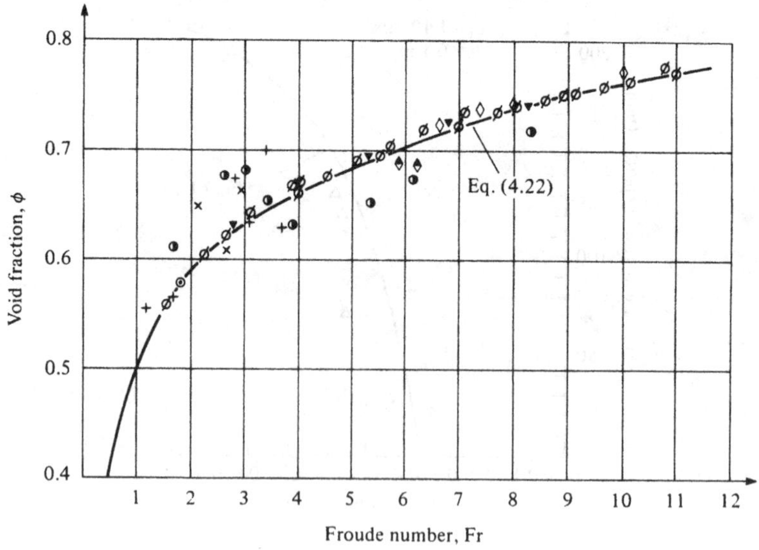

and, finally, substitution of Fr/4 in place of F yields

$$\phi_{av} = \frac{1}{1 + (1/Fr^{1/2})} \qquad (4.22)$$

We obtain the result that the average void fraction in a rapidly bubbling flow, with an ideal liquid, is given by a very simple function of the Froude number alone. Equation (4.22) is illustrated in Figure 4.2 for an air–water system together with data[41] obtained for various orifice and seive configurations. Clearly, equation (4.22) gives very good results irrespective of flow geometries. Figure 4.3 shows that, over a wide range of different gases and liquids (represented by different symbols),[42] that equation (4.22) still holds good for flow systems whose typical dimension is less than about 200 mm (7.9 in.).

The average relative density of the gas–liquid mixture (the *specific gravity of the foam*) is determined by the formula

$$\psi = \frac{h}{x_1} \qquad (4.23)$$

where ψ is now the average relative density. If we replace x_1 according to equation (4.19), we have

$$\psi = \frac{1}{1 + Fr^{1/2}} \qquad (4.24)$$

4.3 Rapid bubbling with real liquid

In this section we obtain a formula for the void fraction ϕ in rapid bubbling by using the same notion of energy minimizing, except that we will now allow for energy dissipation due to the viscosity of the liquid and due to

Figure 4.3 Mean void fraction against Froude number for various mixtures.

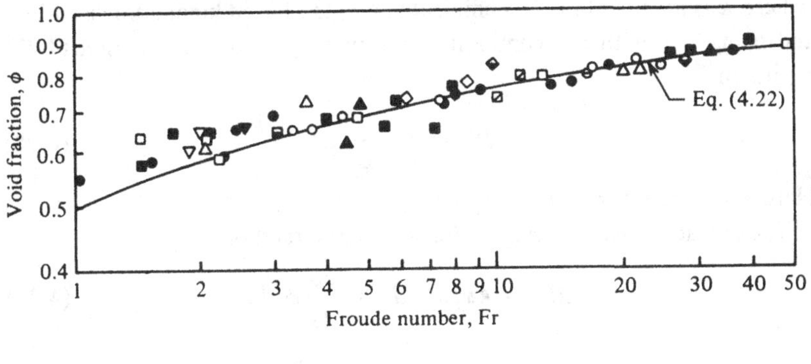

liquid turbulence. We again use the energy balance for the differential layer,

$$dE = dE_1 + dE_2 + dE_3 \tag{4.25}$$

except that dE_2 now represents the energy dissipation in the layer.

The potential energy dE_1 of the liquid can be found from the equation

$$dE_1 = (1 - \phi)\rho_f gx\, dx \tag{4.26}$$

which is simply equation (4.3) restated for convenience.

To find the dissipative energy dE_2 of the differential layer, we shall use the model of Section 3.3, of a uniform bubble distribution in the differential layer with each bubble being located in the center of a "spherical compartment" formed by adjacent bubbles. Calculation of drag based on this model agrees fairly well with experimental data on liquid–liquid and gas–liquid systems.[43–46] With this model the energy dissipation of the bubble is obtained by using equation (3.9),

$$dE_2 = 12\pi\mu_f r_b v^2 \frac{1 - \phi^{5/3}}{(1 - \phi)^2} \tau n\, dx$$

where τ is the time scale of turbulent eddies.

Taking into account that the number of bubbles per unit volume is

$$n = \phi / \tfrac{4}{3}\pi r_b^3$$

we may write the following equation for calculating the energy dissipation produced by motion of the bubbles in the layer,

$$dE_2 = \frac{9\mu_f v^2 \phi (1 - \phi^{5/3})}{r_b^2 (1 - \phi)^2} \tau\, dx \tag{4.27}$$

In Section 3.3 we obtained, for the time scale of the eddies most likely to have an effect on the bubble motion, a value given by equation (3.12):

$$\tau = \frac{k r_b^2 (1 - \phi)^2}{9\nu_f (1 - \phi^{5/3})}$$

(where k is an added mass coefficient and ν_f the liquid kinematic viscosity), and using this with the continuity condition $v_s = v\phi$ in equation (4.27) results in

$$dE_2 = k\rho_f \frac{v_s^2}{\phi}\, dx \tag{4.28}$$

Here k can be taken to be a constant.

The surface tension energy is defined by the relation

$$dE_3 = 4\pi r_b^2 \sigma n\, dx = \frac{3\sigma}{r_b} \phi\, dx \tag{4.29}$$

Substituting the values of dE_1, dE_2, and dE_3 as given by equations (4.26), (4.28), and (4.29) into equation (4.25), we obtain

$$dE = \left[(1 - \phi)\rho_f g x + k\rho_f \frac{v_s^2}{\phi} + \frac{3\sigma}{r_b}\phi\right] dx$$

Consequently, the total energy of a two-phase layer of height x_1 is

$$E = \int_0^{x_1} \left[(1 - \phi)\rho_f g x + k\rho_f \frac{v_s^2}{\phi} + \frac{3\sigma}{r_b}\phi\right] dx \tag{4.30}$$

As in Section 4.2, the equilibrium distribution of the gas void fraction ϕ is established when this energy is a minimum. Therefore, to determine ϕ, we must find the minimum of the integral (4.30).

Following the analysis of Section 4.2, the Euler equation (with the variation with respect to x_1 being neglected as before) in this case is

$$-\rho_f g x - k\rho_f \frac{v_s^2}{\phi^2} + \frac{3\sigma}{r_b} - \lambda = 0$$

Hence

$$\phi(x) = \left[\frac{k\rho_f v_s^2}{(3\sigma/r_b) - \lambda - \rho_f g x}\right]^{1/2} \tag{4.31}$$

With $\phi(x_1) = 1$, where x_1 is the point where the flow is totally gaseous, the value of x_1 may be determined from equation (4.31).

$$1 = \left[\frac{k\rho_f v_s^2}{(3\sigma/r_b) - \lambda - \rho_f g x_1}\right]^{1/2} \tag{4.32}$$

As before, we can eliminate the undetermined Lagrange multiplier by using the constraint equation (4.9).

$$h = \int_0^{x_1} (1 - \phi)\, dx = \int_0^{x_1} \left\{1 - \left[\frac{k\rho_f v_s^2}{(3\sigma/r_b) - \rho_f g x - \lambda}\right]^{1/2}\right\} dx$$

and, integrating, while making allowance for equation (4.32), we obtain

$$h = x_1 + \frac{k v_s^2}{g} - \frac{2}{\rho_f g}(k\rho_f v_s^2)^{1/2}(\rho_f g x_1 + k\rho_f v_s^2)^{1/2}$$

Solving this equation with respect to x_1, we get an equation suitable for calculating the height of a dynamic two-phase flow, allowing for a real liquid,

$$x_1 = h[2(k\mathrm{Fr})^{1/2} + 1] \tag{4.33}$$

It follows from this equation that, as for the ideal liquid, the height of the two-phase flow is a linear function of the superficial gas velocity v_s.

We can now determine the gas void fraction of the flow from equation (4.32). It follows that

$$\lambda = \frac{3\sigma}{r_b} - \rho_f g\left(x_1 + \frac{kv_s^2}{g}\right)$$

Substituting the value of x_1 given by equation (4.33) into this equation, we obtain

$$\lambda = \frac{3\sigma}{r_b} - \rho_f g\left[h(4k\mathrm{Fr})^{1/2} + h + \frac{kv_s^2}{g}\right]$$

Since

$$\frac{kv_s^2}{gh}h = kh\mathrm{Fr}$$

we may write

$$\frac{3\sigma}{r_b} - \lambda = \rho_f gh[(4k\mathrm{Fr})^{1/2} + 1 + k\mathrm{Fr}]$$

and, using this in equation (4.31) yields

$$\phi(x) = \left[\frac{k\rho_f v_s^2}{\rho_f gh[(4k\mathrm{Fr})^{1/2} + 1 + k\mathrm{Fr}] - \rho_f gx}\right]^{1/2}$$

On simplifying, we obtain

$$\phi(x) = \left[\frac{k\mathrm{Fr}}{(k\mathrm{Fr})^{1/2} + k\mathrm{Fr} + 1 - x/h}\right]^{1/2} \tag{4.34}$$

From this equation it follows that the gas content does not depend on the shape and the size of the gas bubbles, because the bubble radius does not occur in the equation. It is noteworthy that the analysis of rapid bubbling for a real liquid gives a formula for the void fraction very similar to that for an ideal liquid [equation (4.20)].

Let us determine the average gas content ϕ_{av} of the flow. This is given by

$$\phi_{av} = \frac{\int_0^{x_1}\phi(x)\,dx}{\int_0^{x_1}dx} = \int_0^{x_1}\frac{(k\mathrm{Fr})^{1/2}\,dx}{x_1[(4k\mathrm{Fr})^{1/2} + k\mathrm{Fr} + 1 - x/h]^{1/2}}$$

$$= -\frac{2h(k\mathrm{Fr})^{1/2}}{x_1}\left\{\left[(4k\mathrm{Fr})^{1/2} + k\mathrm{Fr} + 1 - \frac{x_1}{h}\right]^{1/2} - [(4k\mathrm{Fr})^{1/2} + k\mathrm{Fr} + 1]^{1/2}\right\}$$

Substituting x_1 from equation (4.33) and simplifying, we obtain

$$\phi_{av} = \frac{2(k\mathrm{Fr})^{1/2}}{1 + 2(k\mathrm{Fr})^{1/2}} \tag{4.35}$$

Equation (4.35) shows that for a real liquid (in other words, allowing for dissipative forces), we result with a simple formula for the mean void fraction in terms of the Froude number and a constant. Note the similarity between this equation and equation (4.22) for an ideal liquid.

The relative density of the two-phase flow is

$$\psi = \frac{h}{x_1}$$

and on substituting the expression for x_1 from equation (4.33), we find that

$$\psi = \frac{1}{1 + 2(k\mathrm{Fr})^{1/2}} \tag{4.36}$$

For air–water systems the constant k has been shown to be of the order unity. Hence we can write equations (4.33), (4.35), and (4.36) as

$$x_1 = h(2\mathrm{Fr}^{1/2} + 1) \tag{4.37}$$

for the height of the two-phase mixture,

$$\phi_{av} = \frac{2\mathrm{Fr}^{1/2}}{1 + 2\mathrm{Fr}^{1/2}} \tag{4.38}$$

for the average void fraction, and

$$\psi = \frac{1}{1 + 2\mathrm{Fr}^{1/2}} \tag{4.39}$$

for the relative density.

It follows from equations (4.37), (4.38), and (4.39) that the main hydrodynamic parameters of the flow in rapid bubbling depend neither on the physical properties of the liquid and gas nor on the geometric characteristics of the gas-distributing devices. They are determined by the ratio of the liquid and gas flow rates (where v_s is a characteristic gas velocity) and are characterized by the ratio between the inertia forces and the gravity forces; in other words, they are functions only of the Froude number. Equations (4.38) and (4.39) are illustrated in Figures 4.4 and 4.5, respectively, in comparison to experimental data[41,42] for air–water over a wide range of geometric conditions and flow rates. All the data are for flow systems whose characteristic dimension is greater than 200 mm (7.9 in.). Note the close agreement of experiment and theory.

4.4 Main parameters of the two-phase flow in the slow-bubbling regime

We now turn to a study of the main hydrodynamic parameters of the two-phase flow in the slow-bubbling regime. In this regime, geometrical specifications of the gas-distributing devices are known to have no practical influence on the gas void fraction of the two-phase flow.[11,47–49]

The gas void fraction ϕ by definition is a measure of the amount of gas

Figure 4.4 Mean void fraction against Froude number for air–water mixture.

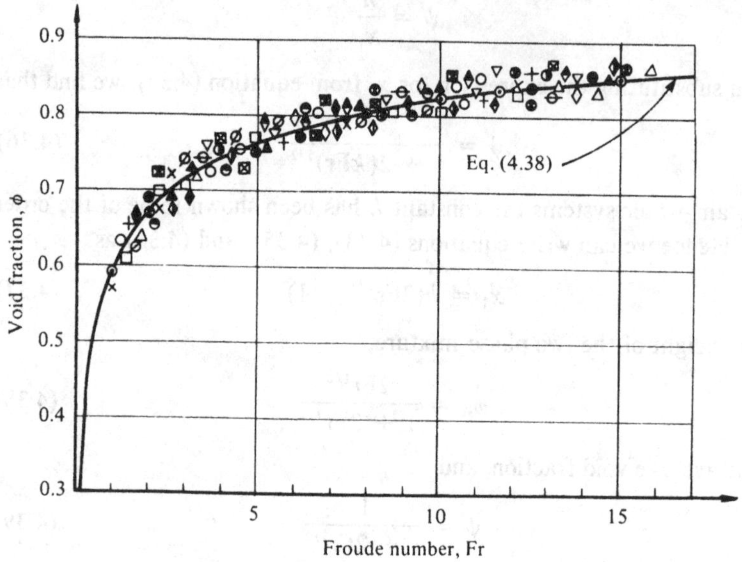

Figure 4.5 Relative density against Froude number for air–water mixture.

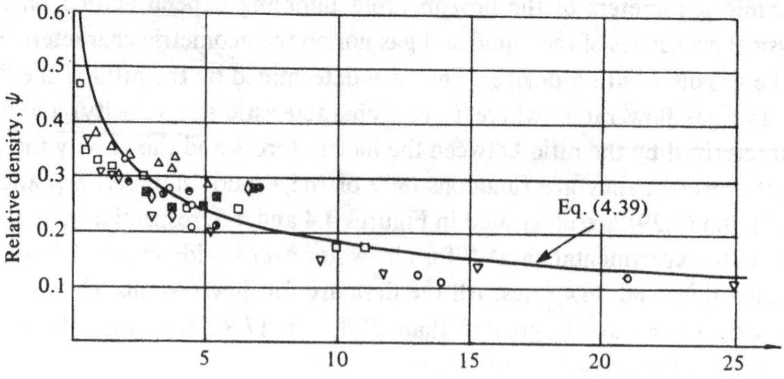

per unit volume, and we can write a continuity equation for the void fraction:

$$\frac{\partial \phi}{\partial t} + \frac{\partial}{\partial x_i}(u_i \phi) = 0 \tag{4.40}$$

where u_i ($i = 1, 2, 3$) is the gas velocity vector at vector coordinate x_i, and the Cartesian tensor convention of summation of repeated indices is implied.

Because of the turbulence of the two-phase flow, the gas content ϕ and velocity u_i change their values in space and time, and we can decompose them into two components, the mean component and the fluctuating component:

$$\phi = \overline{\phi} + \phi'$$

and

$$u_i = \overline{u}_i + u_i'$$

where $\overline{\phi}$ is the mean and ϕ' the fluctuation component of the void fraction, and $\overline{u}_i$ is the mean and u_i' the fluctuation component of the velocity vector.

The turbulent exchange coefficient is defined by

$$\overline{u_i'\phi'} = -D_t \frac{\partial \overline{\phi}}{\partial x_i}$$

where D_t is the coefficient of turbulent diffusion, and substituting the value of $\overline{u'\phi'}$ from the equation above into equation (4.40) and time averaging gives

$$\frac{\partial \overline{\phi}}{\partial t} + \frac{\partial}{\partial x_i}(\overline{u}_i \overline{\phi}) = \frac{\partial}{\partial x_i}\left(D_t \frac{\partial \overline{\phi}}{\partial x_i}\right) \tag{4.41}$$

This modified equation of continuity takes into account gas mass transfer as a result of the interaction of the fluctuating velocity and void fraction.

As a first approximation we may assume that the gas content ϕ can be averaged over the flow cross section so that it depends only on the height, x, in the two-phase flow. Then, for a steady-state flow, equation (4.41) may be rewritten as

$$\frac{d}{dx}\left(\overline{v}\overline{\phi} - D_t \frac{d\phi}{dx}\right) = 0$$

where $\overline{v}$ is now the mean gas velocity in the x direction. The expression in parentheses in this equation is obviously a constant, and in a liquid-free zone $\overline{v} = v_s$, the superficial gas velocity, and $d\phi/dx = 0$ because $\phi \equiv 1$, so that

$$\overline{v}\overline{\phi} - D_t \frac{d\overline{\phi}}{dx} = v_s \tag{4.42}$$

Taking as a boundary condition

$$\overline{\phi}(x_1) = 1$$

the solution of equation (4.42), assuming, as a first approximation, that $\overline{v}$ and D_t are independent of the coordinate x, will be

$$\overline{\phi}(x) = \frac{v_s}{\overline{v}} + \left(1 - \frac{v_s}{\overline{v}}\right) \exp\left[-\frac{\overline{v}}{D_t}(x_1 - x)\right] \tag{4.43}$$

It follows from equation (4.43) that the gas void fraction has only a weak dependency on the position in the two-phase mixture when the liquid height, h, is sufficiently high (~1 m, 3.28 ft) that $(\overline{v}/D_t)(h - x) \gg 1$. Thus this equation implies that if the undisturbed liquid level $h(<x_1)$ is large, then $\phi \simeq$ constant over most of the flow. Local measurements of the gas content[49] confirm this conclusion.

In the lower region[40,48,49] of the two-phase flow, in its initial section, the void fraction is defined by the cross-sectional area of the gas-distributing device (e.g., the orifices). The main region of the two-phase flow, which has a fairly constant void fraction, is upstream of this small initial section, and the void fraction increases in the upper part of the main region, approaching unity at the "interface" of the two-phase flow and the free-gas region. The gas void fraction is higher in the upper part because the surface bubbles burst more slowly than the rate of rise of the newly generated bubbles, but for flow systems where h is large and the superficial gas velocity is small, the influence of the upper part on the space-average gas void fraction of the two-phase flow is not great. However, in flow systems in which h has a moderate value (5 cm $< h <$ 1.2 m, 1.96 in. $< h <$ 3.94 ft), the mean void fraction $\overline{\phi}$ is dependent on the value of h, and there may be a significant effect on the average.

The location at which the void fraction starts increasing rapidly to unity is basically determined by the superficial gas velocity. For $v_s \leq 0.2$ m/s (0.66 ft/s), the transition is at a height of 1 to 2 cm (0.39 to 0.79 in.) and for $v_s \geq 0.2$ m/s (0.66 ft/s) it may be extended to 30 to 35 cm (11.8 to 13.78 in.).

Let us now set up the energy balance of a differential layer, dx (of unit cross section), for a regime of slow bubbling:

$$dE = dE_1 + dE_2 + dE_3 \tag{4.44}$$

where dE is the total energy of the layer, dE_1 the potential energy of the liquid, dE_2 the dissipative energy of the layer, and dE_3 the energy of surface tension.

The values of dE_1 and dE_2 are obtained from the following equations:

$$dE_1 = (1 - \phi)\rho_f g x \, dx \tag{4.45}$$

and

$$dE_2 = C_D \frac{\rho_f v^2}{2} \pi r_b^2 n x \, dx \tag{4.46}$$

where C_D is the drag coefficient, v the local gas velocity, and n the number of bubbles per unit volume.

We have, as before, for gas mass conservation

$$\phi v = v_s$$

and the fraction of the volume of the differential layer occupied by the gas is

$$\phi = \tfrac{4}{3}\rho r_b^3 n$$

Taking these into consideration, equation (4.46) becomes

$$dE_2 = \frac{3}{8} \frac{C_D \rho_f v_s^2}{r_b \phi} x \, dx \tag{4.47}$$

The energy of surface tension is given by

$$dE_3 = 4\pi r_b^2 \sigma n \, dx = \frac{3\sigma}{r_b} \phi \, dx \tag{4.48}$$

Here we have followed the analysis of Sections 4.2 and 4.3. Substituting the values of E_1, E_2, and E_3 from equations (4.45), (4.47), and (4.48) into equation (4.44), we obtain an expression for the total energy of the mixture of height x_1,

$$E = \int_0^{x_1} \left[(1 - \phi)\rho_f g x + \frac{3}{8} \frac{C_D \rho_f v_s^2}{r_b \phi} x + \frac{3\sigma}{r_b} \phi \right] dx \tag{4.49}$$

The kinetic energy of the two-phase flow is ignored in equation (4.49) because its value is usually three orders of magnitude less than the energy of dissipation, for slow bubbling.

As in the analysis of rapid bubbling, the steady-state distribution of the gas content ϕ sets in when the energy given by equation (4.49) is at a minimum.[50] Thus to find the value of ϕ it is necessary to determine the minimum of the integral under the condition that the height of the liquid in the equipment is constant [equation (4.9)]. Ignoring variations due to x_1, Euler's equation in this case assumes the form

$$-\rho_f g x - \frac{3 C_D \rho_f v_s^2 x}{8 r_b \phi^2} + \frac{3\sigma}{r_b} - \lambda = 0$$

where, as in the earlier cases, λ is the Lagrange multiplier. Rearranging, we find that

$$\phi = \left[\frac{3C_D\rho_f v_s^2 x}{8r_b[(3\sigma/r_b) - \rho_f g x - \lambda]}\right]^{1/2} \tag{4.50}$$

The height of the two-phase layer x_1 can be determined from equation (4.9) by substituting into it the value of ϕ from equation (4.50).

$$h = \int_0^{x_1} \left\{1 - \left[\frac{3C_D\rho_f v_s^2 x}{8r_b[(3\sigma/r_b) - \rho_f g x - \lambda]}\right]^{1/2}\right\} dx$$

Integrating, we obtain

$$\begin{aligned} h = x_1 &+ \frac{v_s}{g}\left(\frac{3C_D}{8r_b\rho_f}\right)^{1/2}\left[\left(\frac{3\sigma}{r_b} - \lambda\right)x_1 - \rho_f g x_1^2\right]^{1/2} \\ &- \frac{v_s}{2}\left(\frac{3\sigma}{r_b} - \lambda\right)\left(\frac{3C_D\rho_f}{8r_b}\right)^{1/2}\frac{1}{(\rho_f g)^{3/2}} \\ &\times \sin^{-1}\left\{\frac{2\rho_f g x - [(3\sigma/r_b) - \lambda]}{(3\sigma/r_b) - \lambda}\right\}\Bigg|_0^{x_1} \end{aligned} \tag{4.51}$$

From equation (4.50), for $\phi(x_1) = 1$, it follows that

$$\left(\frac{3\sigma}{r_b} - \rho_f g x_1 - \lambda\right)^{1/2} = \left(\frac{3C_D\rho_f v_s^2 x_1}{8r_b}\right)^{1/2}$$

$$\frac{3\sigma}{r_b} - \lambda = \rho_f g x_1 + \frac{3C_D\rho_f v_s^2 x_1}{8r_b}$$

Figure 4.6 Void fraction against superficial gas velocity.

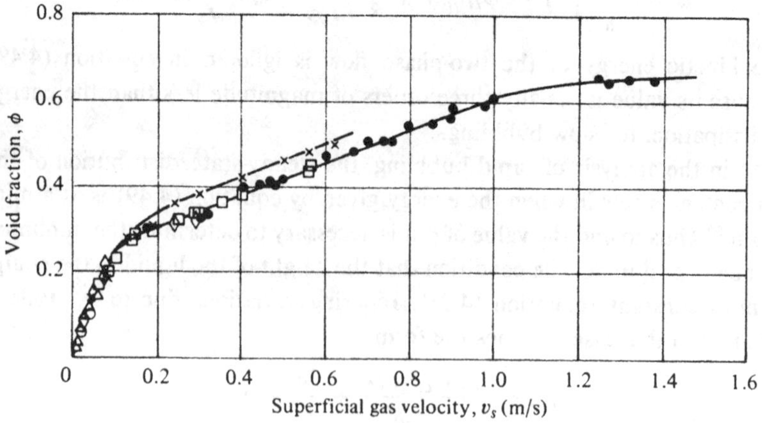

and taking this into account, we can rewrite equation (4.51) in the form

$$h = x_1 + \frac{3C_D v_s^2}{8r_b g} x_1 - \left(\frac{3C_D}{8gr_b}\right)^{1/2} \times \left(1 + \frac{3C_D v_s^2}{8r_b g}\right) \frac{v_s x_1}{2} \left[\sin^{-1}\left(\frac{1 - 3C_D v_s^2/8r_b g}{1 + 3C_D v_s^2/8r_b g}\right) + \frac{\pi}{2}\right]$$

Thus the height of the two-phase mixture is given by

$$x_1 = \frac{h}{(1 + ab)\left\{1 - \frac{(ab)^{1/2}}{2}\left[\sin^{-1}\left(\frac{1 - ab}{1 + ab}\right) + \frac{\pi}{2}\right]\right\}} \tag{4.52}$$

and the average gas void fraction is simply

$$\begin{aligned} \phi_{av} &= \frac{x_1 - h}{x_1} \\ &= 1 - \frac{h}{x_1} \\ &= ab - \frac{(ab)^{1/2}}{2}(1 + ab)\left[\sin^{-1}\left(\frac{1 - ab}{1 + ab}\right) + \frac{\pi}{2}\right] \end{aligned} \tag{4.53}$$

where

$$a = 3C_D/8 \quad \text{and} \quad b = v_s^2/gr_b$$

Equation (4.53) is illustrated in Figure 4.6 in comparison to experimental data.[10,12] Now, for superficial gas velocities $v_s < 0.1$ m/s (0.33 ft/s), for instance, in the bubbling process that usually takes place in chemical reactors, the drag coefficient can be estimated by the formula[51]

$$C_D = 0.82\left(\frac{g\mu_f^4}{\rho_f \sigma^3}\right)^{1/4} \text{Re} \tag{4.54}$$

where σ is the surface tension.

With this formula, equations (4.52) and (4.53) can be used to obtain the height and average void fraction, respectively, of the two-phase flow in the slow bubbling regime.

4.5 Main parameters of the two-phase flow with dissipation and inertia

The influences of the dissipative and inertia forces have been investigated separately in the preceding sections. We now consider the general problem of the calculation of the average void fraction when dissipative and inertia forces are present at the same time.

For this case we use equation (3.16) for the relative motion of a gas bubble in the two-phase flow. Multiplying this equation by v, we obtain an expression for the change of energy of a single bubble during its motion,

$$g(\rho_f - \rho_g)Vv - C_D \frac{\rho_f v^3}{2} A_b - (\rho_g + k\rho_f)V \frac{d}{dt}\frac{v^2}{2} = 0$$

or in the integral form,

$$E_b = (\rho_f - \rho_g)gVx - C_D \frac{\rho_f A_b}{2} \int_0^x v^2\,dx - \frac{(\rho_g + k\rho_f)Vv^2}{2}$$

where E_b is the energy of a single bubble and V and A_b are the bubble volume and cross-sectional area, respectively.

Assuming that the differential layer dx contains n bubbles per unit volume whose volume is a fraction $\phi = \frac{4}{3}\pi r_b^3 n$ of the total volume in the layer, we can express the energy of all the bubbles in a unit cross section of the layer dx as follows:

$$E_b n\,dx = \left[\left(\rho_g + \frac{\rho_f}{2}\right)\frac{v^2}{2}\phi + C_D \frac{\rho_f}{2}\phi \frac{A_b}{V}\int_0^x v^2\,dx - g(\rho_f - \rho_g)\phi x\right] dx$$

Note that, as in equation (4.5), we have set $k = \frac{1}{2}$. If we combine the potential energy of the liquid and the energy of surface tension with the energy of the bubbles, the total energy of the two-phase layer will be

$$E = \int_0^{x_1}\left[(\rho_f - \rho_g)gx(1 - \phi) + \frac{1}{2}\left(\rho_g + \frac{\rho_f}{2}\right)v^2\phi + C_D \frac{3}{4r_b}\phi \frac{\rho_f}{2}\int_0^x v^2\,dx + \frac{3\sigma}{r_b}\phi\right] dx$$

We shall ignore the influence of the transition zones (lower and upper) described in Section 4.4 on the average void fraction of the two-phase flow, because these zones essentially balance each other out. With this condition of negligibly small change in the void fraction with height, the energy of the two-phase flow per unit liquid height h will be

$$\frac{E}{h} = \rho_f g h(1 - \phi)\frac{\eta_1^2}{2} + \frac{\rho_f v_s^2 \eta_1}{4\phi} + \frac{3C_D \rho_f v_s^2 \eta_1^2 h}{16 r_b \phi} + \frac{3\sigma\phi\eta_1}{r_b} \qquad (4.55)$$

where $\eta_1 = x_1/h$ (h is the height of the static liquid and x_1 is the height of the two-phase flow), v_s is the superficial gas velocity, and we have used the relation $v = v_s/\phi$ and the approximation $\rho_f - \rho_g \simeq \rho_f$.

Noting that $\phi_{av} \simeq \phi \simeq$ constant by neglecting the transition regions, we find that

$$\eta_1 \simeq \frac{1}{1-\phi}$$

and equation (4.55) can be rewritten

$$\frac{E}{h} = \frac{\rho_f g h}{2(1-\phi)} + \frac{\rho_f v_s^2}{4\phi(1-\phi)} + \frac{3C_D\rho_f v_s^2 h}{16 r_b \phi(1-\phi)^2} + \frac{3\sigma\phi}{r_b(1-\phi)} \tag{4.56}$$

where in the rest of this section ϕ will denote the average void fraction. The equilibrium distribution of the void fraction ϕ is established when the specific energy E/h of the two-phase flow is at a minimum. Consequently, in order to determine this equilibrium ϕ, we must vary equation (4.56) with respect to ϕ to obtain

$$\frac{\rho_f g h}{2(1-\phi)^2} - \frac{\rho_f v_s^2(1-2\phi)}{4\phi^2(1-\phi)^2} - \frac{3C_D\rho_f v_s^2 h(1-3\phi)}{16 r_b \phi^2(1-\phi)^3} + \frac{3\sigma}{r_b(1-\phi)^2} = 0$$

After algebraic transformations we can rewrite this equation as

$$\phi^3 - B\phi^2 - 3A\phi + A = 0 \tag{4.57}$$

where A and B are the dimensionless quantities

$$A = \frac{\frac{1}{2}v_s^2(1 + 3C_D h/4r_b)}{gh + 6\sigma/\rho_f r_b} \tag{4.58}$$

and

$$B = 1 - \frac{v_s^2}{gh + 6\sigma/\rho_f r_b} \tag{4.59}$$

In many practical applications quantity B approaches unity, and therefore equation (4.57) gives in this case

$$A \simeq \frac{\phi^2(1-\phi)}{1-3\phi} \tag{4.60}$$

In Figure 4.7 a plot is given of $A^{1/2}$ against ϕ according to equation (4.60). This curve can be used to determine the void fraction once the value of A is known, for $B \simeq 1$.

Let us now eliminate from the quantities A and B the part that is dependent only on one flow system parameter (h), using equations (4.58) and (4.59):

$$\frac{A}{1-B} = \frac{1}{2}\left(1 + \frac{3C_D h}{4r_b}\right) = \alpha \tag{4.61}$$

where α is a function of h.

Taking this into account, we can rewrite equation (4.57) in the form

$$\phi^3 - B\phi^2 - 3\alpha(1 - B)\phi + (1 - B)\alpha = 0$$

or

$$B = \frac{\phi^3 - 3\phi\alpha + \alpha}{\phi^2 - 3\phi\alpha + \alpha}$$

From this equation it follows that for values of ϕ which are the roots of the equation

$$\phi^2 - 3\alpha\phi + \alpha = 0 \tag{4.62}$$

$B = -\infty$ (because $0 \leq \phi \leq 1$), which means, as indicated in equation (4.59), that for h being constant, the superficial velocity, v_s, is infinite. Thus equation (4.62) is applicable when we have flows with large superficial velocity, and the roots of this equation will be

$$\phi = \frac{3\alpha}{2}\left[1 \pm \left(1 - \frac{4}{9\alpha}\right)^{1/2}\right]$$

Because ϕ must be less than or equal to unity, we have only one value of ϕ from this equation, which is

$$\phi = \frac{3\alpha}{2}\left[1 - \left(1 - \frac{4}{9\alpha}\right)^{1/2}\right] \tag{4.63}$$

For the equation above, for $\alpha = 1$ [about the smallest possible value of α as is clear from equation (4.61) because $h/r_b \gg 1$], ϕ achieves a maximum

Figure 4.7 Void fraction against factor $(A)^{1/2}$.

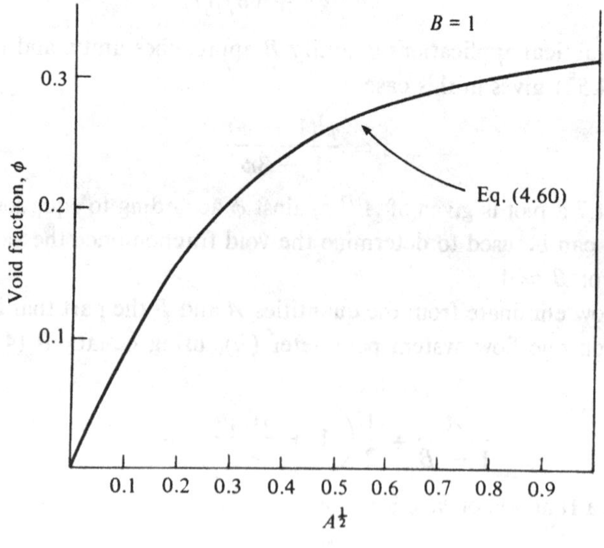

value, equal to $\phi = 0.382$. This, then, is the value of the void fraction in the central zone, which cannot be surpassed either by decreasing the height, h, or by increasing the gas velocity, v_s, beyond a certain (high) value.

Further, for $\alpha \rightarrow \infty$, we find that

$$\phi \simeq \frac{3\alpha}{2}\left(1 - 1 + \frac{2}{9\alpha}\right) = \frac{1}{3}$$

and this is the limit that the void fraction reaches when there are no restrictions on the upper limits of gas velocity and static liquid height.

It should be noted that according to experimental data, the void fraction in this center zone actually achieves higher values than we have estimated because of the effect of the two transition zones (which we neglected in our analysis). Nevertheless, for high gas velocities, we can see that the analysis gives the result that the void fraction ϕ lies in the range 0.3 to 0.4.

Now, in Figure 4.8 is given a graph of quantity B against ϕ, with different values of the parameter A, according to equation (4.57) in the form

$$B = \phi + \frac{A(1 - 3\phi)}{\phi^2}$$

Thus we can see in the figure that the possible values of ϕ and B are limited from above by line $B = 1$, from the left by $\phi = 0$, and from the right by $\phi = 1$. The lines $B = \phi$ and $\phi = \frac{1}{3}$ divide the field into four sections. For $A > 0$ (which is physically necessary), the section limited by the lines B

Figure 4.8 Void fraction against factors A and B.

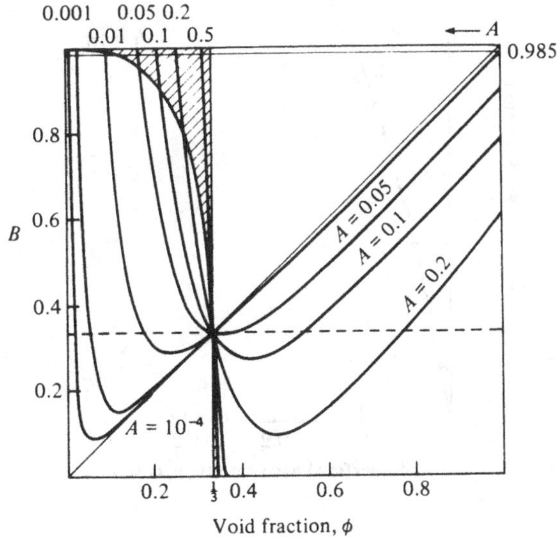

= 1, $B = \phi$, and $\phi = \frac{1}{3}$ and the section limited by the lines $B = 0$, $B = \phi$, and $\phi = \frac{1}{3}$ are eliminated.

Equations (4.57) through (4.63) agree fairly well with experimental data in flow systems where the liquid depth h is large and the superficial gas velocity in sufficiently high regions of the liquid are small. If the extent of the transition zones, created by larger superficial gas velocities (more than 0.5 to 0.6 m/s, 1.64 to 1.97 ft/s) is 35 to 40 cm (1.15 to 1.31 ft) in total, the prediction of ϕ by these equations will still hold good when the liquid depth is large (in excess of 1 m, 3.28 ft). The upper limit for using the equations (divergence from experimental data being no more than ± 10 percent) is for $B \leq 0.985$.

Now, to use this set of equations we must have a means of determining the bubble radius r_b and the drag coefficient C_D. The bubble radius and drag coefficient have been shown[52] in equation (3.18) to be related in the form

$$r_b = \frac{C_D v_b^2}{4g}$$

Figure 4.9 Void fraction against superficial gas velocity for air–water mixture.

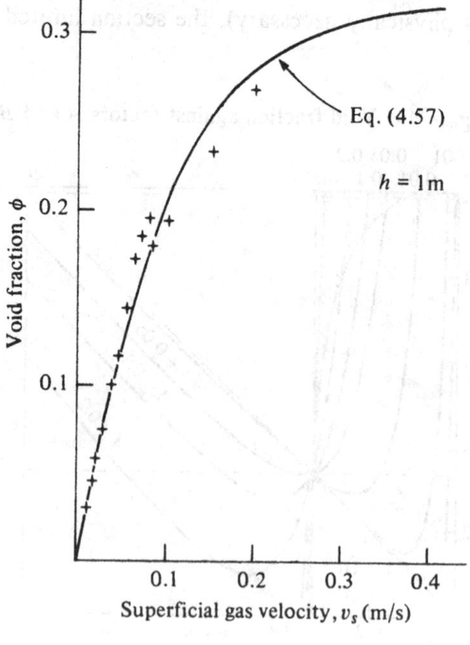

where v_b is the velocity of the bubble. Consequently,

$$\frac{3}{8}\frac{C_D}{r_b} = \frac{3g}{2v_b^2} \tag{4.64}$$

and we can use for v_b the velocity of buoyancy given[7] by

$$v_b = \frac{2}{3}\left(\frac{4\sigma^2 g}{3s\rho_f\mu_f}\right)^{1/5}$$

The factor s is a shape factor, and when the bubbles are spherical, $s \simeq 1$. It is possible to take other values of s. For example, by taking $s = 4$ (the ratio of the major to minor axes of the bubble then being equal to 2), we get better results. Equation (4.64) can then be used in equation (4.61). We still have to determine the factor $6\sigma/\rho_f r_b$ in equations (4.58) and (4.59), and to simplify the calculation, r_b in this term is taken to be 0.2 cm (0.08 in.) (the influence of this term on the final result being very small). Figure 4.9 shows experimental data[37] obtained for an air–water system as well as the curve corresponding to equation (4.57). Note the close correspondence between theory and experiment.

4.6 Effect of static liquid height and equipment diameter on the void fraction

We have seen in equation (4.53) that the average gas void fraction can be defined as

$$\phi_{av} = \frac{x_1 - h}{x_1} = 1 - \frac{h}{x_1} = 1 - \psi$$

where ψ is the average relative density of the two-phase flow and is the ratio of the density of the gas–liquid mixture to that of the liquid. Thus, if ψ (or x_1 and h) is known, ϕ_{av} can be obtained. However, when the diameter D of the equipment is less than 0.2 m (7.88 in.) and the static liquid level h is less than 1 m (3.28 ft), we must introduce correction factors into the foregoing equation to obtain

$$\phi_{av} = 1 - k_d k_h \psi \tag{4.65}$$

where k_d is the correction factor for the equipment diameter and k_h is the correction factor for the static liquid level.

Analyzing experimental data[53,54] with the use of similarity theory yields

$$k_d = 1 - \exp\left\{-1.1\left(\frac{v_s}{v_b}\right)^{1/2}\left[\frac{(gd)^{1/2}}{v_s}\right]^{3/4}\right\} \tag{4.66}$$

and

$$k_h = 1 - \exp\left[-0.405\left(\frac{v_s}{v_b}\right)^{0.7}\frac{(gh)^{1/2}}{v_s}\right] \tag{4.67}$$

where v_b is the rising velocity of a single bubble, given by the formula

$$v_b = 1.18\left[\frac{g\sigma(\rho_f - \rho_g)}{\rho_f^2}\right]^{1/4}$$

References

1. Teletov, S. G., *Dokl. Acad. Nauk SSSR, 4* (1945).
2. Teletov, S. G., *Vestu. Mosk. Ůniv.*, no. 2, *Ser. Mekh.* (1958).
3. Teletov, S. G., Doctoral dissertation, ENIN, AN SSSR, Moscow, 1948. (In Russian.)
4. Teletov, S. G., Candidate's dissertation, Moscow University, 1938. (In Russian.)
5. Levich, V. G., *K teorii poverkhnostnykh yavlenii (Theory of Surface Phenomena)*, Izdatelstvo Sov. Nauka, Moscow, 1941.
6. Landau, L. D., and Lifshitz, E. M., *Fluid Mechanics*, Pergamon Press, Elmsford, N.Y., 1959.
7. Levich, V. G., *Physicochemical Hydrodynamics*, Prentice-Hall, Englewood Cliffs, N.J., 1962.
8. Kochin, N. E., Kibel, I. A., and Roze, N. N., *Teoreticheskaya gidrodinamika* (Theoretical Hydrodynamics), Gostekhizdat, Moscow, 1955.
9. Akselrod, L. S., and Dilman, V. V., *Zh. Prikl. Khim. (Moscow), 29*(12), 1803 (1956).
10. Kutateladze, S. S., and Styrikovich, M. A., *Hydraulics of Gas–Liquid Systems*, Gosudazstvennoe Energeticheskoye Izdatelstvo, Moscow, 1958. Wright Field trans. F-TS-9814v.
11. Kasatkin, A. G., Dyinerskii, Yu. I., and Popov, D. M., *Khim. Prom. (Moscow)*, no. 7, 482 (1961).
12. Aizenbud, M. B., and Dilman, V. V., *Khim. Prom. (Moscow)*, no. 3, 199 (1961).
13. Akselrod, L. S., and Dilman, V. V., *Zh. Prikl. Khim. (Moscow), 27*, 5 (1954).
14. Chhabra, P. S., and Mahajan, S. P., *Indian Chem. Eng., 16*(2), 16 (1974).
15. Bhaga, D., and Weber, M. E., *Can. J. Chem. Eng., 50*(3), 323 (1972).
16. Bhaga, D., and Weber, M. E., *Can. J. Chem. Eng., 50*(3), 329 (1972).
17. Pruden, B. B., and Weber, M. E., *Can. J. Chem. Eng., 48*, 162 (1970).
18. Pozin, M. E., Mukhlenov, I. P., and Tarat, E. Ya., *Zh. Prikl. Khim. (Moscow), 30*(1), 45 (1957).
19. Zuber, N., and Finlay. J. A., *J. Heat Transfer Trans. ASME, C87*, 453 (1965).
20. Chekhov, O. S., Candidate's dissertation, Moscow Institute of Chemical Engineers, 1960. (In Russian.)
21. Solomakha, G. P., Candidate's dissertation, Moscow Institute of Chemical Engineers, 1957. (In Russian.)
22. Artomonov, D. S., Candidate's dissertation, Moscow Institute of Chemical Engineers, 1961. (In Russian.)
23. Brown, R. W., Gomezplata, A., and Price, J. D., *Chem. Eng. Sci., 24*, 1483 (1969).
24. Ribgy, G. R., and Capes, C. E., *Can. J. Chem. Eng., 48*, 343 (1970).
25. Gomezplata, A., Munson, R. E., and Price, J. D., *Can. J. Chem. Eng., 50*, 669 (1972).
26. Gomezplata, A., and Sung, P. T., *Chem. Eng. Sci., 48*, 336 (1970).
27. Koide, K., Hirahara, T., and Kubota, H., *Kogaku Koguku, 5*(1) 38 (1967).
28. Stepanek, J., *Chem. Eng. Sci., 25*, 751 (1970).

29. Pruden, B. B., Hayduk, W., and Laudie, H., *Can. J. Chem. Eng.*, *52*, 64 (1974).
30. Laudie, H. A., M.A.Sc. thesis, University of Ottawa, 1969.
31. Bell, R. L., *AIChE J.*, *18*(3), 498 (1972).
32. Kuz'minykh, I. N., and Koval', G. A., *Zh. Prikl. Khim. (Moscow)*, *28*(1) 21 (1955).
33. Noskov, A. A., and Sokolov, V. N., *Tr. Leningr. Tekhnol. Inst. Lensoveta*, *39*, 110 (1957).
34. Pozin, M. E., *Sb. Vopr. Massoperedachi*, p. 148 (1957).
35. Sherherd, E. B., *Ind. Chem.*, no. 4, 175 (1956).
36. Jackson, R., *Ind. Chem.*, no. 336, 16; no. 338, 109 (1953).
37. Hyghes, R. R., *Chem. Eng. Prog.*, *51*(12), 555 (1955).
38. Loitsyanskii, L. G., *Mekhanika zhidkosti i gaza*, Gostekhizdat, Moscow, 1957.
39. Smirnov, V. I., *Kurs vysshey matematiki*, vol. 4, Gostekhizdat, Moscow, 1941.
40. Vinokur, Ya. G., and Dilman, V. V., *Khim. Prom. (Moscow)*, no. 7, 619 (1959).
41. Kashnikov, A. M., Candidate's dissertation, D. I. Mendeleev Moscow Institute of Chemical Technology, 1965. (In Russian.)
42. Rodionov, A. I., Kashnikov, A. M., and Radikovsky, V. M., *Khim. Prom. (Moscow)*, no. 10, 17 (1964).
43. Akselrod, L. S., and Dilman, V. V., *Khim. Prom. (Moscow)*, no. 1, 8 (1954).
44. Azbel, D. S., *Khim. Prom. (Moscow)*, no. 11, 854 (1962).
45. Gal-Or, B., *AIChE J.*, *12*, 3, 604.
46. Marucci, G., *Ind. Eng. Chem. Fund.*, *2*, 224 (1965).
47. Kurbatov, A. V., *Tr. Mosk. Energ. Inst.*, *2*, 82 (1953).
48. Sterman, L. S., *Zh. Tekh. Fiz.*, *26*(7), 1512 (1956).
49. Sterman, L. S., and Surnov, A. B., *Teploenergetika*, *8*, 39 (1955).
50. Lavrent'ev, M. A., and Luysternak, L. A., *A Course of Variation Calculus*, Gosizdat (state publishing house) of Theoretical and Technical Literature, Moscow, 1950. (In Russian.)
51. Peebles, F. N., and Garber, H. J., *Chem. Eng. Prog.*, *49*(2), 88 (1953).
52. Azbel, D. S., *Khim. Prom. (Moscow)*, no. 1, 43 (1964).
53. Artomonov, D. S., Candidate's dissertation, Moscow Institute of Chemical Engineers, 1961. (In Russian.)
54. Popov, D. M., Candidate's dissertation, D. I. Mendeleev Moscow Institute of Chemical Technology, 1960. (In Russian.)

5

Modes of liquid entrainment

5.1 Introduction

In this chapter we study different aspects of the phenomenon of liquid entrainment, in which gas, being forced into liquid through an array of orifices, carries off droplets of liquid. This study is important because it can illustrate which factors contribute to the maximizing (or minimizing) of the amount of liquid entrained by the flowing gas. As in the other phenomena studied in this text, the complicated nature of two-phase behavior militates against the development of anything approaching a complete analytical description, so several different possible entrainment mechanisms will be described. It should be noted that, although the following analysis is applied to the case of gas passing through an orifice (or orifices) into liquid, the results obtained are in no way restricted to this flow system. In fact, the results obtained are also valid for the two-phase dynamic flows described in Chapter 1.

5.2 Liquid cone stability

In Section 2.6 we discussed the transition from single-bubble formation to gas jet formation, at the orifice, as the gas flow rate is increased. We now propose to determine whether or not the gas jet can be a stable configuration in the flow of gas through an array of orifices submerged in liquid.

Penetration of gas into the liquid raises the potential energy of the liquid in the chamber, which then becomes transformed into the kinetic energy of the descending stream of liquid. The mean velocity of the liquid has a maximum on the chamber axis and is zero at the walls of the chamber.[1] This velocity gradient is responsible for rotation of the gas bubbles,[2] which are thus acted on by forces directed toward the walls of the chamber.[3] Also, circulation of the liquid brings about forces on the bubbles acting in the direction of higher velocities—in other words, from the walls toward the center.[4] As a result of interaction between these effects, the central portion of the bubbling flow is found to be the part most highly concentrated with bubbles and a gas core forms in this part, which rises at a much higher rate than do isolated bubbles.

As the gas flow is increased, trickling of the liquid downward diminishes, and the amount of liquid doing so becomes inadequate to support the formation of new bubbles. In such a case, a layer of foam, a film-cellular system whose individual bubbles are connected by their common interface, accumulates on the free surface of the liquid. The thickness of the cellular foam layer is determined by the average lifetime of the individual bubbles and by the rate of growth of the layer from beneath due to the arrival of new bubbles. Breakdown of the layer of cellular foam sets in at a certain critical flow speed. Some investigators[5–7] have suggested that breakdown of foam in the two-phase flow is followed by the formation of gas channels with liquid walls, called the *mode of open spray cones* by those authors. This is an alternative description of the phenomenon described in Section 2.6.

When we attempt to investigate the possible stable existence of these spray cones (or gas jets), it is convenient to consider as a model an arbitrary array of orifices at the bottom of a liquid-containing chamber, this being the general case of gas being forced through orifices into a liquid. We consider that the gas flow rate is such that gas jets that are formed expand with distance from the orifices, with liquid filling the spaces between the gas jets, as illustrated in Figure 5.1. Following is an analysis to determine whether or not this configuration can be in dynamic equilibrium. If it cannot, the liquid cones (which exist in the spaces between the gas jets) will break up to produce a random-size distribution of droplets, which can then be entrained by the flowing gas, and this possibility of entrainment indicates the importance of this analysis.

For simplicity, we ignore the viscosity of the liquid and the gas, and the liquid motion induced by the gas motion is neglected. The flow is assumed to be one-dimensional; that is, the velocity of the gas is dependent only on the perpendicular distance from the orifices. Later, the analysis will be modified to make some allowance for liquid motion.

Consider Bernoulli's equation for a streamline in the gas:

$$\frac{\rho_g v^2(x)}{2} + p_g(x) = c_1, \qquad c_1 = \text{const.} \tag{5.1}$$

where ρ_g is the gas density (assumed constant), $v(x)$ the gas velocity at height location x, and $p_g(x)$ the gas static pressure at location x. Note that the term $\rho_g g x$ has been neglected as insignificant.

Also for the liquid, assumed stationary, we have

$$p_f(x) + \rho_f g x = c_2, \qquad c_2 = \text{const.} \tag{5.2}$$

where $p_f(x)$ is the liquid pressure at location x and ρ_f the liquid density. Now, at any location x, we require the flowing gas to be in dynamic equilibrium with the liquid, so that, for no lateral motion of the liquid or gas we must have a balance of pressure and surface-tension forces on each elemental interface area,

$$p_g(x) = p_f(x) - \frac{\sigma}{R(x)} \tag{5.3}$$

where σ is the surface tension. The term $R(x)$ is the equivalent radius of the liquid cones at location x. These liquid cones are generated in the spaces between the gas jets, and as the gas jets expand with increasing x, the value of $R(x)$ decreases with x. In general, the liquid cones will have irregular cross sections (determined by the shape of the gas jets, arrangement of the orifices, and so on), and thus some suitably defined equivalent radius is used (see Figure 5.1). The exact nature of this definition is not important to the analysis.

Combining equations (5.1), (5.2), and (5.3), we obtain the condition for dynamic equilibrium at any height location x:

$$c_1 - \frac{\rho_g v^2(x)}{2} = c_2 - \rho_f g x - \frac{\sigma}{R(x)} \tag{5.4}$$

Figure 5.1 Open spray cones.

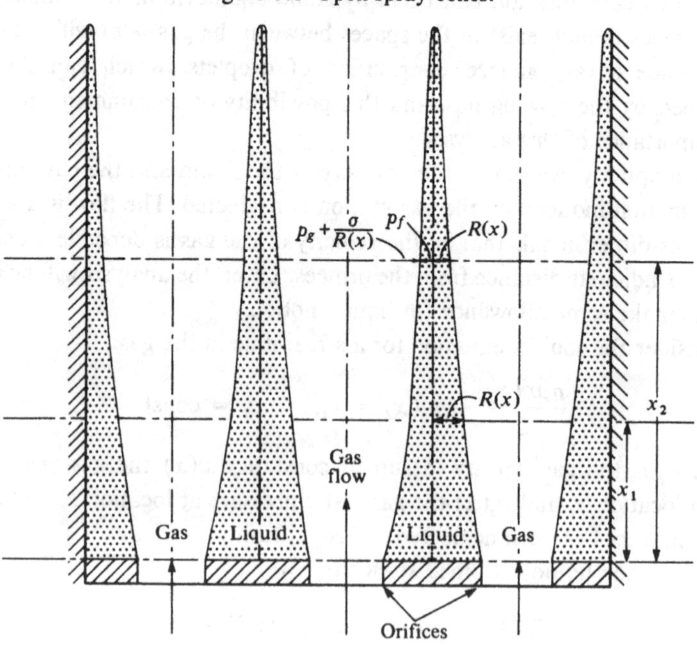

Let us assume that equation (5.4) is fulfilled when $x = x_1$, some arbitrary value of x. We wish to see if, as required for the stable existence of the gas jets, the equation can also be satisfied at any and all other locations $x = x_2 \neq x_1$. Consider, first, $x_2 > x_1$. As the gas leaves the orifices, its velocity $v(x)$ decreases as the distance from the orifices, x, increases. This means, as noted earlier, that the total cross-sectional area of the jets increases as x increases, with the result that $R(x)$ decreases. Hence, in equation (5.4), for x changing from x_1 to $x_2 > x_1$, the left-hand side of the equation becomes more positive. Also, as x goes from x_1 to $x_2 > x_1$, the right-hand side of the equation becomes increasingly negative, and hence, if the two sides of equation (5.4) balance at location x_1, they cannot balance at any location $x_2 > x_1$. A similar argument shows that equation (5.4) cannot be fulfilled for any $x_2 < x_1$. The gas jets cannot be in equilibrium with the liquid at any other location if they are in equilibrium at any one location, and the gas jet–liquid cone system cannot be maintained, with the result that breakup must occur. This means that, under gas flow rates in the range of continuous gas flow at the orifice, we have liquid breakup and liquid entrainment.

Equation (5.4) was derived with the assumption that the pressure in the liquid is hydrostatic, in other words, that we have no liquid motion. To make sure that the conclusion we have just arrived at is correct as far as a more realistic flow system is concerned, we now relax this condition by adding a correction term to the liquid pressure to allow for liquid motion.

We assume a correction to the hydrostatic pressure produced by the induced motion of the liquid to be

$$\delta p = c_3 x / x_1 \qquad c_3 = \text{const.} \tag{5.5}$$

where x_1 is the depth of the liquid in the chamber (the height of the liquid cones). We also assume, for simplicity, that the pressure drop in the gas in the x direction is linear:

$$p_g(x) = \frac{\Delta p}{x_1}(x_1 - x) \tag{5.6}$$

where Δp is the pressure difference, in the gas, between the orifice location and the free surface of the liquid ($x = x_1$). This is a first-order approximation to the actual pressure distribution.

Using equations (5.5) and (5.6) in a force balance per unit area of the liquid–gas interface, for equilibrium we find that

$$\rho_f g(x_1 - x) - \frac{\sigma}{R(x)} + \frac{c_3 x}{x_1} = \frac{\Delta p}{x_1}(x_1 - x)$$

Hence

$$R(x) = \frac{\sigma}{(\rho_f g x_1 - \Delta p) + (x/x_1)[c_3 - (\rho_f g x_1 - \Delta p)]} \tag{5.7}$$

Now, we know that $R(0) = R_{\max}$, and $R(x_1) = R_{\min}$, as described earlier, so that for equation (5.7) to be physically realistic, we require that

$$R_{\max} = R(0) = \frac{\sigma}{g\rho_f x_1 - \Delta p} > R_{\min} = R(x_1) = \frac{\sigma}{c_3} \tag{5.8}$$

or

$$c_3 > \rho_f g x_1 - \Delta p$$

Now, if we introduce the notation

$$A = \rho_f g x_1 - \Delta p$$

$$B = \frac{1}{x_1}[c_3 - (\rho_f g x_1 - \Delta p)]$$

equation (5.7) becomes

$$R(x) = \frac{\sigma}{A + Bx} \tag{5.9}$$

We make the entirely reasonable assumption that the total amount of liquid Q_l in the chamber is a constant, and then, if we have N_c liquid cones produced, we have the constraint that

$$Q_l = N_c \int_0^{x_1} \pi R^2(x)\, dx \tag{5.10}$$

Combining equations (5.9) and (5.10) and carrying out the integration, we find that

$$Q_l = N_c \pi \sigma^2 \frac{1}{B}\left(\frac{1}{A} - \frac{1}{A + Bx_1}\right)$$

or

$$Q_l = N_c \pi \sigma^2 \frac{1}{B}\left[\frac{1}{(\rho_f g x_1 - \Delta p)} - \frac{1}{(\rho_f g x_1 - \Delta p) + c_3 - (\rho_f g x_1 - \Delta p)}\right]$$

and, on using equation (5.8) this becomes

$$Q_l = N_c \pi \sigma \frac{1}{B}(R_{\max} - R_{\min}) \tag{5.11}$$

Provided that $R_{\max} \gg R_{\min}$,

$$B = \frac{1}{x_1}[c_3 - (\rho_f g x_1 - \Delta p)] \simeq \frac{c_3}{x_1} = \frac{\sigma}{x_1 R_{\min}}$$

and equation (5.11) is changed to

$$Q_l = N_c \pi x_1 R_{max} R_{min} \tag{5.12}$$

The cross-sectional area of the liquid chamber is approximately the same as the total cross-sectional area of the liquid cones at location $x = 0$, because at this location the gas jets are just beginning, and have their minimum diameter; in other words,

$$A \simeq N_c \pi R_{max}^2 \tag{5.13}$$

where A is the chamber cross-sectional area. Combining equations (5.12) and (5.13), we find

$$\frac{Q_l}{A} = \frac{x_1 R_{min}}{R_{max}} \tag{5.14}$$

For any given system, Q_l, A, and x_1 are known quantities, and R_{max}, the equivalent radius of the liquid cones at the location $x = 0$ (at the level of the orifices), can be estimated from the orifice configuration. Using the same definition of equivalent radius (the definition is not critically important), a value of R_{min}, the radius of the cones at the location $x = x_1$ (the free surface of the liquid), can be estimated using equation (5.14). When this is done it is found that, for a wide range of conditions, the value of R_{min} obtained is unrealistically small, and this result supports the conclusion implied by equation (5.4); that is, when the gas flow rate is such that the gas is injected into the liquid continuously rather than in the form of bubbles, the gas jet cannot be stable, and breaks down, in doing so entraining the liquid at the unstable interface.

5.3 Entrainment of turbulent eddies

During intensive bubbling (i.e., if the Froude number $\mathrm{Fr} = v_s^2/gh$ exceeds unity, where v_s denotes the superficial gas velocity or the ratio of volumetric flow rate to total cross-sectional area of flow, g is the acceleration due to gravity, and h is the height of the liquid chamber) the gas–liquid flow is markedly turbulent. Disordered motions, during which the various parameters (e.g., pressure, velocity, void fraction) undergo random changes in space and time, are generated in this mixture.

In this section we consider a possible entrainment mechanism in which the flowing gas interacts with the turbulent liquid, such that eddies which are sufficiently small are carried off by the gas.[8] We make the assumption that the eddies most likely to be so entrained are those in the "inertial subrange," because eddies smaller than this are strongly dominated by liquid viscosity and are thus less likely to be separated from the bulk of the liquid, and eddies larger than this are considered too large to be carried off by the gas. We first consider some basic characteristics of turbulent flows.

Turbulization of the bubbling flow results in superposition of motions with decreasing periods, or in other words, of eddies of decreasing size.[9] From experimental data it is known that turbulence may be represented by a succession of three main stages of eddy sizes, or of wavenumbers κ ($= 2\pi n/v$, where n denotes the turbulent frequency and v the time average of the velocity), in which, as a result of inertia interactions, energy is transmitted from a stage with high wavenumbers, through an intermediate stage, to a stage with low wavenumbers.

We should note that turbulent motions differ in systems of different construction and may also depend on the discharge rates of the two separate phases, and for this reason the time-averaged properties of a two-phase (gas–liquid) flow depend on the vertical position in the mixture. Averaged flow rates in the chamber also vary from one lateral point to another (due to the wall effect), so that fluctuating velocities vary from point to point and give rise to an interaction between fluctuating and averaged components of the motion, this interaction being related to the existence of transport effects. Passage of gas through, for example, a sieve plate having regularly arranged orifices, and mounted perpendicular to the flow, superimposes a random velocity distribution on the motion in the bubbling mixture. These random motions degenerate as the distance from the plate increases, and seem, therefore, not to be statistically uniform. However, the rate at which the turbulence degenerates is so small that for practical applications this turbulence can be taken to be uniform, and in this case the turbulence closely resembles uniform isotropic turbulence, whose characteristics do not depend on the position and direction of the coordinate axes. In practice, the turbulence of a gas–liquid mixture may be completely developed, and the energy spectrum then drops steeply at high frequencies n (in other words, at large wavenumbers κ); neglect of this fact may result in too high a predicted value for the entrainment of liquid from the two-phase mixture.

Now, although the wavenumber spectrum of an actual turbulent flow is not continuous, it is possible to assign a definite amount of the total energy to each value of the wavenumber; in other words, we can derive the relationship between the wavenumbers in the form of an energy spectrum.[10] According to Kolmogoroff's first law[11,12] "at sufficiently high Reynolds numbers there exists a range of high wavenumbers in which the turbulence is in static equilibrium, being defined unambiguously by the values of ϵ (energy dissipation) and ν (kinematic viscosity)." The spectral energy density is then given by the equation[10]

$$E(\kappa) = \left(\frac{8\epsilon}{9\gamma}\right)^{2/3} \kappa^{-5/3} \left(1 + \frac{8\nu_f^3\kappa^4}{3\gamma^2\epsilon}\right)^{-4/3} \tag{5.15}$$

where $E(\kappa)$ denotes the spectral density of the energy per unit mass contained in eddies with wavenumbers in the interval between κ and $(\kappa + d\kappa)$, ϵ denotes the energy dissipation per unit mass and unit time, $\gamma = 0.4$ is a universal constant, κ is the wavenumber ($\kappa = r^{-1}$, where r denotes the eddy size), and ν_f is the kinematic viscosity coefficient.

For high Reynolds numbers, in the subrange of wavenumbers of finite magnitude (inertial subrange) defined by the inequality

$$\kappa_e \ll \kappa \ll \kappa_d \tag{5.16}$$

the effect of dissipation is negligibly small compared with energy transmitted under the action of inertia forces, and in this subrange the energy spectrum is determined unambiguously by a single ϵ value and does not depend on ν_f. Here κ_e denotes the wavenumber defining the range of energy containing eddies,

$$\kappa_e = \frac{1}{r_e}$$

(r_e is the mean size of energy containing eddies), and κ_d denotes the wavenumber defining the range of marked effect of viscous forces, being equal to the inverse size of the smallest eddy,

$$\kappa_d = \frac{1}{r_d} = \left(\frac{3\gamma^2\epsilon}{8\nu_f^3}\right)^{1/4} \tag{5.17}$$

At wavenumbers within the inertial subrange of the energy spectrum, the viscous term in equation (5.15) can be neglected and the equation for the spectral energy density then becomes

$$E(\kappa) = \left(\frac{8}{3.6}\right)^{2/3} \epsilon^{2/3}\kappa^{-5/3} = 1.7\epsilon^{2/3}\kappa^{-5/3} \tag{5.18}$$

If $N(\kappa)$ denotes the number of eddies per unit mass of liquid and per unit interval of wavenumbers κ, and

$$e = \rho_f \tfrac{4}{3}\pi r^3 (v_e^2/2) \tag{5.19}$$

denotes the energy of one eddy, where ρ_f denotes the density of the liquid, r the radius of the eddy, and v_e the velocity of a liquid eddy, then the spectral density $E(\kappa)$ can be expressed by the number of eddies $N(\kappa)$ and by the energy e of one eddy by

$$E(\kappa) = N(\kappa)e \tag{5.20}$$

Substituting equation (5.18) for $E(\kappa)$ and equation (5.19) for e into equation (5.20), we get

$$E(\kappa) = 1.7\epsilon^{2/3}\kappa^{-5/3} = N(\kappa)\frac{2\pi\rho_f v_e^2}{3\kappa^3}$$

Hence it follows that the number of eddies is

$$N(\kappa) = 0.812\frac{\epsilon^{2/3}\kappa^{-5/3}\kappa^3}{\rho_f v_e^2} \tag{5.21}$$

According to Kolmogorov's law, for the inertial subrange of the spectrum,

$$v_e^2 = \beta(\epsilon/\kappa)^{2/3} \tag{5.22}$$

(the absolute constant $\beta = 8.2$) and equation (5.21) can be transformed into

$$N(\kappa) = \frac{0.812}{8.2}\frac{\epsilon^{2/3}\kappa^{-5/3}\kappa^3\kappa^{2/3}}{\rho_f\epsilon^{2/3}} = 0.1\frac{\kappa^2}{\rho_f} \tag{5.23}$$

Let us now assume the following mechanism for the entrainment of liquid from a two-phase flow. The gas intrudes between the liquid eddies (intrusion evidently being possible at the moment the eddy forms). If the gas is capable of lifting the eddy formed, liquid is atomized and entrained from the bulk of the liquid; but if the liquid eddy is too large and cannot be lifted by the gas, it remains in the bulk of the liquid.

We can determine the maximum radius r_0 of the liquid eddies that can be entrained by the gas from the liquid, and to do so, we set the force due to gravity on the liquid eddy equal to the lifting force exerted by the gas:

$$\tfrac{4}{3}\pi r_0^3\rho_f g = C_D\rho_g v^2\pi r_0^2$$

where $v(x)$ denotes the velocity of the gas in the liquid, and C_D is a drag coefficient. From this we derive

$$r_0 = \frac{3}{4}C_D\frac{\rho_g v^2}{\rho_f g}$$

or

$$\kappa_0 = \frac{4}{3}\frac{\rho_f g}{C_D\rho_g v^2} \tag{5.24}$$

Note that for $r_0 \leq r_d$ (or $\kappa_0 \geq \kappa_d$), turbulent entrainment is clearly impossible, because r_d denotes the smallest eddy size for which liquid viscosity is not significant. It is obvious that under this condition no turbulence will develop in the two-phase flow, since the mechanism of liquid atomization and entrainment from the system is also the mechanism of development of turbulence. So, for $r_0 \leq r_d$, the regime is a pure bubbling regime without entrainment of liquid. We introduce the dimensionless ratio

$$\eta = \frac{r_0}{r_d} = \frac{\kappa_d}{\kappa_0} \tag{5.25}$$

which characterizes the regime of entrainment of liquid from a two-phase mixture during bubbling, so that for $\eta < 1$ we have no liquid entrainment, and for $\eta > 1$ we have entrainment.

Now, the mass of liquid entrained, Y, per unit liquid mass and time from the system is proportional to the number of eddies formed per unit time and is capable of being entrained by the gas:

$$Y = c_4 \int_{\kappa_0}^{\kappa_d} N(\kappa) \frac{4}{3} \frac{\pi \rho_f}{\kappa^3} \frac{1}{\tau(\kappa)} \, d\kappa$$

where c_4 is a constant of proportionality. The formation time of an eddy of size r is

$$\tau(\kappa) = \frac{r}{v_e(r)} = \frac{1}{\kappa v_e(\kappa)}$$

and substituting the value of $v_e(\kappa)$ given by equation (5.22), we get

$$\tau(\kappa) = 0.35 \left(\frac{1}{\epsilon^{1/3}} \right) \frac{1}{\kappa^{2/3}} \tag{5.26}$$

Substituting the values of $N(\kappa)$ given by equation (5.23) and $\tau(\kappa)$ given by equation (5.26) into the entrainment expression yields

$$Y = c_4 \epsilon^{1/3} \int_{\kappa_0}^{\kappa_d} \frac{d_\kappa}{\kappa^{1/3}}$$

Carrying out the integration, we obtain

$$Y = c\epsilon^{1/3}(\kappa_d^{2/3} - \kappa_0^{2/3}) \tag{5.27}$$

where c is a constant. Utilizing equation (5.25), we transform equation (5.27) as follows:

$$Y = c\epsilon^{1/3}\kappa_d^{2/3}(1 - \eta^{-2/3}) \tag{5.28}$$

Equation (5.28) is the final result of the analysis of the contribution of turbulent motion to liquid entrainment. If we can find for any gas–liquid flow system being considered (e.g., two-phase flow in a pipe, gas injection through an orifice) the quantities κ_d, η (or κ_0), and ϵ, then the equation gives us a value for the mass of liquid per unit liquid mass and time that is entrained by the gas. Equations (5.17) and (5.24), in fact, give κ_d and κ_0 in terms of flow system properties; further, in combination with equation (5.25), they indicate whether or not liquid entrainment can occur for that flow system. Equation (5.25) also defines the region of applicability of equation (5.28); namely, $\eta \geq 1$. Finally, note that the constant c must be determined empirically.

Let us consider as an example a flow system consisting of air being injected into water from below, through an array of orifices. Let the depth of liquid h be 2 cm (0.79 in.), the superficial gas velocity v_s be 25 cm/s (9.84 in./s), and the pressure drop in the air during passage through the water, Δp, be 2 cm water (equivalent to 196 Pa, 0.9 lbf/in.^{2}). The kinematic viscosity of water, ν_f, is 1×10^{-6} m^2/s (1×10^{-5} ft^2/s), the density of water ρ_f is 10^3 kg/m^3 (62.4 lb/ft^3), and the density of air ρ_g is 1.2 kg/m^3 (7.49×10^{-2} lb/ft^3).

Now, instead of the superficial gas velocity v_s, we need the local gas velocity v, and the two are related by the equation

$$v = v_s/\phi \tag{5.29}$$

where ϕ is the void fraction, that is, the ratio of the instantaneous volume of the gas to the total volume. The void fraction can be obtained from [13]

$$\begin{aligned}\phi &= \frac{2\mathrm{Fr}^{1/2}}{1 + 2\mathrm{Fr}^{1/2}} \\ &= \frac{2(v_s^2/gh)^{1/2}}{1 + 2(v_s^2/gh)^{1/2}}\end{aligned} \tag{5.30}$$

where Fr is the Froude number. Equation (5.30) is identical with equation (4.38) derived in Section 4.3. Using this formula, we obtain

$$\phi = 0.53$$

Therefore, from equation (5.29) we find that

$$v = 25/0.53 = 47 \text{ cm/s } (1.54 \text{ ft/s})$$

The energy dissipation per unit mass and time is given by

$$\epsilon = \frac{v\,\Delta p}{\rho_f h} = 4.6 \text{ J/kg}\cdot\text{s } (49 \text{ ft lbf/lb}\cdot\text{s})$$

We can then find the size r_d of the smallest eddy formed in the liquid from equation (5.17):

$$r_d = \left(\frac{8\nu_f^3}{3\gamma^2\epsilon}\right)^{1/4} = 4.36 \times 10^{-5} \text{ m } (1.7 \times 10^{-3} \text{ in.})$$

On the other hand, the largest eddy capable of being carried off by the air flow is given by equation (5.24):

$$r_0 = \frac{3}{4} C_D \frac{\rho_g}{\rho_f} \frac{v^2}{g}$$

and the drag coefficient is usually of the order unity, so we can write

$$r_0 \simeq \frac{3\rho_g v^2}{4\rho_f g} = 2.02 \times 10^{-5}\ \text{m}\ (8 \times 10^{-4}\ \text{in.})$$

Hence, for this example, $r_0 < r_d$ and the smallest eddy present in the turbulent liquid is too large to be entrained by the gas, so we have no entrainment.

5.4 Entrainment at the liquid free surface during bubbling

Sections 5.2 and 5.3 considered mechanisms of liquid entrainment by the gas in interior regions of the liquid. We now consider possible mechanisms of entrainment in the form of liquid droplets at the upper (free) surface of the liquid.

5.4.1 Cellular foam conditions

When gas bubbles through a liquid layer, the bursting of the shells of the bubbles coming to the surface and the simultaneous formation of droplets give rise to a considerable decrease in the total surface area of the bubbles. This decrease is usually many times greater than the increase in the total surface area contributed by the newly formed drops, and therefore drop formation and propulsion during bubbling must be due not only to the kinetic energy of the gas but also to the surface energy released when bubbles burst.

At a moderate rate of bubbling (at a small superficial gas velocity) and with a considerable depth of liquid through which bubbles rise, which is typical of much engineering equipment, the kinetic energy of the gas approaching the surface is relatively small and the surface energy of the bubble shells plays a major role in the total energy balance. For example, the average velocity of gas rising in liquid under conditions of high-pressure steam boilers ($p \simeq$ 100 to 120 atm) does not exceed 0.7 m/s (2.3 ft/s) and in low-pressure distillation columns ($p \simeq$ 1 atm) this velocity has a maximum of about 2 to 3 m/s (6.6 to 9.9 ft/s). Taking into account that the kinetic energy of a bubble is

$$L_k = \tfrac{1}{2} V \rho_g v_b^2$$

where V is the volume and v_b is the velocity of the bubble, and that the surface energy of a bubble is

$$L_\sigma = \pi d_b^2 \sigma$$

where d_b is the bubble diameter and σ is the surface tension, we can make a rough estimate of L_k and L_σ for these two cases. Letting, for illustration, the mean bubble diameter be $d_b = 3$ mm (0.12 in.) in a steam boiler and $d_b = 5$ mm (0.2 in.) in a distillation column, we get:

1. For high-pressure boilers:

$$\sigma \simeq 1.75 \times 10^{-2} \text{ N/m } (1 \times 10^{-4} \text{ lbf/in.})$$

$$L_k \simeq 4.2 \times 10^{-9} \text{ J } (3.3 \times 10^{-9} \text{ lbf·ft})$$

$$L_\sigma \simeq 5 \times 10^{-7} \text{ J } (3.94 \times 10^{-7} \text{ lbf·ft})$$

2. For distillation columns:

$$\sigma \simeq 7.28 \times 10^{-2} \text{ N/m } (4.15 \times 10^{-4} \text{ lbf/in.})$$

$$L_k \simeq 2 \times 10^{-7} \text{ J } (1.6 \times 10^{-7} \text{ lbf·ft})$$

$$L_\sigma \simeq 5.7 \times 10^{-6} \text{ J } (4.5 \times 10^{-6} \text{ lbf·ft})$$

Thus, under these conditions, the total kinetic energy of the gas is much less than released surface energy, and an analysis of the mechanism by which the surface energy affects droplet formation is of clear interest. At low gas flow rates, the gas breaks through the liquid mass, at regular intervals, in the form of bubbles. When the superficial gas velocity is increased, a cellular foam layer, consisting of tightly packed bubbles, much enlarged and deformed, is developed on the surface of the liquid, and at a given ratio of flow rates of gas and liquid, all the liquid may be changed into cellular foam.[14]

A possible mechanism for liquid entrainment under foaming conditions is suggested by experimental investigations[15] of this phenomenon in the *bubble-by-bubble regime,* where it is observed that at low flow rates of gas through individual orifices the bubbles do not coalesce to form jets when coming to the surface. Under these cellular foam conditions, then, drops of liquid are usually generated by bursting of the bubble film, and can be subsequently entrained by the gas escaping from the cellular foam layer at the liquid surface.

5.4.2 Bubble breakup

We now develop a mathematical model to describe entrainment of liquid due to bubble breakup at the free surface of the liquid.[16] When a gas bubble is released in a stagnant liquid, it will rise, under the action of the buoyancy force F_b to the liquid surface, where it is arrested by the surface tension force F_σ exerted by the liquid membrane, with base radius r_m, protruding out of the liquid surface and forming the top part of the bubble boundary, as shown in Figure 5.2. We assume, for simplicity, that the bubble remains spherical in shape, with radius r_b, and with its center 0 located at a distance s, below the liquid surface, under equilibrium conditions. We further assume that near the edge of the liquid membrane the surface-tension force, F_σ, acts in a direction tangential to the boundary of the bubble. Now, the buoyancy force, F_b, acting in the vertical direction is

$$F_b = V_d(\rho_f - \rho_g)g \tag{5.31}$$

where V_d is the displacement volume of the bubble, or the volume of that portion of the bubble which lies below the liquid surface. The vertical component of the surface-tension force acting on the edge of the membrane is

$$F_{\sigma_v} = 2\pi r_m \sigma \cos\theta \tag{5.32}$$

where θ is the angle between the horizontal and the bubble radius ending at the edge of the membrane, r_m the base radius of the membrane, and σ the interface surface tension.

Figure 5.2 Sketch of equilibrium bubble configuration.

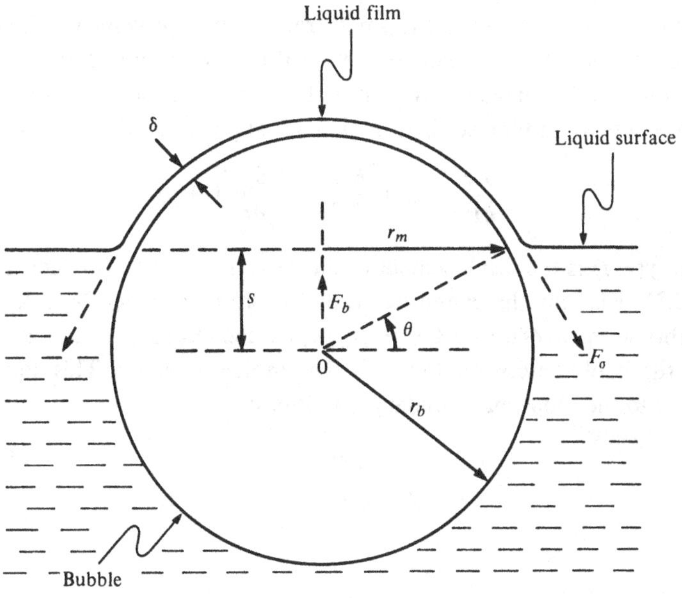

For use in equations (5.31) and (5.32), the following relations can be obtained from geometric considerations:

$$V_d = \tfrac{2}{3}\pi r_b^3 + \pi s r_b^2 - \tfrac{1}{3}\pi s^3 \tag{5.33}$$

$$\cos\theta = \left(1 - \frac{s^2}{r_b^2}\right)^{1/2} \tag{5.34}$$

$$r_m = \left(r_b^2 - s^2\right)^{1/2} \tag{5.35}$$

At equilibrium, $F_b = F_{\sigma_v}$, and therefore we have from equations (5.31), (5.32), (5.33), (5.34), and (5.35),

$$\eta^3 - \beta\eta^2 - 3\eta + (\beta - 2) = 0$$

where

$$\eta = \frac{s}{r_b} \quad \text{and} \quad \beta = \frac{6\sigma}{(\rho_f - \rho_g) g r_b^2}$$

This cubic equation has a root at $\eta = -1$, but the particular root we are concerned with will be given by

$$\eta^2 - (\beta + 1)\eta + (\beta - 2) = 0$$

or

$$\eta = \tfrac{1}{2}[(\beta + 1) - (\beta^2 - 2\beta + 9)^{1/2}] \tag{5.36}$$

Now, in a manner similar to the case of a thin, planar elastic membrane of uniform thickness subject to small transverse displacements, the oscillation of the bubble membrane is governed by the wave equation, with the addition of an inhomogeneous term that accounts for the static pressure difference across the film, so that, in cylindrical polar coordinates,[17] we have

$$\frac{\partial^2 y}{\partial t^2} = a^2\left(\frac{\partial^2 y}{\partial r^2} + \frac{1}{r}\frac{\partial y}{\partial r}\right) + p \tag{5.37}$$

where $y(r, t)$ is the displacement in the vertical direction (as shown in Figure 5.3), r is now the cylindrical radial coordinate, $a^2 = 2\sigma/\rho_f\delta$, δ is the membrane thickness, $p = (p_g - p_a)/\rho_f\delta$ is the bubble pressure increment, p_g is the bubble pressure, and p_a is the ambient pressure. This equation is subject to the following boundary conditions:

$$y(r, t) = 0 \tag{5.38a}$$

$$\frac{\partial y(0, t)}{\partial r} = 0 \tag{5.38b}$$

We will let the membrane displacement $y(r, t)$ be made up of two parts, one steady, y_0, the other time-dependent, $y'(r, t)$:

$$y(r, t) = y_0(r) + y'(r, t) \tag{5.39}$$

Introducing equation (5.39) into equations (5.37) and (5.38), we have for each of the new dependent variables, $y_0(r)$ and $y'(r, t)$, a simpler governing equation. For the steady displacement $y_0(r)$, we have the governing equation

$$\frac{d^2y_0}{dr^2} + \frac{1}{r}\frac{dy_0}{dr} = -\frac{p}{a^2} \tag{5.40}$$

subject to the boundary conditions

$$y_0(r_m) = 0$$

$$\frac{dy_0(0)}{dr} = 0$$

and for the time-dependent displacement, $y'(r, t)$, we have from equations (5.37) and (5.40),

$$\frac{\partial^2 y'}{\partial t^2} = a^2\left(\frac{\partial^2 y'}{\partial r^2} + \frac{1}{r}\frac{\partial y'}{\partial r}\right) \tag{5.41}$$

subject to the boundary conditions

$$y'(r_m, t) = 0 \tag{5.42a}$$

$$\frac{\partial y'(0, t)}{\partial r} = 0 \tag{5.42b}$$

A solution of equation (5.40) for the steady displacement $y_0(r)$, satisfying the boundary conditions, can then be readily obtained as

$$y_0(r) = c\left(1 - \frac{r^2}{r_m^2}\right)$$

where $c = pr_m^2/4a^2$.

Figure 5.3 Sketch of film cap in oscillation.

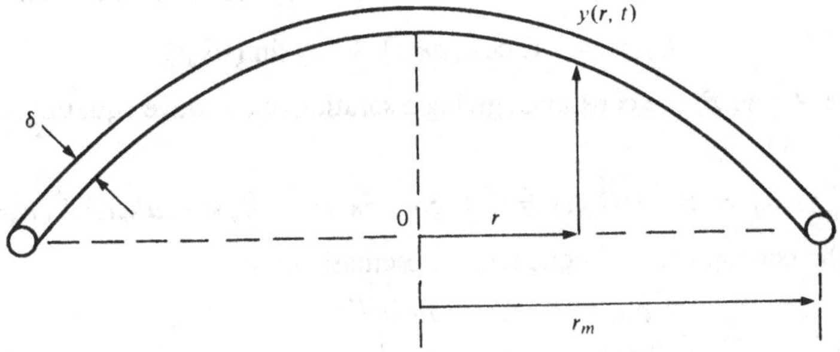

Now, the time-dependent displacement $y'(r, t)$ is governed by the wave equation, equation (5.41), subject to the homogeneous boundary conditions, equations (5.42), similar to the case of oscillations of a flat circular elastic membrane. Therefore, a solution for $y'(r, t)$ is readily obtained by the method of separation of variables in the form

$$y'(r, t) = W(r)G(t)$$

and substituting this into equation (5.41), we have

$$\ddot{G} + (ak)^2 = 0, \qquad k = \text{const.} \tag{5.43}$$

$\ddot{G}$ denotes d^2G/dr^2, and

$$W'' + \frac{1}{r} W' + k^2 W = 0 \tag{5.44}$$

where W' and W'' denote dW/dr and d^2W/dr^2.

Introducing the substitution $S = kr$, we obtain from equation (5.44)

$$\frac{d^2W}{dS^2} + \frac{1}{S}\frac{dW}{dS} + W = 0$$

This is a form of Bessel's equation, and for our application has the solution

$$W(r) = J_0(S) = J_0(kr)$$

where J_0 is the Bessel function of the first kind and order zero. Using the boundary conditions, equations (5.42), we find that

$$J_0(k_n r_m) = 0$$

where $k_n = \alpha_n/r_m$ and $\alpha_n(n = 1, 2, 3, \ldots)$ are the positive zeros of the Bessel function J_0. Hence

$$W_n(r) = J_0\left(\frac{\alpha_n r}{r_m}\right), \qquad n = 1, 2, 3, \ldots$$

is the solution of equation (5.44) with boundary conditions given by equations (5.42).

The corresponding solutions of equation (5.43) are easily obtained as

$$G_n(t) = A_n \cos(ak_n t) + B_n \sin(ak_n t)$$

where A_n and B_n are constants, giving a solution of the wave equation, equation (5.41),

$$y'_n(r, t) = W_n(r)G_n(t) = [A_n \cos(ak_n t) + B_n \sin(ak_n t)]\, J_0(k_n r)$$

and the corresponding frequencies of oscillation are

$$\Omega_n = \frac{a\alpha_n}{2\pi r_m} = \frac{\alpha_n}{2\pi r_m}\left(\frac{2\sigma}{\rho_f \delta}\right)^{1/2}, \qquad n = 1, 2, \ldots$$

the first of which is the fundamental normal mode,

$$\Omega_1 = \frac{\alpha_1}{2\pi r_m}\left(\frac{2\sigma}{\rho_f \delta}\right)^{1/2} \tag{5.45}$$

where $\alpha_1 \simeq 2.404$.

We thus have a formula specifying the natural frequency of oscillations of the membrane; we now need to obtain a formula for the oscillation frequency of the gas in the bubble. It is well known that, for a compressible fluid (the gas), the velocity potential, ϕ, of the motion is described by the wave equation:[18,19]

$$\frac{\partial^2 \phi}{\partial t^2} = c_s \left(\frac{\partial^2 \phi}{\partial r^2} + \frac{2}{r}\frac{\partial \phi}{\partial r}\right) \tag{5.46}$$

where c_s is the speed of sound in the gas and r is now the radial distance from the bubble center, in spherical coordinates. Note that in equation (5.46) we have assumed spherical symmetry for the flow in the bubble interior. We need boundary and initial conditions to obtain a solution of this equation, so we consider the gas to be constrained in a rigid sphere (the bubble surface). This is, of course, not the case for the bubble, but our intention here is only to obtain an approximate value for the natural frequencies of the gas motion, so we take as the boundary condition

$$\left.\frac{\partial \phi}{\partial r}\right|_{r=r_b} = 0 \tag{5.47}$$

which states that the normal velocity of the gas at the bubble surface is zero. As for the analysis of the membrane motion, we use the method of separation of variables,

$$\phi(r, t) = T(t)R(r)$$

which, in conjunction with equation (5.46), produces

$$\ddot{T}(t) + \lambda^2 c_s T(t) = 0 \tag{5.48}$$

and

$$R''(r) + \frac{2}{r}R'(r) + \lambda^2 R(r) = 0 \tag{5.49}$$

where λ is a constant and $\ddot{X}$, X', and X'' denote differentiations, as in equations (5.43) and (5.44), with the boundary condition, equation (5.47), being

$$\left.\frac{dR}{dr}\right|_{r=r_b} = 0 \tag{5.50}$$

The general solution of equation (5.47) is of the form

$$R(r) = C_1 \frac{\sin \lambda r}{r} + C_2 \frac{\cos \lambda r}{r}$$

where C_1 and C_2 are constants. Clearly, $C_2 = 0$, because otherwise the solution would diverge as r approaches zero and C_1 can be set equal to unity without loss of generality. Hence we can write

$$R(r) = \frac{\sin \lambda r}{r}$$

and using this in the boundary condition, equation (5.50), we obtain

$$\tan \lambda r_b = \lambda r_b \tag{5.51}$$

If we let the eigenvalues of equation (5.51) be λ_n, $n = 1, 2, \ldots$, the solution of equation (5.48) will be of the form

$$T_n(t) = C_3 \cos(\lambda_n c_s t) + C_4 \sin(\lambda_n c_s t)$$

and the solution of equation (5.46) is of the form

$$\phi_n(r, t) = \frac{\sin \lambda_n r}{r} [C_3 \cos(\lambda_n c_s t) + C_4 \sin(\lambda_n c_s t)]$$

where C_3 and C_4 are constants. The natural frequencies of the gas motion are then

$$\omega_n = \frac{\lambda_n c_s}{2\pi}, \qquad n = 1, 2, \ldots$$

the first of which is the fundamental normal mode:

$$\omega_1 = \frac{\lambda_1 c_s}{2\pi} \tag{5.52}$$

where $\lambda_1 \simeq 4.493/r_b$, this being obtained by solving equation (5.51).

We can now postulate that the amplification of the membrane oscillations, fed by background acoustic noise in the liquid, is mainly responsible for the rupture, and we consider that the direct coupling of the membrane to the bulk liquid at its rim is generally not sufficient for this feeding, but that the gas bubble cavity plays the central role of an intermediary in the process. When a bubble is stable, coupling via the cavity is weak, with the membrane-cavity system oscillating passively in the manner of a kettledrum. When the natural frequencies of the membrane and the cavity are close, however, this coupling becomes significant, and we propose that the latter condition is achieved at some point in the course of membrane draining or runoff. Since the speed of sound in the bulk liquid far exceeds both that of the gas cavity and the speed of the associated membrane wave, only fundamental modes of motion of the cavity and the membrane can be expected to receive efficient excitation, and then the onset of rupture is interpreted as the state in which this strong coupling allows ambient energy to preferentially enter the membrane oscillation in its fundamental mode.

Subsequently, with the amplitude of motion increasing progressively beyond the linear range, secondary pumping, through mode coupling, from the fundamental to the higher modes then becomes possible, and this stage is just prior to the final disintegration of the membrane into droplets.

We thus arrive at the condition for rupture of the membrane by equating the fundamental frequencies of oscillation of the membrane, Ω_1, and of the gas cavity, ω_1, from equations (5.45) and (5.52), respectively, to get

$$\frac{r_m^2}{r_b^2} = \frac{0.5726\sigma}{\rho_f \delta c_s^2}$$

which, by use of equation (5.35), can be written as

$$\eta = (1 - \gamma)^{1/2} \tag{5.53}$$

where

$$\gamma = \frac{0.5726\sigma}{\rho_f \delta c_s^2} \quad \text{and} \quad \eta = \frac{s}{r_b}$$

as defined previously.

The combination of equations (5.36) and (5.53) provides a relationship between the bubble size, r_b, and the critical membrane thickness, δ_c, at which the condition of rupture is established:

$$\gamma = \tfrac{1}{2}[(\beta + 1)(\beta^2 - 2\beta + 9)^{1/2} - (\beta^2 + 3)]$$

A plot of this in terms of the nondimensional bubble radius

$$r_b \left[\frac{(\rho_f - \rho_g)g}{6\sigma}\right]^{1/2} = \frac{1}{\beta^{1/2}}$$

and the nondimensional critical film thickness

$$\delta_c \left(\frac{\rho_f c_s^2}{0.5726\sigma}\right) = \frac{1}{\gamma}$$

is shown in Figure 5.4.

The mass of liquid entrained due to the bursting of a single gas bubble can then be computed from the total mass of the liquid in the membrane at the onset of rupture:

$$M_b = \rho_f \delta_c A_m \tag{5.54}$$

where $A_m = 2\pi r_m (r_b - s)$ is the surface area of the membrane. Introducing equations (5.36) and (5.53) into equation (5.54), we have, after simplification, the liquid entrainment from one gas bubble:

$$M_b = k_1 r_b^2 \left\{\frac{(1 - 2r_b^2/k_2 + 9r_b^4/k_2^2)^{1/2} + [(r_b^2/k_2) - 1]}{(1 + 3r_b^2/k) - (1 - 2r_b^2/k_2 + 9r_b^4/k_2^2)^{1/2}}\right\}^{1/2} \tag{5.55}$$

where

$$k_1 = \frac{1.15\pi\sigma}{c_s^2} \quad \text{and} \quad k_2 = \frac{6\sigma}{(\rho_f - \rho_g)g}$$

We are now in a position to compare this theory, of liquid entrainment due to bubble rupture, with experimental results. A study[15] of droplet formation due to bubbles collapsing at the gas–liquid interface indicates that droplets of two distinctly different size ranges are generated; small droplets, due to the disintegration of the bubble membranes, and large droplets, which appear only for bubbles of small enough size and which are due to the breaking up of a liquid jet induced from the bubble cavity. Owing to a limitation on the measuring technique, droplets smaller than 20 μm, which were expected to show up in large numbers, were not measured. However, the number of droplets from one single bubble of a given size was reported for each of these two distinctly separated droplet size ranges for a number of bubble sizes; the corresponding sizes of droplets were evaluated on the basis of the Sauter mean diameter, $d_{s.m.} = \Sigma\, d_b^3/\Sigma\, d_b^2$. The use of the Sauter mean diameter is probably more justifiable for the larger droplets than for the smaller droplets, because the total surface area per unit mass of the smaller droplets is significantly greater than that of the larger droplets. For these reasons, no quantitative comparison can be made between the theoretical result we have obtained and the experimental results for the small-droplet-size range, which relates only to the disintegration of the membrane.

Figure 5.4 Critical film thickness versus bubble size.

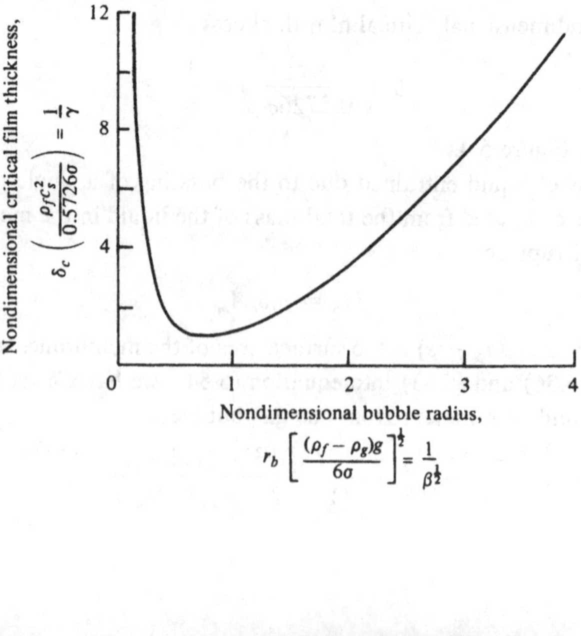

However, it might not be unreasonable to expect that the mean size of droplets that are generated as a result of the disintegration of the bubble membrane will be related to the critical thickness of the membrane at which disintegration first takes place, and in fact, the dependence of the experimental mean droplet size for small droplets on bubble size has been found to agree quite well qualitatively with the dependence on the theoretical critical film thickness, as shown in Table 5.1.

A similar experiment,[20] over a much wider bubble-size range, measured the size distributions of both the larger and the smaller droplets, which were created by the bursting air bubbles, down to a size of 4.9 μm; it yielded results in terms of the liquid mass entrainment per unit mass of air as a function of the bubble size. Knowing the bubble size, we can readily obtain the mass of air in the bubble by first computing the static pressure inside the bubble, with the surface-tension effect included, and then the total liquid mass entrainment per air bubble is simply the product of the specific (per unit mass of air) liquid mass entrainment and the mass of air so computed. By this method, the specific mass entrainment data can be compared with the corresponding theoretical results of equation (5.55), as shown in Figure 5.5. It should be noted that only results for bubbles with radii greater than 0.25 cm have been included in this comparison, because droplets from bubbles above 0.25 cm radius were found to be produced entirely from the collapse of the bubble dome.[20] Close agreement on liquid entrainment is found between theory and experiment for bubbles of these larger sizes, with the exception of a much lower measured liquid entrainment than that predicted

Table 5.1. *Comparison of dependence of experimental mean droplet size for the small-droplet-size range and theoretical film thickness on bubble size*

Bubble radius, r_b (cm)	Experimental Sauter mean diameter of small droplets,[15] $d_{s.m.}$ ($\times 10^6$ m)	Theoretical nondimensional critical film thickness, $\delta_c\left(\frac{\rho_f c_s^2}{0.5726\sigma}\right)$	The ratio $\frac{d_{s.m.}}{\delta_c(\rho_f c_s^2/0.5726\sigma)}$ ($\times 10^6$ m)
0.265	22	1.80	12.2
0.233	26	2.16	12.0
0.205	34	2.70	12.6
0.180	45	3.60	12.5
0.156	58	4.70	12.3

by the theory for the case of the largest bubble. This discrepancy could be due to some extent to shortcomings in the measuring technique employed to determine the size and number distributions of droplets.

5.5 Droplet dynamics

In Section 5.4 we discussed a mechanism for the production of liquid droplets during the bubbling process, and we now consider the behavior of droplets in the gas flow. After formation, the drops move with a considerable initial velocity, which can cause them to rise (in a practically motionless gas) to a great height. For example, it has been[21] observed that drops were ejected in air to more than 2 m (6.56 ft) above a bubbling air–water mixture. However, such a height is reached only by individual drops that have the greatest velocity and nearly vertical direction of flight at the instant of their breakoff, and the bulk of the drops are ejected to a considerably lesser height.

For purposes of illustration, we consider a flow system consisting of an orifice (or orifices) through which gas is being passed into liquid of finite depth. As the gas reaches the liquid free surface, droplets are entrained by one or more of the mechanisms described earlier in this chapter, and possibly by other mechanisms. We define h_s to be the perpendicular distance

Figure 5.5 Comparison of theory with experimental data on liquid entrainment from film cap disintegration.

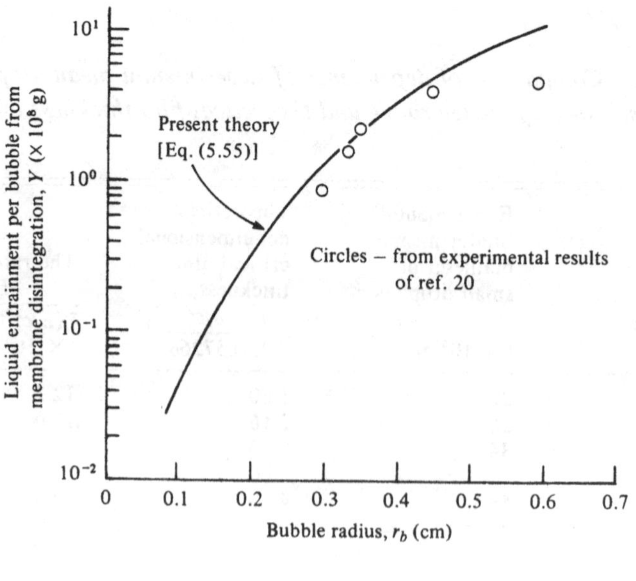

from the liquid free surface to the height at which the liquid entrainment is being evaluated, and obviously the liquid entrainment measured will be a function of the location h_s of the observation point. From experimental data[22] it is apparent that an increase in the height of the gas space h_s from 0.28 to 0.7 m (11 to 27.5 in.) decreases the entrainment coefficient $Y = \Sigma M_f/M_g$ (where ΣM_f is the total mass of entrained droplets per unit time and M_g is the mass of gas per unit time) from 0.16 to 0.0007, or over two orders of magnitude. In other words, at the height $h_s = 0.7$ m (27.5 in.), less than 0.5 percent of all the droplets rising from the surface of the liquid to a height of 0.28 m (11 in.) were entrained. It would be of interest to determine whether the droplet concentration becomes zero at some finite value of h_s, or if, for a given superficial gas velocity v_s, some droplets will always be transported by the gas flow to any value of h_s.

It is useful to consider under what conditions the liquid droplets will tend to rise or settle in the gas space above the liquid level, and at what rate. Whether settling or rising occurs depends upon the size, shape, viscosity, and density of the drops, and the density of the gas, and as a model, we consider the dispersion of droplets in gas where they are sufficiently separated that there are no collisions or other interactions between droplets. Let us assume spherical droplets of density ρ_f and diameter d in a gas of density ρ_g and viscosity μ_g. In this case the gravitational force causing a droplet to fall ($\rho_f > \rho_g$) will be

$$F_{\text{grav}} = \frac{\pi d^3}{6}(\rho_f - \rho_g)g$$

Motion of the droplet in gas results in the drag force

$$F_D = C_D \frac{\rho_g v_r^2}{2} \frac{\pi d^2}{4}$$

where C_D is the drag coefficient and v_r the velocity of the droplet relative to the gas.

After an initial period of acceleration or deceleration, the droplet attains a uniform velocity (called the *terminal settling velocity*) under which the drag force just balances the gravitational force, and this settling velocity may then be obtained from

$$\frac{\pi d^3}{6}(\rho_f - \rho_g)g = C_D \frac{1}{2} v_r^2 \frac{\pi d^2}{4} \rho_g \tag{5.56}$$

and

$$v_r = 1.155 \left[\frac{g(\rho_f - \rho_g)d}{\rho_g C_D}\right]^{1/2}$$

For turbulent flow $C_D \simeq 0.45$ ($\mathrm{Re} = \rho_g v_r d/\mu_g > 800$) and then

$$v_r = 1.72\left[\frac{g(\rho_f - \rho_g)d}{\rho_g}\right]^{1/2} \tag{5.57}$$

Equation (5.57) indicates that the terminal relative velocity of a liquid droplet in the gas increases with droplet size and decreases with gas density. Thus for gases of high density, or for very small droplets, resistance of the gas begins to play a noticeable role in droplet trajectories. For example, for droplets 0.2 mm (0.008 in.) in diameter flying up vertically at an initial velocity of 2 m/s (6.56 ft/s) in motionless air, the height of ascent decreases from 0.2 m (8 in.) at $p = 1$ atm ($\rho_g = 1.2$ kg/m^3, 7.49×10^{-2} lb/ft^3) to 10 mm at $p = 110$ atm ($\rho_g = 100$ kg/m^3, 7.49 lb/ft^3).

Under considerable gas velocities we must take into account the effect of gas velocity on the maximum height of droplet ascent. The presence of a high superficial gas velocity, v_s, increases the total rise height of all the droplets, and this circumstance becomes of significance for entrainment when the quantity v_s becomes comparable with the settling velocity of the droplets. When $v_s \geq v_r$ (where we are considering absolute values), the height of ascent of droplets becomes unlimited, and they will be carried off with the gas stream. Liquid droplets, torn away from the liquid at the initial stage of their trajectory, may rise with a velocity considerably exceeding the upward velocity of the gas, and in this case, droplet motion is noticeably slowed by drag of the gas, this slowing down beiug especially pronounced for small droplets and at high gas density. As the upward velocity of the droplets decreases, the drag of the gas decreases rapidly, and for $v_r \leq v_s$ it becomes negative, so that the gas stream begins to entrain the drops upward.

Thus we see that if the gas superficial velocity v_s exceeds in magnitude the relative settling velocity of the droplets, they will be carried upward by the gas to an unlimited height (completely transportable droplets), but if the settling velocity of the droplets is greater than the velocity of the gas, the droplets, after losing their initial energy, begin to fall with a velocity equal to $(v_r - v_s)$. The maximum height to which such droplets rise depends on the vertical component of the initial velocity of the drops, on the settling velocity v_r, and on the upward velocity of the gas v_s. When $v_r \gg v_s$, the rise height of the drops depends only slightly on v_s, and this is a flow of nearly complete ejection. On the other hand, when the quantities v_r and v_s are approximately equal, the height of ascent will be determined almost entirely by the value v_s, and thus this is a flow of droplet transportation.

We now consider the effect of the walls of the enclosing chamber on droplet motion. Formulas for calculation of liquid entrainment are usually obtained assuming wall-free conditions, when the geometrical configuration (cross-sectional area) of the chamber in which the gas–liquid system moves is not taken into account. Evaluation of the effect of the finite dimensions of the chamber (using a suitably defined equivalent diameter) on the ascent velocity of a single droplet is the first step in a solution of the problem of a cloud of droplets in the same chamber. Consider the steady flow of gas–liquid droplet dispersion in a vertical duct of uniform cross section, where the diameter of a drop is comparable with the diameter of the duct (Figure 5.6). In this case the drag of the drop is determined by the relative velocity

$$v_r = v_d - v_s$$

Figure 5.6 Flow of bubbles in a narrow duct.

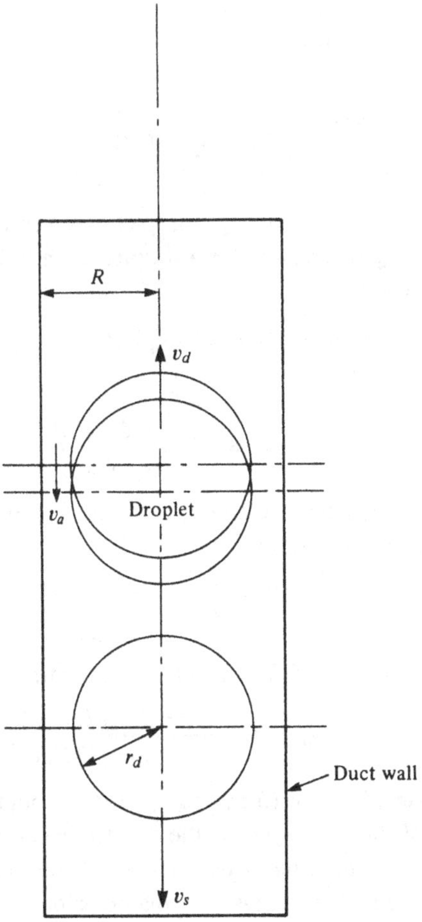

in which v_d is the absolute velocity of the drop and v_s the superficial velocity of the gas.

Let us first evaluate the gas velocity v_a in the annular space between the drop and the wall by considering the equivalence of the volume of the liquid moving up and the volume of gas moving down in the annular space, in a coordinate system moving at the superficial gas velocity:

$$(v_d - v_s)\pi r_d^2 = -(v_a - v_s)\pi(R^2 - r_d^2)$$

or

$$v_d \pi r_d^2 = -v_a \pi(R^2 - r_d^2) + v_s \pi R^2$$

where r_d is the droplet radius and $R = D/2$ is the equivalent radius of the chamber. After manipulation we obtain

$$\frac{v_d r_d^2}{v_s R^2} = \frac{-v_a}{v_s}\left(1 - \frac{r_d^2}{R^2}\right) + 1$$

and

$$\frac{v_a}{v_s} = \frac{1 - v_d r_d^2/v_s R^2}{1 - r_d^2/R^2} \tag{5.58}$$

The terminal velocity of a single droplet is determined by balancing gravitational and drag forces, and for a droplet in an infinite medium we have, as in equation (5.56),

$$\tfrac{4}{3}\pi r_d^3 \rho_f g = C_D \pi r_d^2 \tfrac{1}{2}\rho_f (v - v_{d\infty})^2$$

or

$$v_{d\infty} = v - \left(\frac{8 r_d g}{3 C_D}\right)^{1/2} \tag{5.59}$$

where $v_{d\infty}$ is the droplet absolute velocity in an infinite medium, and for a single drop in a finite medium

$$v_f = v_a - \left(\frac{8 r_d g}{3 C_D}\right)^{1/2} \tag{5.60}$$

Combining equations (5.59) and (5.60), we obtain

$$v_d = v_{d\infty} \frac{v_a - (8 r_d g/3C_D)^{1/2}}{v - (8 r_d g/3C_D)^{1/2}} \tag{5.61}$$

By combining equations (5.58) and (5.61), we obtain a formula illustrating the effect of the ratio r_d/R on the droplet velocity, and we now consider empirical evidence for such a dependence. According to measurements,[23] the minimum size of the equivalent diameter of a chamber in which this wall correction is to be negligible is at least 10 (and preferably 15) times the diameter of the largest droplet occurring in the two-phase flow. Experiments disclose a more or less substantial influence of the ratio r_d/R (or d/D) on v_d. For example, below[24] are given the empirical correction factors K by which the free ascent velocity should be multiplied to obtain the actual velocity of the droplet:

1. For laminar flow,

$$K = 1 - \left(\frac{d}{D}\right)^2 \tag{5.62}$$

2. For turbulent flow,

$$K = 1 - \left(\frac{d}{D}\right)^{1/5} \tag{5.63}$$

These equations give reasonable order-of-magnitude results that are in close accord with the elementary calculation given above in equations (5.58) and (5.61). It should also be noted that for small sizes, droplets and solid particles behave dynamically in a very similar fashion, and it is therefore possible to use the theoretical and experimental results obtained for solid spheres in analyzing droplet motion.

For large values of d/D, we can assume[25] that the surface of the falling sphere near its equator and of the nearby cylindrical container can be approximated by two flat plates, and an analysis based on this assumption leads to the expression

$$K = 0.6\left(1 - \frac{d}{D}\right)^{2.5} \tag{5.64}$$

This equation is valid when inertia effects are negligible, so under these conditions, we obtain

$$\frac{v_d}{v_{d\infty}} = K = 0.6\left(1 - \frac{d}{D}\right)^{2.5} \tag{5.65}$$

Analysis of the wall effect on the rate of fall of solid spheres, assuming negligible inertial effects and a small d/D ratio, has been carried out.[26–29] The resulting equation for the terminal-velocity correction factor is

$$\frac{v_d}{v_{d\infty}} = K = 1 - 2.105\left(\frac{d}{D}\right) + 2.087\left(\frac{d}{D}\right)^3$$

A binomial of $[1 - (d/D)]^{2.1}$ leads to

$$K = 1 - 2.1\left(\frac{d}{D}\right) + 1.155\left(\frac{d}{D}\right)^2 - 0.0385\left(\frac{d}{D}\right)^3$$

and the difference in the two expressions is small in the range over which they may be applied, so for the case of negligible inertia and small d/D, we have

$$\frac{v_d}{v_{d\infty}} \simeq \left[1 - \left(\frac{d}{D}\right)\right]^{2.1} \tag{5.66}$$

All the foregoing correction factors suggest that the correction factor may be expressed in a general form:

$$K = \phi\left(1 - \frac{d}{D}\right)^n$$

In a different approach, if we consider that section of the cylindrical chamber near the equator of the droplet to be approximated by a section of a large sphere, we can use spherical coordinates and after analysis obtain[25] the following equation for the range of negligible inertial effects and low d/D ratio:

$$K = \frac{1}{1 + \frac{9}{4}(d/D) + [\frac{9}{4}(d/D)]^2}$$

Now, a binomial expansion of $[1 - (d/D]^{-2.25}$ is

$$1 + 2.25\left(\frac{d}{D}\right) + 3.66\left(\frac{d}{D}\right)^2 + 5.18\left(\frac{d}{D}\right)^3 + \cdots$$

which, in the range of variables over which it is applicable, does not differ greatly from the previous expression, and therefore for the case of negligible inertia and small d/D, an alternative formula to equation (5.66) is

$$\frac{v_d}{v_{d\infty}} \simeq \left[1 - \left(\frac{d}{D}\right)\right]^{2.25} \tag{5.67}$$

It has been shown[30] that for the case of large inertial effects,

$$\frac{v_d}{v_{d\infty}} = K = \left[1 - \left(\frac{d}{D}\right)^2\right]\left[1 - \frac{1}{2}\left(\frac{d}{D}\right)^2\right]^{1/2} \tag{5.68}$$

Finally, the equation of momentum has been used[30] to arrive at an equation for the case of d/D nearly equal to unity and negligible viscous forces. The result is

$$K = \frac{[1 - (d/D)]^2}{d/D} \tag{5.69}$$

Equations (5.62) to (5.69) furnish formulas that can be used to correct the droplet velocity when allowing for the finite dimensions of the enclosing chamber.

5.6 Droplet statistics

When determining quantitative relationships of the droplet entrainment process with a large number of factors (physical, structural, and operational) that effect entrainment, it is advantageous to isolate what seem to be the two principal ones: the superficial velocity of the gas, v_s, and the gas space above the level of the liquid, h_s. To analyze this extremely complex phenomenon, a simplified model is examined, in which we assume that, for given values of v_s and h_s, entrainment takes place in a unique manner. Obviously, the accuracy of such a model would be increased considerably by increasing the number of determining factors in the process, but such an attempt would lead to an impractical solution by virtue of its greatly increased complexity.

In the gas space above the liquid many droplets of different sizes are moving at various velocities, but the time-averaged entrainment behavior of this large number is found to be independent of the individual random features of the flight of the drops, and the entrainment is found to be virtually nonrandom. The stability of this time-averaged entrainment, repeatedly confirmed by experiment,[31–34] enables probability methods to be used to establish quantitative relations for predicting entrainment.

Now, owing to breakdown of bulk liquid in the bubbling mixture, droplets of various radii r_d are continually being produced, and these droplets have initial escape velocities w from the liquid surface having random values scattered around a certain value w_0, which depends on the escape velocity of the gas from the bubbling layer. The subsequent motion of each drop is determined by gravity and drag of the gas, as we outlined in Section 5.5. We make the following two assumptions (which have experimental support[35]):

1. The random quantities w and r_d are independent of one another.
2. The probability distribution law for the random quantities is close to normal.

Indeed, if it is assumed that the deviation of the individual drop velocities from the mean value w_0 is due to the effect of the large number of independently acting random factors, each of which alters its velocity a little, then in this case the distribution law for the drop velocities will be close to normal.[36]

Based on assumption 1, the probability of an event M, which means an ejected droplet having a radius in the range r_d to $(r_d + dr_d)$ and an escape velocity in the range w to $(w + dw)$, is given by

$$P\{M\} = \int_{r_d}^{r_d+dr_d} f(r_d)\, dr_d \int_{w}^{w+dw} f(w)\, dw$$

where $f(w)$ is the probability density distribution of the velocity w and $f(r_d)$ the probability density distribution of the droplet radii.

The total mass of all such drops is then

$$Y = \int_{r_d}^{r_d+dr_d} f(r_d)(\tfrac{4}{3}\pi r_d^3\rho_f)\, dr_d \int_w^{w+dw} f(w)\, dw$$

where $\frac{4}{3}\pi r_d^3\rho_f$ is the mass of a drop of radius r_d. The entrainment at height h_s corresponding to an escape velocity w is then

$$Y = \int_{r_{d_{\min}}}^{r_{d_{\max}}} \tfrac{4}{3}\pi r_d^3\rho_f f(r_d)\, dr_d \int_w^{w_{\max}} f(w)\, dw \tag{5.70}$$

Now it is clear that the total entrainment of liquid from the surface of a two-phase mixture is

$$Y_0 = \int_{r_{d_{\min}}}^{r_{d_{\max}}} f(r_d)(\rho_f \tfrac{4}{3}\pi r_d^3)\, dr_d$$

and using this, equation (5.70) becomes

$$Y(w) = Y_0 \int_w^{w_{\max}} P(w)\, dw \tag{5.71}$$

The probability density distribution of the drop initial (or escape) velocity, according to assumption 2, is

$$f(w) = \frac{\dfrac{1}{\sigma_w(2\pi)^{1/2}} \exp\left[-\dfrac{1}{2}\left(\dfrac{w-w_0}{\sigma_w}\right)^2\right]}{P(w>0)} \tag{5.72}$$

where σ_w is the standard deviation of the velocity w and $P(w > 0)$ is the probability that the velocity $w > 0$. The denominator of equation (5.72) is a normalization factor, so that

$$\int_0^\infty P(w) = 1$$

and the probability that w is negative is zero,

$$P(w < 0) = 0$$

Therefore, the probability that the velocity $w > 0$ is

$$P(w>0) = \frac{1}{(2\pi)^{1/2}\sigma_w} \int_0^\infty \frac{\exp\left[-\dfrac{1}{2}\left(\dfrac{w-w_0}{\sigma_w}\right)^2\right]}{P(w>0)}\, dw = 1$$

and equation (5.71) takes the form

$$Y(w) = \frac{Y_0}{(2\pi)^{1/2}\sigma_w} \int_w^{w_{\max}} \exp\left[-\frac{1}{2}\left(\frac{w - w_0}{\sigma_w}\right)^2\right] dw \qquad (5.73)$$

We now determine entrainment as a function of the gas space height, h_s; in other words, we replace the limits of integration in equation (5.73):

$$Y(h_s) = \frac{Y_0}{(2\pi)^{1/2}\sigma_w} \int_{(2gh_s)^{1/2}}^{(2gh_s)^{1/2}_{\max}} \exp\left[-\frac{1}{2}\left(\frac{w - w_0}{\sigma_w}\right)\right]^2 dw$$

$$\simeq \frac{Y_0}{(2\pi)^{1/2}\sigma_w} \int_{(2gh_s)^{1/2}}^{\infty} \exp\left[-\frac{1}{2}\left(\frac{w - w_0}{\sigma_w}\right)\right]^2 dw$$

and, using the substitution $(w - w_0)/\sigma_w = x$, we obtain the final equation for calculating the entrainment of liquid as

$$Y(h_s) = \frac{Y_0}{(2\pi)^{1/2}\sigma_w}\sigma_w \int_{[(2gh_s)^{1/2} - w_0]/\sigma_w}^{\infty} e^{-x^2/2} dx$$

$$= \frac{Y_0}{(2\pi)^{1/2}} \int_{[(2gh_s)^{1/2} - w_0]/\sigma_w}^{\infty} e^{-x^2/2}\, dx$$

or

$$Y = Y_0 \left\{\frac{1}{(2\pi)^{1/2}} \int_{[(2gh_s)^{1/2} - w_0]/\sigma_w}^{0} e^{-x^2/e^2}\, dx + \frac{1}{(2\pi)^{1/2}} \int_0^{\infty} e^{-x^2/2}\, dx\right\}$$

Finally,

$$Y = Y_0 \left\{\frac{1}{2} - \frac{1}{(2\pi)^{1/2}} \int_0^{[(2gh_s)^{1/2} - w_0]/\sigma_w} e^{-x^2/2}\, dx\right\} + Y_c \qquad (5.74)$$

where Y_c is an entrainment component that is independent of the spacing height.

To calculate liquid droplet entrainment by equation (5.74), it is first necessary to determine the parameters Y_0, w_0, σ_w, and Y_c. For this we use experimental entrainment values, Y_e, which have been obtained for various values of the superficial gas velocity, v_s, and gas space height, h_s. For this we formulate the function

$$\psi(h_s, v_s, Y_0, w_0, \sigma_w, Y_c) = \sum_{i=1}^{n} \frac{(Y_i - Y_{e_i})^2}{Y_{e_i}} \qquad (5.75)$$

where Y_i is determined from equation (5.74), Y_{e_i} the experimental values, and n the number of experimental points. The unknown quantities Y_0, w_0, σ_w, and Y_c are found from the conditions for a minimum of equation (5.75).

This can be done on an electronic computer, and for an air–water system with a fixed value of h_s and various values of v_s, we find that

$$Y_0 = 34.29v_s - 9.24 \text{ kg/m}^2\cdot\text{min}$$

$$w_0 = 0.3342v_s - 0.0179 \text{ m/s}$$

$$\sigma_w = 0.1261v_s + 0.8997 \text{ m/s}$$

$$Y_c = 6 \times 10^{-4} \text{ kg/m}^2\cdot\text{min}$$

Using these values in equation (5.74), we obtain

$$Y \simeq (34v_s - 9)\left[0.5 - \frac{1}{(2\pi)^{1/2}}\int_0^s e^{-x^2/2}\,dx\right] + 6 \times 10^{-4} \text{ kg/m}^2\cdot\text{min} \quad (5.76)$$

where

$$s \simeq \frac{4.42(h_s)^{1/2} - 0.334v_s + 0.018}{0.126v_s + 0.9}$$

and v_s is in meters per second and h_s is in meters.

For a given value of s, the probability integral

$$\frac{1}{(2\pi)^{1/2}}\int_0^s e^{-x^2/2}\,dx$$

can be found from tables, and equation (5.76) then gives us a value for the liquid entrainment from an air–water flow solely in terms of the height above the water level, h_s, and the superficial air velocity, v_s. Equations similar to equation (5.76) can be obtained from equation (5.75) by the same method for any gas–liquid combination (e.g., methane–water, air–kerosene, Freon 12–water), if experimental data are available.[37] It has been found that the use of equation (5.76) gives predicted results accurate to ±10 percent with experimental results. The usefulness of equation (5.74) is diminished, in that experimental data are needed to obtain the final equation (5.76), and also the analysis of these data in minimizing equation (5.75) is quite laborious.

5.7 Similarity analysis

It is clear from the discussion in earlier sections of this chapter that a total analytical treatment of the phenomenon of liquid entrainment in a two-phase flow is very difficult. Nevertheless, we can approach the problem by considering the equations of motion of the liquid and the gas, then using the method of similarity analysis to obtain dimensionless groups, which provide a convenient way of collecting and interpreting data.

We make the assumption that the gas phase is turbulent and that the "apparent" turbulent stresses dominate viscous stresses in the gas. It is assumed that the liquid phase is dominated by molecular viscosity and that the two phases are coupled by the stresses occurring at the interface. Now the equation of motion for the (incompressible) liquid is

$$\frac{\partial u_{f_i}}{\partial t} + u_{f_k}\frac{\partial u_{f_i}}{\partial x_k} = -\frac{1}{\rho_f}\frac{\partial p_f}{\partial x_i} + \frac{\mu_f}{\rho_f}\frac{\partial^2 u_{f_i}}{\partial x_k\,\partial x_k} + g \tag{5.77}$$

where u_{f_i} is the velocity vector of the liquid, x_i the Cartesian coordinate vector, ρ_f the liquid density, p_f the liquid pressure, μ_f the liquid dynamic viscosity, and g the acceleration due to gravity. Note that equation (5.77) is simply the Navier–Stokes equations written in cartesian tensor notation, with the summation of index notation being used (so: $\partial^2/\partial x_k\,\partial x_k \equiv \partial^2/\partial x^2 + \partial^2/\partial y^2 + \partial^2/\partial z^2$). Also, the continuity equation for the liquid is

$$\frac{\partial u_{f_i}}{\partial x_i} = 0 \tag{5.78}$$

For the gas phase, again assumed incompressible, we have an equation similar to equation (5.77) except that the viscous term is replaced by the apparent turbulent stress term, and the gravitational force is neglected:

$$\frac{\partial \bar{u}_i}{\partial t} + \bar{u}_k\frac{\partial \bar{u}_i}{\partial x_k} = -\frac{1}{\rho_g}\frac{\partial \bar{p}_g}{\partial x_i} - \frac{\partial \overline{u'_i u'_k}}{\partial x_k} \tag{5.79}$$

Here $\bar{u}_i$ is the time-averaged gas velocity vector and u'_i the fluctuating gas velocity vector (where the gas velocity vector $u_i = \bar{u}_i + u'_i$). The term $\bar{p}_g$ is the time-averaged gas pressure. The continuity equation for the gas is

$$\frac{\partial \bar{u}_i}{\partial x_i} = 0 \tag{5.80}$$

Now we need to couple the liquid and gas equations, and these are coupled by the interface boundary conditions. We have a balance of normal stresses at the interface,

$$-p_f + 2\mu_f\frac{\partial u_{f_3}}{\partial x_3} = -\bar{p}_g - \rho_g\overline{u_3'^2} - \sigma\left(\frac{1}{R_1} + \frac{1}{R_2}\right) \tag{5.81}$$

where the index 3 refers to the coordinate normal to the interface, indices 1 and 2 refer to coordinates parallel to the interface, the fundamental radii of curvature of the interface are given by R_1 and R_2, and σ is the surface tension. The second terms on each side of the equation are the normal viscous and turbulent stresses, respectively. For the shear stresses at the interface, we find that

$$\mu_f\left(\frac{\partial u_{f_1}}{\partial x_2}+\frac{\partial u_{f_2}}{\partial x_1}\right)=-\rho_g\overline{u_2'u_3'} \tag{5.82}$$

where the left-hand side of the equation represents liquid viscous shear and the right-hand side represents gas turbulent shear.

Equations (5.77) through (5.82) provide a complete description of the two-phase motion, but this group of equations present great analytical difficulties, and instead of attempting a solution, we will introduce dimensionless functions into the equations, thereby generating the dimensionless groups useful in understanding experimental data (e.g., the Reynolds number). The dimensionless functions are

$$U_{f_i}=\frac{u_{f_i}}{v_s},\qquad \overline{U}_i=\frac{\overline{u}_i}{v_s},\qquad U_i'=\frac{u_i'}{v_s}$$

$$P_f=\frac{p_f}{\Delta p},\qquad \overline{P}_g=\frac{\overline{p}_g}{\Delta p}$$

$$T=\frac{t}{\tau},\qquad X=\frac{x}{h_s},\qquad R_{o_i}=\frac{R_i}{h_s}$$

Here v_s is the gas superficial velocity, Δp a typical flow system pressure difference (e.g., between the walls of a chamber), τ a characteristic time, and h_s the distance above the free liquid level.

By using these in the equations above, we obtain

$$\begin{aligned}\frac{v_s}{\tau}\left(\frac{\partial U_{f_i}}{\partial T}\right)+\frac{v_s^2}{h_s}\left(U_{f_k}\frac{\partial U_{f_i}}{\partial X_k}\right)&=-\frac{\Delta p}{\rho_f h_s}\left(\frac{\partial P_f}{\partial X_i}\right)\\&\quad+\frac{\mu_f v_s}{\rho_f h_s^2}\left(\frac{\partial^2 U_{f_i}}{\partial X_k\,\partial X_k}\right)+g\end{aligned} \tag{5.83}$$

$$\frac{v_s}{h_s}\left(\frac{\partial U_{f_i}}{\partial X_i}\right)=0$$

$$\begin{aligned}\frac{v_s}{\tau}\left(\frac{\partial \overline{U}_i}{\partial T}\right)+\frac{v_s^2}{h_s}\left(U_{gk}\frac{\partial U_i}{\partial X_k}\right)&=-\frac{\Delta p}{\rho_g h_s}\left(\frac{\partial \overline{P}_g}{\partial X_i}\right)\\&\quad-\frac{v_s^2}{h_s}\left(\frac{\overline{\partial U_i'\,U_k'}}{\partial X_k}\right)\end{aligned} \tag{5.84}$$

$$\frac{v_s}{h_s}\left(\frac{\partial \overline{U}_i}{\partial X_i}\right)=0$$

$$-\Delta p(P_f) + \frac{\mu_f v_s}{h_s}\left(2\frac{\partial U_3}{\partial X_3}\right) = -\Delta p\,(\overline{P}_g) - \rho_g v_s^2\left(\overline{U_3'^2}\right) - \frac{\sigma}{h_s}\left(\frac{1}{R_{o_1}} + \frac{1}{R_{o_3}}\right)$$

In the foregoing equations all the terms in parentheses are of the order unity, so in any study of the relative influence of pressure, viscosity, surface tension, gravity, and time fluctuations, we need only consider the coefficients in front of the parentheses. It turns out that, for this procedure, only equations (5.83) and (5.84) are required. Hence we find that

$$\frac{v_s}{\tau}[0(1)] + \frac{v_s^2}{h_s}[0(1)] = -\frac{\Delta p}{\rho_f h_s}[0(1)] + \frac{\mu_f v_s}{\rho_f h_s^2}[0(1)] + g$$

or

$$\frac{h_s}{v_s \tau}[0(1)] + [0(1)] = -\frac{\Delta p}{\rho_f v_s^2}[0(1)] + \frac{\mu_f}{\rho_f h_s v_s}[0(1)] + \frac{g h_s}{v_s^2} \quad (5.85)$$

and

$$-\Delta p[0(1)] + \frac{\mu_f v_s}{h_s}[0(1)] = -\Delta p[0(1)] - \rho_g v_s^2[0(1)] - \frac{\sigma}{h_s}[0(1)]$$

or

$$\frac{-\Delta p}{\rho_f v_s^2}[0(1)] + \frac{\mu_f}{\rho_g h_s v_s}[0(1)] = -\frac{\Delta p}{\rho_f v_s^2}[0(1)] - \frac{\rho_g}{\rho_f}[0(1)] - \frac{\sigma}{\rho_g h_s v_s^2}[0(1)] \quad (5.86)$$

In the above, 0(1) signifies that the term is of the order unity. It is clear from equations (5.85) and (5.86) that we have several important nondimensional groups in the two-phase flow that is described by the original equations:

$$\left(\frac{\Delta p}{\rho_f v_s^2}, \frac{\rho_f v_s h_s}{\mu_f}, \frac{v_s^2}{g h_s}, \frac{\rho_g v_s^2 h_s}{\sigma}, \frac{v_s \tau}{h_s}, \frac{\rho_g}{\rho_f}\right)$$

We can write these as

$$\left(\text{Eu, Re, Fr, We, Ho}, \frac{\rho_g}{\rho_f}\right)$$

where

$\text{Eu} = \dfrac{\Delta p}{\rho_f v_s^2}$ the Euler number, ratio of pressure to inertia forces

$\mathrm{Re} = \dfrac{\rho_f v_s h_s}{\mu_f}$ the Reynolds number, ratio of inertia to viscous forces

$\mathrm{Fr} = \dfrac{v_s^2}{g h_s}$ the Froude number, ratio of inertia to gravitational forces

$\mathrm{We} = \dfrac{\rho_g v_s^2 h_s}{\sigma}$ the Weber number, ratio of inertia to surface tension forces

$\mathrm{Ho} = \dfrac{v_s T}{h_s}$ the homochronity number, ratio of spatial to temporal inertia forces

All of these are independent except for the Euler number. This is dependent because it expresses the relationship between the pressure and velocity fields, and once the velocity field is given the pressure field is fully determined, and vice versa. Hence we can say that

$$\mathrm{Eu} = \mathrm{Eu}\left(\mathrm{Re}, \mathrm{Fr}, \mathrm{We}, \mathrm{Ho}, \frac{\rho_g}{\rho_f}\right)$$

This relationship holds not only for the Euler number but for any dependent quantity, so, for example, for the entrainment coefficient $\mathbf{Y} = \Sigma M_f / M_g$, where ΣM_f denotes the total liquid (droplet) entrainment per unit cross-sectional area and time and M_g is the mass flow rate of gas, we will have

$$\mathbf{Y} = \mathbf{Y}\left(\mathrm{Re}, \mathrm{Fr}, \mathrm{We}, \mathrm{Ho}, \frac{\rho_g}{\rho_f}\right) \tag{5.87}$$

Experimental determination of this relationship presents serious difficulties, but experience indicates that this set of dimensionless numbers may, to some advantage, be combined into groups. Thus we can write

$$A = \frac{\mathrm{Fr}^{1/2}\mathrm{We}^{3/2}\rho_g}{\mathrm{Re}^2(\rho_f - \rho_g)} = \frac{\left(\dfrac{v_s^2}{g h_s}\right)^{1/2}\left[\dfrac{(\rho_f - \rho_g)v_s^2}{\sigma} h_s\right]^{3/2}\rho_g}{(v_s h_s/\nu_f)^2(\rho_f - \rho_g)}$$

or

$$A = \frac{\nu_f^2 v_s^2 \rho_g}{g^2\left(\dfrac{\sigma}{g(\rho_f - \rho_g)}\right)^{3/2} h_s(\rho_f - \rho_g)} \tag{5.88}$$

Note that we have replaced ρ_g by $(\rho_f - \rho_g)$ in the Weber number. This is simply an alternative definition of We. Also,

$$B = \frac{\mathrm{HoFr}^{1/2}}{\mathrm{We}^{1/2}} = \frac{(v_s\tau/h_s)(v_s^2/gh_s)^{1/2}}{\left[\dfrac{v_s^2(\rho_f - \rho_g)h_s}{\sigma}\right]^{1/2}} = \frac{v_s\tau\sigma^{1/2}}{h_s^2[(\rho_f - \rho_g)g]^{1/2}} \tag{5.89}$$

In most applications the homochronity number is approximately equal to unity, and hence

$$\tau \simeq \frac{h_s}{v_s}$$

and equation (5.89) may be rewritten as

$$B = \frac{\{\sigma/[g(\rho_f - \rho_g)]\}^{1/2}}{h_s}$$

The value of factor A in equation (5.88) for any system is determined by the magnitudes of gravitational, viscous, and surface-tension forces during droplet generation, and also the gravitational and inertia forces during droplet motion in the gaseous region. Factor B in equation (5.89) is determined by the relative magnitude of gravitational, inertia, and surface tension forces alone. Now, the simplest possible expression of equation (5.87) in terms of factors A and B is

$$\mathrm{Y} = CA^mB^n \tag{5.90}$$

Figure 5.7 Entrainment coefficient against factor B.

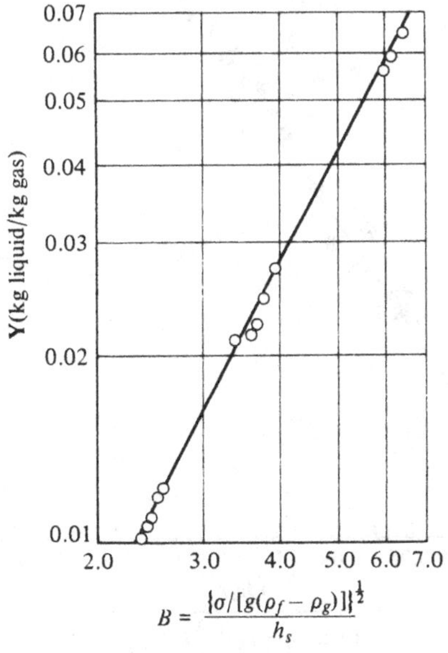

where C, m, and n are constants to be determined experimentally. It now remains to be seen if equation (5.90) adequately "condenses" experimental data. Experiments have been conducted[31] using an air–water system to test this equation. The straight line in Figure 5.7 represents the least-squares fit to the data and has a slope of 1.8. Hence constant n is seen to have the value of 1.8. The effect of factor A on Y has also been determined experimentally, as seen in Figure 5.8, for various liquid–gas flow systems.

The least-squares fitted straight line through the data represents the expression

Figure 5.8 $(\mathbf{Y}B^{-1.8})$ against factor A.

$$A = \frac{\nu_f^2 v_s \rho_g}{g^2 \{\sigma/[g(\rho_f - \rho_g)]\}^{\frac{3}{2}} h_s (\rho_f - \rho_g)}$$

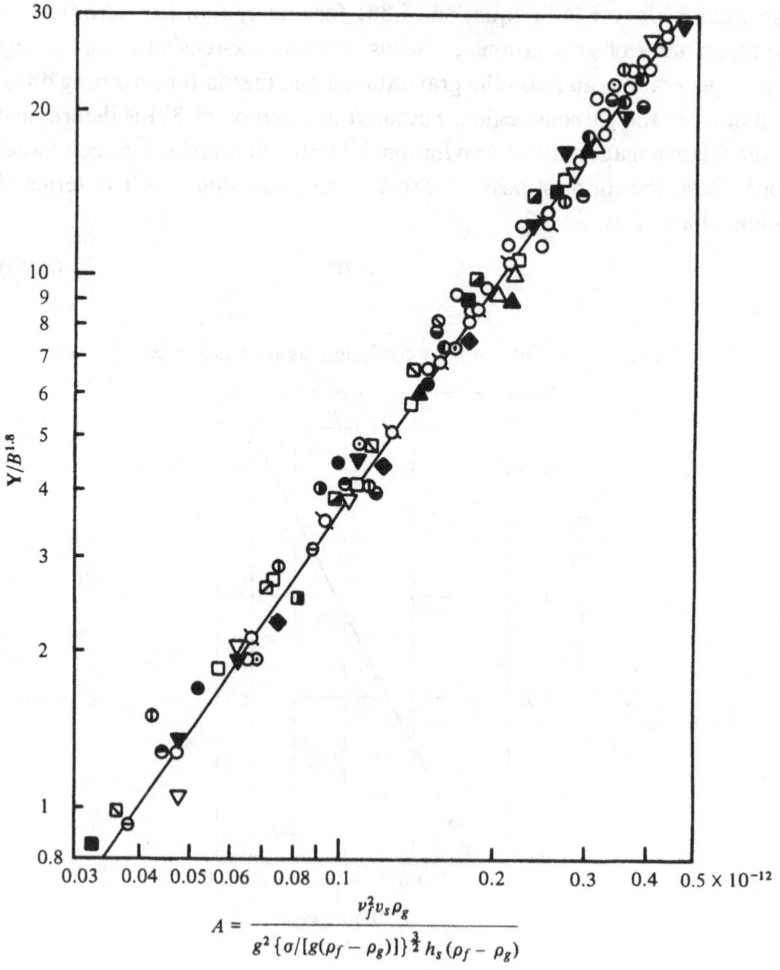

$$\frac{\mathbf{Y}}{B^{1.8}} = f(A)$$

and from this line we find that $C = 3.17 \times 10^{13}$ and $m = 1.4$. Hence equation (5.90) is verified and takes the form

$$\mathbf{Y} = 3.17 \times 10^{13} A^{1.4} B^{1.8}$$

or

$$\mathbf{Y} = 3.17 \times 10^{13} \frac{\left[\dfrac{\nu_f^2 v_s^2 \rho_g}{g^2(\rho_f - \rho_g)}\right]^{1.4}}{\left[\dfrac{\sigma}{g(\rho_f - \rho_g)}\right]^{1.2} h_s^{3.2}} \tag{5.91}$$

Equation (5.91) is compared to data from two-phase systems[37] over a range of physical properties (e.g., Freon 12–H_2O, air–CCl_4) in Figure 5.9. Clearly, there is a good correlation between the equation and the experimental data.

Figure 5.9 Equation (5.91) compared with experimental data.

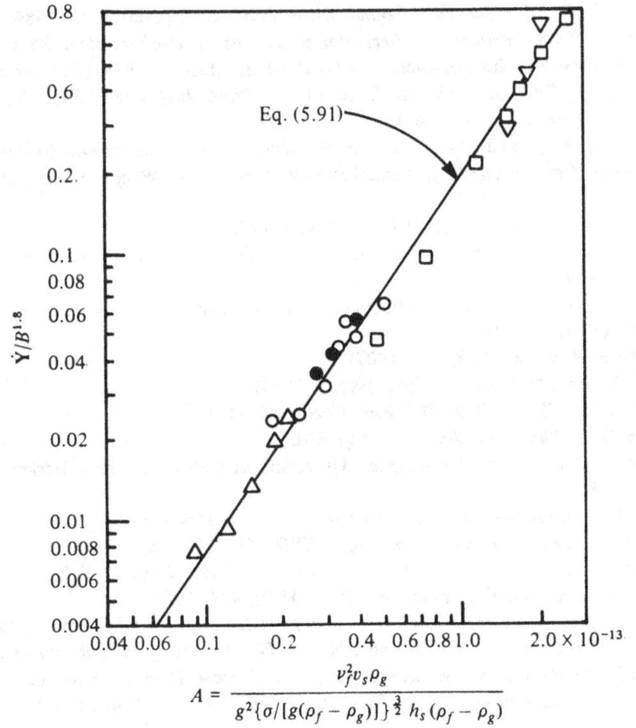

References

1. Pavlov, V. P., Dissertation, Moscow Institute of Chemical Engineers, 1963. (In Russian.)
2. Einstein, A., *Ann. Phys., 19,* 289 (1906).
3. Rubinow, S. I., and Keller, T. B., *Fluid Mech., 11,* 3 (1961).
4. Kurbatov, P. V., *Tr. Mosk. Energ. Inst., 2,* 82 (1953).
5. Chekhov, O. S., and Matrosov, V. I., *Tr. Mosk. Inst. Khim. Mashinostr., 13,* 78, (1957).
6. Salomakha, G. P., Dissertation, Moscow Institute of Chemical Technology, 1957. (In Russian.)
7. Stabnikov, V. N., *Muravskay, Isv. Vusov, Pichshevay Technol., 5,* 108, Moscow (1959).
8. Azbel, D. S., *Theor. Found. Chem. Eng. (USSR), 5*(5), 335, (1971).
9. Hinze, I. O., *Turbulence,* McGraw-Hill, New York (1959).
10. Batchelor, G. K., *Theory of Homogeneous Turbulence,* Cambridge University Press, Cambridge, 1970.
11. Kolmogoroff, A. N., *Dokl. Akad. Nauk SSSR, 30,* 229 (1941).
12. Kolmogoroff, A. N., *Dokl. Akad. Nauk SSSR, 31,* 538 (1941).
13. Azbel, D. S., Doctoral dissertation, D. I. Mendeleev Moscow Institute of Chemical Technology, 1966. (In Russian.)
14. Azbel, D. S., *Khim. Mashinostr. (Moscow),* no. 5, 14 (1960).
15. Newitt, D. M., Dombrowski, N., and Kneeman, F. H., *Trans. Inst. Chem. Eng., 32,* 4 (1954).
16. Azbel, D. S., Lee, L. L., and Lee, T. S., *Proceedings of 1978, International Seminar on Momentum, Heat and Mass Transfer in Two-Phase Energy and Chemical Systems, Dubrovnik, Yugoslavia, September 1978.*
17. Kreyszig, E., *Advanced Engineering Mathematics,* Wiley, New York, 1967.
18. Rayleigh, I. W., *The Theory of Sound,* Dover Publications, New York, 1954.
19. Koshliakov, N. S., *Principal Differential Equations of Mathematical Physics.* ONTI (Obyedinennoe Naucho Technicheskoe Izdatelstvo), Moscow, 1936. (In Russian.)
20. Garner, F. H., Ellis, S. R. M., and Lacey, T. A., *Trans. Inst. Chem. Eng., 32,* 4 (1954).
21. Blinov, V. I., and Feynberg, Ye. L., *Zh. Tekh. Fiz., 3,* 5 (1933).
22. Kutateladze, S. S., and Styrikovich, M. A., *Hydraulics of Gas–Liquid Systems,* Gosudazstvennoe Energeticheskoye Izdatelstvo, Moscow, 1958. Wright Field trans. F-TS-9814v.
23. Uno, S., and Kintner, R. C., *AIChE J., 2,* 420 (1956).
24. Govier, G. W., and Aziz, K., *The Flow of Complex Mixtures in Pipes,* Van Nostrand Reinhold, New York, 1972.
25. McNown, J. S., Lee, H. M., MacPherson, M. B., and Engez, S. M., *Proc. VII Int. Congr. Appl. Mech., Lond., 1948.*
26. Ladenburg, *Ann. Phys., 23,* 447 (1907).
27. Faxen, H., *Ark. Mat. Astron. Fys., 19*(13) (1925).
28. Happel, J., and Byrne, B. J., *Ind. Eng. Chem., 46,* 1181 (1954).
29. Wakiya, S. J., *Phys. Soc. Jap., 8,* 254 (1953).
30. Newton, I., *Mathematical Principles,* University of California Press, Berkeley, Calif., 1934, p. 348.
31. Azbel, D. S., *Khim. Mashinostr. (Moscow),* no. 6, 14 (1960).
32. Azbel, D. S., *Theor. Found. Chem. Eng. USSR, 1*(1), 91 (1967).
33. Holbrook, G. E., and Baker, E. M., *Trans. Am. Inst. Chem. Eng., 30,* S. 20 (1933–1934).
34. Jones, J. B., and Pyle, C., *Chem. Eng. Prog. 51*(9), 424 (1955).
35. Cumo, M., and Farello, G., *Tenth Int. Heat Transfer Conf., AIChE–ASME 1968.*
36. Dunin-Barkovskii, I. V., and Smirnov, N. V., *The Theory of Probability and Mathematical Statistics in Engineering,* Gostekhizdat, Moscow, 1955. (In Russian.)
37. Hunt, Ch. A., Hanson, D. H., and Wilke, C. R., *AIChE J., 1*(4), 441 (1955).

6

The motion of solid particles in a liquid

6.1 Introduction

Many operations in chemical technology involve the simultaneous flow of liquid and solid particles. The field of liquid–solid particle flows encompasses a tremendous range of practical problems, but we will deal here only with dilute liquid–solid particle systems in which the particles are far enough removed from one another that we may consider the motion of individual particles independently of the motion of the other particles in the system.

The motion of solid particles in such systems is often of primary concern when dealing with problems of mass transfer, heat transfer, and kinetics in chemical technology, and therefore liquid–solid particle mixtures have been extensively studied, both experimentally and analytically.[1–11] In spite of these studies, however, our knowledge of such flows is limited, because most of the useful correlations are of empirical origin and have application only to the specific systems from which they were developed, and sometimes these correlations result in limited accuracy when they are applied to other systems. Thus we will attempt to characterize dispersion of a low concentration of solid particles in a liquid flow by means of the volume fraction of solid particles, and also the superficial velocities υ_{sf} and υ_s of the liquid and solid phases, respectively, in an attempt to develop formulas having wider applicability.

6.2 Average particle flow velocity and average fraction of particles in a liquid flow

In this section we present some basic terms we will use in describing liquid–solid particle flows. The average velocity of solid particles in a liquid stream, or, in other words, the superficial particle velocity, υ_s, when velocities of the two phases are low and the duct is large enough (so that we have no wall effect) may be defined if we imagine some control plane to pass through the flow perpendicular to the direction of the time-averaged flow velocity vector and we write the volumetric flow rate of solid particles per unit area of this control plane as

$$v_s = \frac{Q}{A} = \frac{1}{A\,\Delta\tau} \int_{\Delta\tau} \int_A v_p \, dA \, d\tau \qquad (6.1)$$

where Q (m^3/s) is the volumetric flow rate of particles, A (square meters) is the flow system cross-sectional area, $\Delta\tau$ (seconds) is a period of time substantially greater than the quantity $1/f$ [where f (second^{-1}) is the frequency of movement of solid particles through the cross section in question], and v_p is the actual velocity of the particles. At a given instant of time the particles do not occupy the whole cross section of the duct and, in general, a part of every cross section of the duct is occupied by the liquid. Therefore, equation (6.1) must be written in the form of a sum of integrals:

$$v_s = \frac{1}{A\,\Delta\tau} \int_{\Delta\tau} \left(\sum_i \int_{A_i} v_p \, dA \right) d\tau \qquad (6.2)$$

where i is the number of particle formations, or groups, in a given cross section at a given instant of time. The ratio

$$\phi = (1/A) \sum_i A_i$$

is defined as the instantaneous value of that fraction of the section of the dispersed flow which is occupied by particles (this section being of thickness dn, where n is normal to the surface A). The true average velocity of particles in the liquid flow is then

$$\bar{v}_p = \frac{1}{\Delta\tau \, \Sigma_i \, A_i} \int_{\Delta\tau} \left(\sum_i \int_{A_i} v_p \, dA \right) d\tau$$

and this equation, with equation (6.2), indicates that the true average velocity and superficial velocity of the particles are related by

$$\bar{v}_p = v_s/\phi$$

Now, the ratio of the total mass flow rate of the phases to the product of the cross-sectional area of the channel and the density of the liquid is usually designated as the flow velocity of the two-component flow:

$$v = \frac{\rho_p v_s + \rho_f v_{sf}}{\rho_f}$$

where ρ_p and ρ_f are the particle and fluid densities, respectively, and v_{s_f} is the liquid superficial velocity. It provides a measure of the velocity of a liquid flow with mass flow rate equal to that of the mixture.

The average relative velocity of the solid phase is defined by

$$\bar{v}_r = \bar{v}_f - \bar{v}_p = \frac{v_{s_f}}{1-\phi} - \frac{v_s}{\phi}$$

and hence

$$\bar{v}_r\phi^2 - (\bar{v}_r - v_{s_f} - v_s)\phi - v_s = 0 \tag{6.3}$$

Solving this equation for ϕ, taking into account that when $v_s = 0$ we have $\phi = 0$, we obtain

$$\phi = \frac{\bar{v}_r - v_{s_f} - v_s}{2\bar{v}_r} - \left[\left(\frac{\bar{v}_r - v_{s_f} - v_s}{2\bar{v}_r}\right)^2 + \frac{v_s}{\bar{v}_r}\right]^{1/2}$$

When the average relative velocity of the particles, $\bar{v}_r$, is equal to zero, we see from equation (6.3) that the volumetric flow fraction, defined below, becomes equal to ϕ,

$$\beta \equiv \frac{v_s}{v_{s_f} + v_s} = \phi$$

and when the particle relative velocity is positive, the true fraction of the particles in the two-phase flow is less than the flow fraction β.

The inequality of the averaged phase velocities, which is a measure of the "slip" of the particles relative to the liquid flow, changes the energy losses in the two-phase system as compared to the case where the relative phase velocity is zero. The mass flow rate of the mixture per unit area is

$$M = \rho_p v_s + \rho_f v_{s_f}$$

and introducing the concept of average density of the mixture, we have

$$\rho = \frac{M}{v_s + v_{s_f}}$$

or

$$\rho = (1-\phi)\rho_f + \phi\rho_p$$

leading to

$$\rho = \rho_f - (\rho_f - \rho_p)\phi$$

Note that ϕ is then the equivalent, for solid particles, of the void fraction ϕ used in describing gas–liquid flows. The relative density of the mixture is then given by

$$\psi = \frac{\rho}{\rho_f} = 1 - \frac{\rho_f - \rho_p}{\rho_f}\phi$$

6.3 Steady motion of solid particles in a laminar viscous flow

When a spherical particle moves through a viscous liquid in which the ratio of inertia to viscous forces is very small (the Reynolds number $\mathrm{Re} = 2\rho_f v_r r_p/\mu_f \ll 1$), the inertia forces may be neglected. Viscosity is then the controlling factor,[12] and the resistance of a liquid to the motion of a particle is expressed by the well-known Stokes law:

$$F_D = 6\pi\mu_f r_p v_r \tag{6.4}$$

where μ_f is the liquid dynamic viscosity, r_p the particle radius, and v_r the particle relative velocity. In the derivation of equation (6.4), the inertia terms were omitted from the equations of motion, so that the equation may be only considered as a first approximation, and a second approximation, which partly takes into account inertia forces, was obtained by Oseen[13] as

$$F_D = 6\pi\mu_f r_p v_r\left(1 + \frac{3}{16}\mathrm{Re}\right)$$

and somewhat more accurately by Goldstein,[14]

$$F_D = 6\pi\mu_f r_p v_r\left(1 + \frac{3}{16}\mathrm{Re} - \frac{19}{1280}\mathrm{Re}^2 + \cdots\right)$$

In this *creeping flow* ($\mathrm{Re} \ll 1$), the resistance is thus proportional to the particle velocity. However, the region of large Reynolds numbers, where inertia forces cannot be ignored, is characterized by the resistance varying as v_r^n, where n steadily increases with Re, and so, in general, this nonlinear dependency of resistance has to be expressed in two variables (size and velocity of the particle) instead of one. Introduction of the dimensionless quantity

$$C_D = \frac{F_D}{\frac{1}{2}\rho_f v_r^2 \pi r_p^2}$$

called the *drag coefficient,* which is a unique function of the Reynolds number, enables this complex problem to be simplified considerably. Where Stokes's law applies, the drag coefficient becomes

$$C_D = \frac{24}{\mathrm{Re}}$$

and Oseen's formula is expressed by

$$C_D = \frac{24}{\mathrm{Re}} + 4.5$$

A large number of empirical formulas relating C_D and Re have been obtained[15] and the most successful from the standpoint of simplicity and accuracy seems to be the one given by[16]

$$C_D = \frac{24}{\mathrm{Re}} + \frac{4}{\mathrm{Re}^{1/3}}$$

in the range $3 < \mathrm{Re} < 500$.

6.4 Nonuniform motion of solid particles

The motion of a particle with constant velocity, which was examined in Section 6.3, must be regarded as ideal, because in reality the velocity is always changing, both in magnitude and direction. The nonuniform motion of solid particles is, certainly, more complex than motion at constant velocity, and the relevant differential equations can be solved in comparatively few cases, and in most cases one must resort to numerical methods. Only a few of the more important cases of nonuniform motion will be discussed here, assuming for simplicity that the particles are spherical.

The differential equation for nonsteady rectilinear motion of a particle in a liquid has been obtained and modified by several authors.[13,17–23] Neglecting external forces, the final form of the equation for the motion of a spherical particle in a turbulent flow, with nonzero mean velocity, is usually obtained as

$$\begin{aligned} \frac{\pi d_p^3 \rho_f}{6} \frac{dv_p}{dt} = {} & \frac{\pi d_p^3 \rho_f}{6} \frac{dv_f}{dt} + \frac{1}{2} \frac{\pi d_p^3 \rho_f}{6} \left(\frac{dv_f}{dt} - \frac{dv_p}{dt} \right) \\ & + 3\pi \nu_f \rho_f d_p (v_f - v_p) \\ & + \frac{3}{2} d_p^2 \rho_f \pi^{1/2} \nu_f \int_{t_0}^{t} \left(\frac{dv_f}{dx} - \frac{dv_p}{dx} \right) \frac{dx}{(t-x)^{1/2}} \end{aligned} \tag{6.5}$$

where d_p is the diameter of the particle, ρ_p the particle density, ρ_f the density of the liquid, v_p the velocity of a particle, v_f the absolute velocity of a liquid, ν_f the liquid kinematic viscosity, and t the time.

The first term on the right-hand side of the equation is the surface force acting on the particle resulting from the pressure variation on the surface, and this variation in turn is a function of the pressure field in the entraining liquid. The second term describes the inertia of the added mass (entrained with the relative motion of the particle in the entraining liquid) and is equivalent to an increase in mass of the spherical particle equal to half the mass of the liquid displaced. The third term describes the viscous force due to the relative motion of the particle in liquid, and the fourth term is the resistance that is due to the energy expended in setting the liquid itself in motion, called the *history term.*

Equation (6.5) is valid when the following relations hold:[24,25]

$$\frac{d_p^2}{\nu_f}\frac{\partial v_f}{\partial x} \ll 1$$

$$\frac{v_f}{d_p^2(\partial^2 v_f/\partial x^2)} \gg 1$$

where x is here a linear dimension. Following an approach described elsewhere,[25] we write equation (6.5) in the form

$$\frac{dv_p}{dt} + av_p + c\int_{t_0}^{t} \frac{dv_p/dx}{(t-x)^{1/2}}\,dx = av_f + b\frac{dv_f}{dt} + c\int_{t_0}^{t}\frac{dv_f/dx}{(t-x)^{1/2}}\,dx \tag{6.6}$$

where

$$a = \frac{18\nu_f}{[(\rho_p/\rho_f) + 0.5]d_p^2}$$

$$b = \frac{3}{2[(\rho_p/\rho_f) + 0.5]}$$

$$c = \frac{9\nu_f^{1/2}}{[(\rho_p/\rho_f) + 0.5]\pi^{1/2}}$$

Now expressing v_f and v_p in terms of Fourier integrals, we obtain

$$\begin{aligned} v_f &= \int_0^\infty (\zeta\cos\omega t + \lambda\sin\omega t)\,d\omega \\ v_p &= \int_0^\infty (\sigma\cos\omega t + \psi\sin\omega t)\,d\omega \end{aligned} \tag{6.7}$$

where ζ, λ, σ, and ψ are Fourier amplitudes and ω is the frequency, and substituting equations (6.7) into equation (6.6) yields, after straightforward manipulation,

$$\sigma = (1 + f_1)\zeta + f_2\lambda$$

and

$$\psi = -f_2\zeta + (1 + f_1)\lambda$$

in which

$$f_1 = \frac{\omega[\omega + c(\pi\omega/2)^{1/2}](b - 1)}{[a + c(\pi\omega/2)^{1/2}]^2 + [\omega + c(\pi\omega/2)^{1/2}]^2}$$

and

$$f_2 = \frac{\omega[a + c(\pi\omega/2)^{1/2}](b - 1)}{[a + c(\pi\omega/2)^{1/2}]^2 + [(\omega + c(\pi\omega/2)^{1/2}]^2}$$

In developing the theory to deal with the influence of turbulence on the motion of a particle in a very dilute flow, certain assumptions and modifications of equation (6.5) are necessary. For example, we may neglect the integral (or history term) in equation (6.5); we may modify it by neglecting the added mass and the integral; and we may ignore the pressure-gradient effect of the fluid acceleration, in addition to neglecting the added mass and the integral. In fact, deviation of the particle motion from the liquid motion has been investigated[26] for these three types of approximations, as follows:

Type I: When ignoring the integral or history term in equation (6.6):

$$a = \frac{18\nu_f}{[(\rho_p/\rho_f) + 0.5]d_p^2}, \qquad b = \frac{3}{2[(\rho_p/\rho_f) + 0.5]}, \qquad c = 0$$

Type II: When ignoring the integral and added mass terms:

$$a = \frac{18\nu_f}{d_p^2(\rho_p/\rho_f)}, \qquad b = \frac{1}{\rho_p/\rho_f}, \qquad c = 0$$

Type III: When in addition to the conditions of Type II, the pressure-gradient effect is neglected:

$$a = \frac{18\nu_f}{d_p^2(\rho_p/\rho_f)}, \qquad b = c = 0$$

We now introduce the concepts of amplitude ratio and phase angle between the particle and liquid motions and, expressing the velocity as a Fourier integral, we have

$$v_p = \int_0^\infty \eta[\zeta \cos(\omega t + \beta) + \lambda \sin(\omega t + \beta)]\, d\omega$$

where the *amplitude ratio* is given by

$$\eta = [(1 + f_1)^2 + f_2^2]^{1/2}$$

and the *phase angle* is given by

$$\beta = \tan^{-1}\left(\frac{f_2}{1 + f_1}\right)$$

These formulas allow the various approximations to be compared by contrasting the respective phase angles and amplitude ratios, and for convenience, f_1 and f_2 can be expressed in dimensionless form for the general case, so we have

$$f_1 = \frac{\left[1 + \dfrac{9}{2^{1/2}(s+\frac{1}{2})} N_s\right]\left(\dfrac{1-s}{s+\frac{1}{2}}\right)}{\dfrac{81}{(s+\frac{1}{2})^2}\left(2N_s^2 + \dfrac{N_s}{2^{1/2}}\right)^2 + \left[1 + \dfrac{9}{2^{1/2}(s+\frac{1}{2})} N_s\right]^2} \tag{6.8}$$

and

$$f_2 = \frac{\dfrac{9(1-s)}{(s+\frac{1}{2})^2}\left(2N_s^2 + \dfrac{N_s}{2^{1/2}}\right)}{\dfrac{81}{(s+\frac{1}{2})^2}\left(2N_s^2 + \dfrac{N_s}{2^{1/2}}\right)^2 + \left[1 + \dfrac{9}{2^{1/2}(s+\frac{1}{2})}\right]^2} \tag{6.9}$$

where

$$s = \rho_p/\rho_f$$

and

$$N_s = [\nu_f/\omega d_p^2]^{1/2}$$

N_s is the Stokes number. Hence for the Type I approximation, we find that

$$f_1 = \frac{(1-s)/(s+\frac{1}{2})}{[18N_s^2/(s+\frac{1}{2})]^2 + 1} \tag{6.10a}$$

and

$$f_2 = \frac{[18N_s^2/(s+\frac{1}{2})^2](1-s)}{[18N_s^2/(s+\frac{1}{2})]^2 + 1} \tag{6.10b}$$

For the Type II approximation,

$$f_1 = \frac{(1-s)/s}{(18N_s^2/s)^2 + 1} \tag{6.11a}$$

and

$$f_2 = \frac{18N_s^2(1-s)/s^2}{(18N_s^2/s)^2 + 1} \tag{6.11b}$$

and for the Type III approximation,

$$f_1 = \frac{-1}{(18N_s^2/s)^2 + 1} \tag{6.12a}$$

and

$$f_2 = \frac{-18N_s^2/s}{(18N_s^2/s)^2 + 1} \tag{6.12b}$$

The importance of added mass, the integral (or history) term, and the pressure gradient due to fluid acceleration, in their effects on particle motion in turbulent fluid, is indicated by analyzing the approximations of equation (6.5) that we have given above. Only for high density ratios and small particles were these terms found to be unimportant, and it turns out that for liquid–solid particle systems the greatest deviations from the general theory occur within the frequency range of general interest, and therefore other approaches are needed.

However, computations of amplitude ratio and phase angle for various density ratios based on the general case [using equations (6.5), (6.8), and (6.9)] and the three types described above [equations (6.10), (6.11), and (6.12)] have been made,[26] for density ratios of 8.6×10^{-5}, 2.65, and 1000. It was found that the approximations are good for high Stokes numbers (low frequencies), although at higher frequencies a considerable error arises for Type I, neglecting the integral or history term, the difference between this and the general case being greater than that between it and Type II, neglecting additional mass. For Type III, neglecting the pressure term, we have for the small-density-ratio case a negative phase angle, which is an incorrect prediction. These results may, however, be of some use in indicating the circumstances in which each approximation may be made.

6.5 Effect of system walls on particle velocities in a dilute suspension

In Section 6.4 we did not consider the possible effects on the liquid–solid particle flow of the finite dimensions of the flow system. It is obvious that there will, in general, be a nontrivial modification to the flow due to this wall effect, and in fact it is found that, for dilute suspension of solid particles in a liquid, the effect is the same as that on a gas–liquid flow, as described in Chapter 5. We refer the reader to Section 5.5, where this wall effect is discussed. The conclusions drawn there are found to be applicable to liquid–solid particle flow.

6.6 Basic equations for the motion of solid particles in a turbulent flow

In studying the dynamics of solid particles in a turbulent flow it is possible to apply precisely the same procedure as that used when developing the momentum and energy equations for a single liquid, and it is also possible to consider the effect of turbulence by applying the well-known Reynolds procedure.[25] The main physical assumption needed to realize these possibilities is that the size of the suspended particles be small in comparison with the characteristic scale of turbulence, because this provision makes it possible to treat the particles as an admixture distributed continuously in the entraining liquid.

The motion of the entraining liquid and suspended solid particles may be regarded as the flow of a homogeneous medium, if we write the mass and momentum equations for the solid and liquid components separately and then sum these to derive the corresponding equations for the mixture as a whole. We can introduce the notions of the density and velocity of the mixture, and for an elementary volume of space we will have a continuity equation:

$$\frac{\partial \rho}{\partial t} + \frac{\partial(\rho u_i)}{\partial x_i} = 0 \qquad (6.13)$$

where i is the Einstein summation index, indicating a summation over all the values of the repeated index, and where

$$\rho = \rho_f(1 - \phi) + \rho_p\phi = \rho_f + (\rho_p - \rho_f)\phi \qquad (6.14)$$

is the mixture density and

$$u_i = \frac{\rho_f(1 - \phi)u_{f_i} + \rho_p\phi u_{p_i}}{\rho} \qquad (6.15)$$

is the mixture velocity vector. Here ϕ is the solid content of the mixture (by volume), u_{f_i} the average liquid velocity vector in the interparticle space, u_{p_i} the average particle velocity vector, and x_i the position vector.

Now, the liquid and the solid particles are incompressible, so we may eliminate the densities ρ_f and ρ_p as constant factors from the mass conservation equations and obtain conservation equations for the liquid and the solid particles, as follows:

$$\frac{\partial}{\partial t}(1 - \phi) + \frac{\partial}{\partial x_i}(1 - \phi)u_{f_i} = 0 \qquad (6.16)$$

$$\frac{\partial \phi}{\partial t} + \frac{\partial}{\partial x_i}(\phi u_{p_i}) = 0 \qquad (6.17)$$

If we add equations (6.16) and (6.17), we obtain another mass conservation or continuity equation for the liquid–solid particles mixture:

$$\frac{\partial}{\partial x_i}[(1 - \phi)u_{f_i} + \phi u_{p_i}] = 0 \qquad (6.18)$$

The momentum equations for the liquid and the solid particles may be determined by making appropriate allowances for stresses in the liquid phase due to the motion of the solid particles and the reciprocal stresses on the particles due to liquid motion. These equations may be written as[23]

$$\frac{\partial}{\partial t}[\rho_f(1-\phi)u_{f_i}] + \frac{\partial}{\partial x_j}[\rho_f(1-\phi)u_{f_i}u_{f_j}]$$
$$= \rho(1-\phi)g_i - (1-\phi)\frac{\partial p}{\partial x_i} + \frac{\partial \tau_{ij}^{(f)}}{\partial x_j} - F_i \quad (6.19)$$

and

$$\frac{\partial}{\partial t}(\rho_p u_{p_i}) + \frac{\partial}{\partial x_j}(\rho_p \phi u_{p_i} u_{p_j}) = \rho_p \phi g_i - \phi\frac{\partial p}{\partial x_i} + \frac{\partial \tau_{ij}^{(p)}}{\partial x_j} + F_i \quad (6.20)$$

where g_i is the acceleration due to gravity, p the pressure, $\tau_{ij}^{(f)}$ a tensor representing viscous stresses in the liquid, $\tau_{ij}^{(p)}$ a stress tensor resulting from the direct interaction of solid particles, and F_i the interaction force per unit volume between the liquid and the solid particles.

Adding equations (6.19) and (6.20) gives a momentum equation for the mixture:

$$\frac{\partial(\rho u_i)}{\partial t} + \frac{\partial M_{ij}}{\partial x_j} = \rho g_i - \frac{\partial p}{\partial x_i} + \frac{\partial \tau_{ij}}{\partial x_j} \quad (6.21)$$

where the tensors for momentum flux and shear stresses[27,28] are, respectively,

$$M_{ij} = \rho_f(1-\phi)u_{f_i}u_{f_j} + \phi\rho_p u_{p_i}u_{p_j} \quad (6.22)$$

$$\tau_{ij} = \tau_{ij}^{(f)} + \tau_{ij}^{(p)} \quad (6.23)$$

We now have a system of equations [equations (6.13), (6.14), (6.15), (6.18), (6.21), (6.22), and (6.23)] that provide a basis for further analysis of the problem. Two further simplifying assumptions are now required in order to close this system of equations:

1. The concentration of particles in the liquid is small, $\phi \ll 1$.
2. The accelerations of the liquid and the particles are small in comparison with the acceleration due to gravity.

Since the concentration and sizes of the particles are small, the direct interaction of particles may be ignored, or, in other words, we may assume that

$$\tau_{ij}^{(p)} \ll \tau_{ij}^{(f)}$$

so the tensor of shear stresses τ_{ij} may be written in the same form as that for a homogeneous, viscous, incompressible liquid,

$$\tau_{ij} = \mu\left(\frac{\partial u_i}{\partial x_j} + \frac{\partial u_j}{\partial x_i}\right)$$

where, for the mixture viscosity μ may be made of the Einstein formula for the effective dynamic viscosity of a liquid with suspended particles:

$$\mu = \mu_f(1 + C\phi)$$

Here μ_f is the viscosity of the corresponding homogeneous liquid, and the constant $C \simeq 2.5$.

Additionally, in a Cartesian coordinate system, let the x_3 axis be directed vertically upward so that

$$g_1 = g_2 = 0, \qquad g_3 = -g$$

If we have the x_3 axis forming a small angle θ with the vertical then

$$g_1 = g \sin\theta, \qquad g_2 = 0, \qquad g_3 = -g\cos\theta \simeq -g$$

As we stated in our assumptions, the size of the particles is small, and the substantial or total accelerations of the particles is small in comparison with that due to gravity, so that the components of the particle velocities and those of the liquid in planes normal to the x_3 axis must coincide, and the components u_{p_3} and u_{f_3} must differ by the value of the gravitational settling velocity of the particles (their terminal velocity, u_r). This means that

$$u_{p_i} = u_{f_i} - u_r\,\delta_{i3} \tag{6.24}$$

where δ_{ij} is the Kronecker delta function ($\delta_{ij} = 1,\ i = j;\ \delta_{ij} = 0,\ i \neq j$). This assumption means that the additional inertia force on a particle in its accelerated relative motion is neglected, and this makes it possible to obtain equations of motion for the liquid–solid mixture in which the terminal velocity of the particles appears in place of the interaction forces between the suspended particles and the liquid.

Equation (6.24) then yields the following relationships for the components of the total mass velocity and momentum flux:

$$u_i = u_{f_i} - \frac{\rho_p}{\rho}\phi u_r\,\delta_{i3} = u_{p_i} + \frac{\rho_f}{\rho}(1-\phi)u_r\,\delta_{i3} \tag{6.25}$$

$$\begin{aligned} M_{ij} &= \rho_f(1-\phi)u_{f_i}u_{f_j} + \rho_p\phi u_{p_i}u_{p_j} \\ &= \rho u_i u_j + \frac{\rho_f\rho_p\phi}{\rho}(1-\phi)u_r^2\,\delta_{i3}\,\delta_{j3} \end{aligned} \tag{6.26}$$

When we use equations (6.25) and (6.26) in the momentum equation (6.21) and the continuity equation (6.18), we find that

$$\frac{\partial(\rho u_i)}{\partial t} + \frac{\partial(\rho u_i u_j)}{\partial x_j} = \rho g_i - \frac{\partial p}{\partial x_i} + \frac{\partial \tau_{ij}}{\partial x_j} - \rho_f\rho_p\frac{\partial}{\partial x_3}\left[\frac{\phi(1-\phi)u_r^2}{\rho}\right]\delta_{i3} \tag{6.27}$$

and

$$\frac{\partial u_i}{\partial x_i} = -(\rho_p - \rho_f)\frac{\partial}{\partial x_3}\left[\frac{\phi(1-\phi)u_r}{\rho}\right] \tag{6.28}$$

Equations (6.24) through (6.28) and (6.13) now form a closed system of five equations corresponding to the five unknown quantities, ρ, u_i, p. Because we are considering the case of a relatively small suspension concentration, we can use the conditions

$$\phi \ll 1, \qquad \Delta\rho\phi \ll 1, \qquad \Delta\rho = \frac{\rho_p - \rho_f}{\rho_f}$$

and the terminal velocity, u_r, of the particles may as an approximation be regarded as a constant quantity equal to the uniform settling velocity of an individual particle in an unbounded liquid. Consequently, equation (6.28) may be written

$$\frac{\partial u_i}{\partial x_i} = -\Delta\rho u_r \frac{\partial \phi}{\partial x_3}$$

and the equation of mass conservation will thus take the form of a "balance" equation for the suspended particles:

$$\frac{\partial \phi}{\partial t} + u_i \frac{\partial \phi}{\partial x_i} = u_r \frac{\partial \phi}{\partial x_3}$$

Hence the right-hand side of equation (6.28) may be disregarded in all cases (except that of the suspension balance calculation), so the equation may be written

$$\frac{\partial u_i}{\partial x_i} = 0 \tag{6.29}$$

The momentum equation may also be simplified into the following:

$$\frac{\partial u_i}{\partial t} + \frac{\partial}{\partial x_j}(u_i u_j) = (1 + \Delta\rho\phi)g_i - \frac{1}{\rho_f}\frac{\partial p}{\partial x_i} + \frac{1}{\rho_f}\frac{\partial \tau_{ij}}{\partial x_j} - \frac{\partial(u_i u_j)}{\partial x_j} \tag{6.30}$$

This analysis shows that the term $\Delta\rho\phi$ in the coefficient of g_i is dynamically quite essential when applying this equation to the x_3 ($i = 3$) direction, because in this case $g_3 = -g$ is a large factor. Thus the equations of momentum (6.30) and of continuity (6.29) are identical with analogous equations for the motion of liquids of nonhomogeneous density.

The closed system of basic equations we have derived above, which are available for determination of u_i, ϕ, and p, provides a possibility of analyzing the dynamics of solid particles in a turbulent liquid flow, and we consider next a particle suspension in a flow in a horizontal duct, the suspension being caused by turbulent fluctuations of the flow velocity. In this case the flow characteristic quantities may be represented as the sum of their mean and fluctuating quantities:

$$u_i = \bar{u}_i + u'_i; \qquad p = \bar{p} + p'; \qquad \rho = \bar{\rho} + \rho'$$

Fluctuation of the mixture density ρ is entirely the result of the fluctuation of concentration, ϕ (both components of the mixture being incompressible), and hence from equation (6.14), we obtain

$$\rho' = (\rho_p - \rho_f)\phi' \tag{6.31}$$

By averaging we obtain the following equations:

1. For volume conservation [equation (6.28)]:

$$\frac{\partial \bar{u}_i}{\partial x_i} = -\Delta\rho u_r \frac{\partial \bar{\phi}}{\partial x_3} \tag{6.32}$$

2. For mass conservation [equation (6.13) with equations (6.31) and (6.32)]:

$$\frac{\partial \bar{\phi}}{dt} + \bar{u}_i \frac{\partial \bar{\phi}}{\partial x_i} + \frac{\partial \overline{\phi' u'_i}}{\partial x_i} = u_r \frac{\partial \bar{\phi}}{\partial x_3}$$

3. For momentum [equation (6.30)]:

$$\frac{\partial \bar{u}_i}{\partial t} + \frac{\partial \overline{u_i u_j}}{\partial x_j} = (1 + \overline{\Delta\rho\phi}) g_i - \frac{1}{\rho_f}\frac{\partial \bar{p}}{\partial x_i} + \frac{1}{\rho_f}\frac{\partial \bar{\tau}_{ij}}{\partial x_j} - \frac{\partial \overline{u'_i u'_j}}{\partial x_j} \tag{6.33}$$

As the suspension balance has already been calculated, the averaged equation of volume conservation, equation (6.32), takes the simple form

$$\frac{\partial \bar{u}_i}{\partial x_i} = 0 \tag{6.34}$$

In order to study turbulent flow, these averaged equations of motion must be supplemented with a turbulent energy equation, and this may be obtained by multiplying the momentum equations with their corresponding velocity component, u_i, and summing up the equations obtained in this manner over the index i. Applying this method to equation (6.33) and using the continuity equation (6.34), we obtain, after some manipulation, the following equation for the kinetic-energy balance for the mean flow:

$$\frac{\partial E}{\partial t} + \bar{u}_i \frac{\partial E}{\partial x_i} = \underbrace{-\frac{\partial}{\partial x_i}\left(\bar{u}_i \frac{\bar{p}}{\rho_f}\right)}_{\text{I}} + \underbrace{\frac{\partial}{\partial x_i}\left[\bar{u}_j\left(\frac{\tau_{ij}}{\rho_f} - \overline{u'_i u'_j}\right)\right]}_{\text{II}} + \underbrace{\bar{u}_i g_i}_{\text{III}} - \underbrace{\bar{\epsilon}}_{\text{IV}} + \underbrace{\bar{u}_i \bar{u}_j \frac{\partial \bar{u}_i}{\partial x_j}}_{\text{V}} \tag{6.35}$$

with

$$E = \tfrac{1}{2}\bar{u}_i\bar{u}_i, \qquad \bar{\epsilon} = \frac{1}{\rho_f}\bar{\tau}_{ij}\frac{\partial \bar{u}_i}{\partial x_j}$$

The right-hand side of equation (6.35) represents the following factors affecting the loss of mean flow energy, per unit mass and time:

I. Work performed against the pressure field.
II. Work performed against viscous and turbulent stresses.
III. Work against gravity force.
IV. Dissipation of mean flow energy under the action of viscosity.
V. Transition of mean flow energy into that of turbulence (in other words, the generation of turbulent energy).

Applying the same operation to equations (6.30) and (6.31), we obtain an equation for the turbulent kinetic-energy time dependence in a liquid flow transporting solid particles:

$$\frac{\partial e}{\partial t} + \bar{u}_i\frac{\partial e}{\partial x_1} = \underbrace{\frac{\partial D_i}{\partial x_i}}_{\text{I}} + \underbrace{\frac{\partial}{\partial x_i}\left(\frac{\overline{u'_j\tau'_{ij}}}{\rho_f}\right)}_{\text{II}} \underbrace{- \epsilon_t}_{\text{III}} + \underbrace{B}_{\text{IV}} \underbrace{- \overline{u'_iu'_j}\frac{\partial \bar{u}_i}{\partial x_j}}_{\text{V}}$$

$$e = \tfrac{1}{2}\overline{u'_iu'_i}, \qquad D_i = -u_i\left(\frac{\rho'}{\rho_f} + \frac{\overline{u'_ju'_j}}{2}\right), \qquad \epsilon_t = \frac{1}{\rho_f}\left(\overline{\tau'_{ij}\frac{\partial u'_i}{\partial x_j}}\right) \tag{6.36}$$

$$B = \Delta\rho g_i\overline{\phi' u'_i} \simeq -\Delta\rho g\overline{(\phi' u'_3)} \qquad \left(\Delta\rho = \frac{\rho_p - \rho_f}{\rho_f}\right)$$

Now when we compare the energy-balance equation (6.35) for the mean flow with the analogous equation for a suspensionless liquid, we realize that their structures are identical, the difference being a sign change for term V. Also, in the turbulent energy-balance equation (6.36), all the terms (except IV) are present in the analogous equation for a suspensionless liquid. The terms on the right-hand side of equation (6.36) signify:

I. Turbulent diffusion of "total" turbulent energy.
II. Fluctuation work of viscous stresses in turbulent motion.
III. Dissipation of turbulent energy into heat.
IV. Consumption of turbulent fluctuation energy in suspending the solid particles: in other words, the work of suspension.
V. Transition of the energy of mean flow into that of turbulence: in other words, the generation of turbulent energy.

The equations of energy balance (6.35) and (6.36) are found to be identical with analogous equations for a density-stratified homogeneous liquid, and in both cases factor B, which may be written in the form

$$B = \frac{g}{\rho}\overline{p'u_3'} \simeq -\Delta\rho g\overline{\phi' u_3'} \tag{6.37}$$

can be interpreted as work spent in overcoming the buoyancy force arising from the turbulent displacement of liquid elements.

In fully developed turbulence the stresses due to viscosity are negligibly small in comparison with the Reynolds turbulent stresses. Further, the first term in equation (6.36), which is the contribution of pressure fluctuations to the diffusional flux of turbulent energy, is very small and may be neglected. For the same reason the second term, the work due to viscous stress fluctuations, may be neglected. This last assumption becomes valid when we compare this term with the main part of the first term (where this part may be interpreted as the work of the Reynolds stresses in the field of turbulent fluctuations).

Now, if we add the equation of turbulent energy balance to the system of basic equations of the problem, making the simplifications referred to above, we obtain

$$\left.\begin{aligned}
&\frac{\partial \bar{u}_i}{\partial t} + \bar{u}_j\frac{\partial \bar{u}_i}{\partial x_j} = (1 + \Delta\rho\bar{\phi})g_i - \frac{1}{\rho_f}\frac{\partial \bar{p}}{\partial x_i} - \frac{\partial \overline{u_i'u_j'}}{\partial x_j}\\
&\frac{\partial \bar{u}_j}{\partial x_j} = 0\\
&\frac{\partial \bar{\phi}}{\partial t} + \bar{u}_i\frac{\partial \bar{\phi}}{\partial x_i} = u_r\frac{\partial \bar{\phi}}{\partial x_3} - \frac{\partial \overline{\phi' u_i'}}{\partial x_i}\\
&\frac{\partial e}{\partial t} + \bar{u}_i\frac{\partial e}{\partial x_i} = -\frac{1}{2}\frac{\partial}{\partial x_i}\overline{u_i'u_j'u_j'} - \overline{u_i'u_j'}\frac{\partial \bar{u}_i}{\partial x_j} - \Delta\rho g\overline{\phi' u_3'} - \epsilon_t
\end{aligned}\right\} \tag{6.38}$$

This system of equations (6.38) shows that the influence of the particles on the dynamics of the entraining flow will be negligibly small where the flow work done to suspend the particles is sufficiently small or, in other words, when term B of the equation for the turbulent energy balance is small in comparison with the other terms of this equation (note that in this case the term $\Delta\rho\bar{\phi}$ in the equation of momentum flux is negligibly small). When this is the case, the system of equations (6.38) breaks up into a set of independent equations, the first, second, and fourth equations constituting the equations of motion and the third equation being that of suspension balance.

Using equations (6.38), we follow an analysis (developed elsewhere)[29,30] for the special case of steady, plane, uniform flow in an open channel with small slope of angle θ with the horizontal and depth h. In this case the system of basic equations may be considerably simplified, because all mean flow characteristics depend on coordinate x_3 only.

The equation of volume conservation (6.32) may then be rewritten as

$$\frac{d\bar{u}_3}{dx_3} = -\Delta\rho u_r \frac{d\bar{\phi}}{dx_3}$$

and after integration we find that

$$u_3 = -\Delta\rho u_r \bar{\phi}$$

This equation indicates that the velocity of mean flow of the mixture has a small component in the negative x_3 direction, and this is due to the fact that the overall mean flux of each substance (which in the present case is zero in the x_3 direction) includes fluctuating components. In other words,

$$\overline{\phi u_{p_3}} = \bar{\phi}\bar{u}_{p_3} + \overline{\phi' u'_{2p_3}} = 0, \qquad \overline{(1-\phi)u_{f_3}} = (1-\bar{\phi})\bar{u}_{f_3} - \overline{\phi' u'_{f_3}} = 0 \tag{6.39}$$

It may be also shown that

$$\bar{u}_{f_3} = \bar{\phi} u_r, \qquad \bar{u}_{p_3} = -(1-\bar{\phi})u_r$$

Since the velocity component, $\bar{u}_3$, is small, the equation of mass balance (6.34) may be written in the approximate form

$$\frac{d\overline{\phi' u'_3}}{dx_3} = u_r \frac{d\bar{\phi}}{dx_3} - \bar{u}_3 \frac{d\bar{\phi}}{dx_3}$$

If we substitute equations (6.39) into this equation, the last term turns out to be a small quantity of second order (compared to $\bar{\phi}$) and therefore negligible, and integrating yields

$$\overline{\phi' u'_3} = u_r \bar{\phi} \tag{6.40}$$

This is the classic equation of suspension balance (mass conservation) for a uniform flow.

Now, again neglecting terms of small magnitude, we may write the momentum equations for the x_1 and x_3 directions as

$$\frac{d\overline{u'_1 u'_3}}{dx_3} = g\sin\theta, \qquad -\frac{1}{\rho_f}\frac{d\bar{p}}{dx_3} = -(1+\Delta\rho\phi)g + \frac{d(\overline{u_3'^2})}{dx_3} \tag{6.41}$$

Also, by using equation (6.40) the expression for the work of suspension, equation (6.37) may be written in the form

$$B = -\Delta g\overline{\phi' u'_3} = -\Delta\rho g u_r \bar{\phi} \tag{6.42}$$

Further, if we again ignore terms of small magnitude, the equation of energy balance may be written as

$$\frac{d}{dx_3}\overline{u'_3 u'_i u'_i} - \overline{u'_1 u'_3}\frac{d\bar{u}_1}{dx_3} - \Delta\rho g u_r \bar{\phi} - \epsilon_t = 0 \tag{6.43}$$

The first momentum equation (6.41) can be integrated, the constant of integration being determined from the condition that the shear stress is zero on the free surface, and we obtain the usual expression for the distribution of turbulent shear stress, at any position x_3:

$$-\overline{u_1'u_3'} = (v^*)^2\left(1 - \frac{x_3}{h}\right) \qquad [v^* = (gh\sin\theta)^{1/2}] \tag{6.44}$$

Here v^* is, by convention, the *shear velocity*. Using this in equation (6.43), we find

$$v^{*2}\left(1 - \frac{x_3}{h}\right)\frac{d\bar{u}_1}{dx_3} + \frac{d}{dx_3}\overline{u_3'u_i'u_i'} - \Delta\rho g u_r\bar{\phi} - \epsilon_t = 0$$

and these equations, together with the equation of suspension balance,

$$\overline{\phi u_3'} = u_r\bar{\phi} \tag{6.45}$$

are the basic system of equations for the solution of the problem. However, this system is not complete because we need to simplify the turbulent energy equation, and to achieve this we ignore the second term, expressing the diffusional transfer of turbulent energy, on the grounds that its influence is small for most of the depth of the flow.[29,30] We can express the correlations in equations (6.44) and (6.45) in the usual manner, through coefficients of turbulent exchange:

$$\overline{u_1'u_3'} = -k_1\frac{d\bar{u}_1}{dx_3}, \qquad \overline{\phi' u_3'} = -k_2\frac{d\bar{\phi}}{dx_3}$$

For determination of the coefficients k_1 and k_2 of turbulent exchange, as well as the dissipation ϵ_t of turbulent energy, we use Kolmogoroff's hypothesis, according to which the coefficients of exchange and dissipation are uniquely determined by the magnitude of turbulent energy, and by some system linear value l (which may be regarded as a scale of turbulence or, for example, the Prandtl mixing length). Hence we are led to

$$k_1 = le^{1/2}, \qquad k_2 = ale^{1/2}, \qquad \epsilon_t = cl^{-1}e^{3/2}$$

where a and c are nondimensional constant parameters. Substituting these expressions into the equations above, we obtain

$$le^{1/2}\frac{d\bar{u}_1}{dx_3} = v^{*2}\left(1 - \frac{x_3}{h}\right)$$

$$ale^{1/2}\frac{d\bar{\phi}}{dx_3} + u_r\bar{\phi} = 0 \tag{6.46}$$

$$v^{*2}\left(1-\frac{x_3}{h}\right)\frac{d\bar{u}_3}{dx_3}=\Delta\rho g u_r\bar{\phi}+c\frac{\phi^{3/2}}{l} \tag{6.47}$$

Now, this new system of equations will be closed if the dependence of the linear parameter l on coordinate x_3 (or upon the unknown quantities) is known. Using equation (6.46) in equation (6.42), we obtain

$$B = a\,\Delta\rho g l e^{1/2}\frac{d\bar{\phi}}{dx_3}$$

and using this, we find from the equation of turbulent energy (6.47) an explicit expression for the turbulent energy:

$$e=\frac{1}{c^{1/2}}v^{*2}\left(1-\frac{x_3}{h}\right)(1-ak)^{1/2} \tag{6.48}$$

where

$$k=-\Delta\rho g\frac{d\bar{\phi}/dx_3}{(d\bar{u}_1/dx_3)^2}\qquad\left(\Delta\rho=\frac{\rho_p-\rho_f}{\rho_p}\right)$$

The nondimensional factor k is analogous to the Richardson number occurring in the theory of stratified flows. This should not be a surprising conclusion as the method outlined above does in fact describe mathematically the flow of a discrete two-component flow as if it was a continuum with variable (but continuously distributed) density. If suspended particles are absent, $k = 0$ and equation (6.48) gives

$$e=\frac{1}{c^{1/2}}v^{*2}\left(1-\frac{x_3}{h}\right)$$

Comparison of this equation with equation (6.48) shows that the presence of suspended particles in the flow leads to a reduction of turbulent energy, this reduction increasing with k. In the present case k gives the ratio of work produced by the flow for the suspending of particles in the flow to the overall expenditure of turbulent energy, and if the work of suspension is small in comparison with the dissipative term in the turbulent energy equation, we may apply the usual diffusion theory for suspensions (which neglects the reactive influence of suspended particles on the flow dynamics). Thus from the theory[29] we have outlined above, we conclude that the diffusion theory is a reasonable approximation if

$$k \ll 1$$

Note that the work of suspension is proportional to the particle terminal velocity and also to the particle concentration, so that k is not small when only the terminal velocity is small.

It remains to define the linear parameter l and, again, following the analysis given elsewhere,[29,30] we use a modified form of the well-known von Kármán formula, giving

$$l = -K \frac{d\bar{u}_1}{dx_3} \left(\frac{d^2 u_1}{dx_3^2} \right)^{-1} f(ak)$$

Here K is called the von Kármán constant and $f(ak)$ is a function that must decrease with increasing k (or increase of the work of suspension) and become unity at $k = 0$.

These considerations lead us, after some analysis, to conclude that the character of the flow depends not only on k but also on another nondimensional parameter, ω, which occurs in various versions of the diffusion theory:

$$\omega = \frac{u_r}{aKv^*} \simeq \frac{u_r}{Kv^*}$$

In this equation u_r is the terminal velocity of the particles, K the von Kármán constant, and v^* the dynamic (shear) velocity.

The character of the flow appears to differ considerably for $\omega < 1$ and $\omega > 1$. For $\omega > 1$ (when the flow velocity is low and the particles are large), the transport of particles occurs mainly near the bottom surface of the flow, and in the upper region of the flow the influence of suspended particles on the dynamics of the transporting flow is small. However, in the region near the bottom, the theory (based on the assumption of small concentration $\bar{\phi}$) is inapplicable and must be replaced by a more general and exact theory. Far from the bottom region $k \to 0$, and the distribution of suspended particles asymptotically approaches the distribution obtained from the diffusion theory.

For $\omega < 1$ (when the flow velocity is high and the particles are small), the transport of particles occurs within the main body of the flow. The particle distribution at some distance from the bottom surface asymptotically approaches a certain finite self-similar distribution ($k \to$ const.) for which, in the case of very great depth ($h \to \infty$), the concentration $\bar{\phi}$ is inversely proportional to x_3 and the velocity parallel to the surface is

$$\bar{u}_1 = \frac{1}{Ku_r} \ln x_3 + \text{const.}$$

Under these conditions there occurs a certain limiting saturation for the suspension flow, and for concentrations beyond that point the theory becomes inapplicable in the region near the bottom.

As early as 1944, in investigations of the sediment-transporting flow in an open channel, it was discovered that the presence of suspended sand particles in a flow in an open channel caused a decrease of the von Kármán universal constant and a reduction of the resistance coefficient in comparison with the corresponding flow of a homogeneous fluid, and it was concluded that the particles acted to dampen the turbulence. These experimental results have been confirmed by many subsequent experiments. The theory of turbulent particle or sediment-transporting flow outlined above provides a theoretical explanation of the causes of this reduction of both the von Kármán constant and the resistance coefficient. We relate the latter phenomenon to the reduction of velocity fluctuations in any flow transporting suspended particles.

6.7 Microparticles suspended in a turbulent flow

As discussed in Chapter 5, a turbulent flow can be considered to be composed of three main ranges of eddy size (or wavenumber) – small, medium and large – in which, as a result of inertia interactions, energy is transmitted from higher wavenumbers ($\kappa = 2\pi n/v$, where n denotes the turbulence frequency and v the time-averaged velocity) to lower wavenumbers. Turbulent motions differ in apparatus of different construction, and in general, the motion occurring depends on the position in the flow. Although the wavenumber spectrum of any actual turbulent flow is not continuous, it is possible to assign a definite amount of the total energy to certain ranges of wavenumbers, or, in other words, it is possible to derive a relationship between wavenumbers in the form of an energy spectrum.

According to Kolmogoroff's first law,[31–33] at sufficiently high Reynolds numbers there exists a range of high wavenumbers in which the turbulence is in static equilibrium, being defined unambiguously by the values of ϵ (energy dissipation) and ν (kinematic viscosity). Let us consider the theory of uniform, isotropic, turbulence developed by Kolmogoroff (this obviously being a theory for a flow field whose characteristics do not depend on the position or direction of the coordinate axes[25,31–34]) because this theory has become well developed and because it fairly closely approximates flow regimes occurring in commercial equipment.

The largest eddy scale corresponds to the most rapid eddy motions, and the order of magnitude of the velocities of these eddies is given by the characteristic velocity of the system, say U, the maximum mean flow velocity. So, for example, the turbulent flow in a pipe has a largest eddy scale approximately equal to the pipe diameter, and the eddy velocity will approximate the variation of the average velocity over the pipe diameter, or, in this case, it will be equal to the maximum velocity, which occurs in the center of the pipe.[35]

Eddies of any scale can be expressed in terms of a Reynolds number defined as

$$\mathrm{Re}_\lambda = v_e \lambda / \nu_f \tag{6.49}$$

where v_e is the velocity of eddies of scale λ and ν_f the coefficient of kinematic viscosity. Now, for some $\lambda = \lambda_0$, the Reynolds number $\mathrm{Re}_\lambda \simeq 1$, and the eddy motion is accompanied by energy dissipation. Using dimensional considerations,[36] it can be shown that for $\lambda > \lambda_0$ the eddy velocity may be written as

$$v_e = (\epsilon\lambda)^{1/3}; \qquad \lambda > \lambda_0$$

where ϵ is the energy dissipation per unit mass, and when $\lambda \simeq \lambda_0$, by definition [from equation (6.49)],

$$v_{e0} = \nu_f/\lambda_0 \tag{6.50}$$

The acceleration is

$$a_e \simeq (\epsilon^3/\nu_f)^{1/4}; \qquad \lambda > \lambda_0 \tag{6.51}$$

Next, for $\lambda < \lambda_0$, the motion acquires a quasiviscous character, interaction between separate eddies ceases, and eddy frequencies become independent of scale, and the eddy velocity can then be derived from the following relationship:

$$v_e \simeq \lambda/\tau; \qquad \lambda < \lambda_0$$

where τ is the eddy period and the acceleration term assumes the form

$$a_e \simeq (v_{e0}/\lambda_0)^2\lambda; \qquad \lambda < \lambda_0$$

Now, for eddies of scale $\lambda > \lambda_0$, there corresponds a period τ of the motion which is dependent on the scale λ and the turbulent dissipation ϵ only, because for this range the influence of viscosity is ignorable (Kolmogorov's first law) and τ is given by

$$\tau = (\lambda^2/\epsilon)^{1/3}; \qquad \lambda > \lambda_0$$

By comparison, for eddies of scale $\lambda \simeq \lambda_0$, τ is given by

$$\tau_0 \simeq \lambda_0/v_e \simeq (\nu_f/\epsilon)^{1/2}$$

Employing the expressions given above, let us now consider the motion of particles in a homogeneous isotropic turbulent flow. We will use a theory developed elsewhere in which we consider the solid particles and liquid as one fluid, and in using such an approach, we note that the problem of calculating liquid pressure on the surfaces of the particles no longer arises, this being an internal system constraint.[12]

For the motion of a particle in a homogeneous isotropic turbulent flow, let the liquid surrounding the particle have an acceleration, a_f. We would like to determine, in this case, what the acceleration a_p of the particle will be. Neglecting gravitational forces and perturbation of the flow caused by the particle itself, all the elements of the liquid in the neighborhood of the particle have identical accelerations a_f, and to this acceleration there corresponds a pressure drop in the direction of the acceleration vector given by[37]

$$\frac{\partial p}{\partial x} = \rho_f a_f$$

As a result of this pressure drop, on a particle with volume V located in the liquid there acts a force $V\rho_f a_f$ in the direction of the acceleration vector, but since the density ρ_p of the particle is not equal to the density ρ_f of the liquid, the acceleration a_p of the particle is also not equal to the acceleration of the liquid, and the particle thus moves relative to the liquid with an acceleration equal to $(a_p - a_f)$. For relative motion of a particle in the liquid with a nonconstant velocity, the effect exerted on the particle by the liquid is given by an additional term $\rho_f V_a$ added to the mass of the particle.[12] Here V_a can be envisioned as a volume of liquid considered attached to the particle, which moves with the particle. This additional term means that the total influence of liquid motion on the particle can be represented simply as an increase in its mass and, consequently, as an increase of the total system inertia. So, as a result of the relative motion of the particle in the liquid, a resistance is set up directed opposite the driving force and proportional to the "added mass" $\rho_f V_a$, that is, equal to $\rho_f(a_p - a_f)V_a$, and then the equation of relative motion of a particle in a turbulent flow can be represented in the form

$$\rho_p V a_p = \rho_f[a_f V - (a_p - a_f)V_a] \tag{6.52}$$

From equation (6.52), we can find the particle acceleration as a function of the total liquid acceleration:

$$a_p = a_f \frac{V + V_a}{V(\rho_p/\rho_p) + V_a} \tag{6.53}$$

Taking account of the added-mass effect thus makes a noticeable contribution to the value of a_p and, for example, for particles approaching a spherical form, the added mass is equal to half the mass of the displaced liquid.[38]

The motion of particles which are considerably smaller than the scale of turbulence is determined by the small-scale eddies, and the maximum value for the acceleration of these liquid elements is given[36] by equation (6.51). Now, it follows from equation (6.53) that for $\rho_p > \rho_f$, the acceleration a_p of the particle will be less than the acceleration a_f of the flow, and therefore we can write

$$a_p = a_e - a_r \tag{6.54}$$

where we have used the relative acceleration of the motion in the form

$$a_r \simeq v_r/\tau_p \tag{6.55}$$

where v_r is the particle relative velocity and τ_p is the period of particle motion, given by[35]

$$\tau_p \simeq \lambda_0/v_r \tag{6.56}$$

Note that for the smaller turbulent eddies, we have[36]

$$\lambda_0 \simeq (\nu_f^3/\epsilon)^{1/4} \tag{6.57}$$

Using equations (6.51), (6.55), (6.56), and (6.57) in equation (6.54), we obtain

$$a_p \simeq \left(\frac{\epsilon^3}{\nu_f}\right)^{1/4} - \frac{v_r^2}{(\nu_f^3/\epsilon)^{1/4}}$$

and using this equation for the particle acceleration a_p and also equation (6.51) for the acceleration a_f of the liquid, we find that

$$v_r \simeq (\epsilon\nu_f)^{1/4}\left[1 - \frac{V + V_a}{V(\rho_p/\rho_f) + V_a}\right]^{1/2} \tag{6.58}$$

Clearly, for $\rho_p \gg \rho_f$, in equation (6.58) the factor

$$\left[1 - \frac{V + V_a}{V(\rho_p/\rho_f) + V_a}\right]^{1/2}$$

approaches unity and in this case

$$v_r \simeq (\epsilon\nu_f)^{1/4} \tag{6.59}$$

Now, taking into account that the velocity of small-scale eddies[12] is given by equation (6.50) with λ_0 given by equation (6.57), we find that

$$v_{e_0} \simeq (\epsilon\nu_f)^{1/4} \tag{6.60}$$

Consequently, when the density of the particles is significantly greater than the density of the liquid ($\rho_p \gg \rho_f$), the average value of the relative velocity of a particle [given by equation (6.59)] is equal to the velocity of eddies of scale λ_0 [equation (6.60)] and is seen to be uniquely determined by the viscosity of the liquid and the energy dissipation per unit mass, for constant-density fluids.

In actual industrial processes, the turbulence created in the liquid by use of, for example agitators, or that occurring in a bubbling apparatus, usually has an extremely anisotropic character, this being determined by large-scale eddies, and it is therefore of interest to investigate to what degree equation (6.58) is valid for the case of anisotropic turbulence. It is well known[36] that large-scale eddies, whose size (or scale) is determined by a characteristic system dimension (e.g., the diameter of an agitator, or the height of a two-phase mixture), contain most of the energy of the flow, this energy being obtained directly from the system generating the turbulence. However, transfer of energy to small-scale eddies is accompanied by increased disorder of the motion and by a diminishing of the effect of the large-scale eddies on the motion. In addition, the anisotropy of the large-scale eddies has no effect on eddies of the order of the internal scale of the turbulence, and therefore application of the rigorous condition of homogeneity of the turbulence to all the turbulence scales is not required, because the motion of these small-scale eddies can be assumed to take place as in purely homogeneous and isotropic turbulence.

By means of the theory of dimensional analysis, the energy dissipated per unit mass and time is given by

$$\epsilon \simeq U^3/l \tag{6.61}$$

where U is some average velocity of the flow and l a characteristic length scale (equal to the size of largest eddies). Substituting into equation (6.58) the value of ϵ from equation (6.61), we determine the relative velocity of a microparticle suspended in turbulent flow to be

$$v_r \simeq \text{const.}\left[1 - \frac{V + V_a}{V(\rho_p/\rho_f) + V_a}\right]^{1/2}\left(\frac{\nu_f U^3}{l}\right)^{1/4} \tag{6.62}$$

Equation (6.62) shows that the relative velocity, v_r, of a particle increases with the mean velocity U of the flow in the form

$$v_r \propto U^{3/4}$$

Equation (6.62) may be used for calculation of the particle relative velocity in tubular apparatus, because in this application it is possible to measure the average flow velocity, the tube diameter can be used as the characteristic dimension, and the equation has been correlated with experimental data[39] for a liquid droplets–air system; the flow conditions were inferred from the system pressure and from assumed saturation conditions, as shown in Table 6.1.

Droplet diameters for each run were measured and the corresponding velocites v_d were calculated by a distance/time correlation. The drops for each run were of relatively uniform size and were assumed to be moving one-dimensionally and at constant velocity (or, in other words, with zero absolute acceleration).

The vapor velocity was inferred from a vertical force balance of drag against gravity,

$$F_D = F_g$$

or

$$\tfrac{1}{8}C_D \pi d^2 \rho_g (v_g - v_d)^2 = \tfrac{1}{6}\pi d^3 \rho_f g$$

where C_D is the drag coefficient, assumed to have the value 0.45; v_g the vapor velocity; and d the drop diameter. This reduces to

$$v_g = v_d + (4 d \rho_f g / 3 C_d \rho_g)^{1/2}$$

Now for comparison of the theory with experimental data, we can use equation (6.62) if the size of droplets is actually less than the size of small-scale eddies. The size of small-scale eddies is given in equation (5.17) as

$$r_d = (8\nu_f^3 / 3\gamma^2 \epsilon)^{1/4} \tag{6.63}$$

where $\gamma = 0.4$ is a universal constant and ϵ the energy dissipation, obtained from equation (6.61) in the form

$$\epsilon = v_g^3 / l$$

where l is the height of the duct (l = 4.88 m or 16 ft). According to the experimental data, $\nu_f \simeq 0.025$ m²/s (0.27 ft²/s), and $v_g \simeq 10.0$ m/s (32.8 ft/s), so from equation (6.63) we find that the typical size (or radius) of

Table 6.1. *Experimental data for computing the particle relative velocity*

Run number	p ($\times 10^5$ Pa)	σ ($\times 10^{-3}$ N/m)	ρ_f (kg/m³)	ρ_g (kg/m³)	Flood rate (cm/s)
1	1.73	5.79	955.08	0.99	5.08
2	1.24	6.00	936.68	0.73	2.03
3	1.03	6.11	967.56	0.61	1.52
4	1.03	6.11	967.56	0.61	1.52
5	3.78	5.24	933.91	2.07	14.99
6	1.30	5.97	962.55	0.76	2.54

small-scale eddies is $r_d \simeq 3.3$ cm (1.3 in.), which is much greater than any droplet sizes occurring in these experiments, and therefore we may use, for comparison of the theory with experimental data, equation (6.62).

Experimental[39] and calculated data for particle relative velocities indicate that the average variance with experimental data is 3.8 percent, which indicates that the theory is a good model for the flow.

6.8 Motion of microparticles in the intensive-bubbling regime

When gas is injected into a liquid, intensive motion of bubbles can be observed under certain flow conditions, and this motion causes a significant mixing of particle-carrying liquid. As described in Chapter 3, it is usually the practice[40] to distinguish three principal bubbling regimes (unsteady motion, steady motion, and the regime of gas sprays and liquid jets), the transition from one regime to another being determined by gas and liquid rates and by gas and liquid distribution devices. Investigation of the steady regime is of practical importance as well as being the regime most accessible to analysis.

When there is intensive bubbling, it is best to consider the gas–liquid mixture from the point of view of homogeneous isotropic turbulence,[41] because it is known that fully developed turbulence is in fact created as a result of the liquid agitation by the moving bubbles. We make the assumption that all the gas kinetic energy is transferred to the liquid–particle mixture, eventually being dissipated by the turbulent motion, and we follow the analysis of Section 3.2.

The pressure drop in turbulent flow is determined by[36]

$$\Delta p \simeq \rho_f U^2 \tag{6.64}$$

Thus from equations (6.62) and (6.64), we obtain

$$v_r = \text{const.} \left(\frac{\nu_f}{l}\right)^{1/4} \left(1 - \frac{V + V_a}{V(\rho_p/\rho_f) + V_a}\right)^{1/2} \left(\frac{\Delta p}{\rho_f}\right)^{3/8}$$

where l could be, for example, the depth of the bubbling mixture.

6.9 Motion of microparticles in a bubble apparatus with agitators

It is known that the energy E consumed during liquid mixing in a baffled apparatus can be calculated by the following equation:[42]

$$\frac{E}{\rho_f n^3 D_m^5} = \text{const.} \left(\frac{\rho_f n D_m^2}{\mu_f}\right)^{\alpha} \left(\frac{n^2 D_m}{g}\right)^{\beta} \left(\frac{T}{D_m}\right)^{\gamma} \tag{6.65}$$

where $\mathrm{Fr} = n^2 D_m/g$ is the centrifugal Froude number, $\mathrm{Re} = \rho_f n D_m^2/\mu_f$ is the centrifugal Reynolds number (μ_f is the liquid dynamic viscosity), $\Gamma = T/D_m$ is a geometrical similarity parameter (D_m is the diameter of the mixer and T is the diameter of the chamber), and n is the revolution rate of the mixer. Gas is uniformly dispersed in the whole liquid volume when intensively mixed, and because bubble sizes are small in comparison with the size of the apparatus and mixer, the gas–liquid system may be considered as a homogeneous mixture with a uniform average density. For this case, the energy consumption is equivalent to that occurring in homogeneous liquid as described in Section 6.8, except that the "liquid" density is equal to the average density of the mixture. In other words,

$$\rho_{av} = \rho_g \phi + \rho_f (1 - \phi) \tag{6.66}$$

where ϕ (the gas void fraction) may be expressed[43] as a product of the relative gas flow rate q and its residence time τ in the system:

$$\phi = q\tau = q x_1 / v_b$$

where x_1 is the two-phase mixture height and v_b the average rise velocity of a bubble.

This velocity, for a bubble whose size is greater than the scale of internal turbulence, is independent of the viscosity of the liquid, and the bubble motion is determined by the "automodeling" zone of the turbulent regime, and we find that[44]

$$v_b = 1.76 \left[\frac{d_b g(\rho_f - \rho_g)}{\rho_f} \right]$$

where d_b is the bubble diameter. The gas content in an apparatus with a mixer, having calculated the bubble diameter on the basis of homogeneous turbulence, can then be shown to be of the form[45]

$$\phi = \text{const.} \frac{q^3 T^{0.4} \rho_g^{0.3} E^{0.2}}{\sigma^{0.3} \rho_f^{0.2} g^{0.5}} \tag{6.67}$$

where σ is the surface tension. Combining equations (6.66) and (6.67), we obtain a value for ρ_{av} that we can substitute in equation (6.65) in place of term ρ_f.

Experimental coefficients for equation (6.65) have been obtained, describing power consumption during mixing with a turbine-type mixer in a gas–liquid system, leading to

$$K_E = 1.08 \mathrm{Re}^{0.1} \mathrm{Fr}^{-0.03} \mathrm{We}^{-0.12} K_v^{-0.2} s_g \Gamma^{-0.4}$$

where $\mathrm{We} = \rho n^2 D_m^3/\sigma$ is the centrifugal Weber number, $s_g = \rho_g/\rho_f$ a dimensionless density, and $K_v = U/n$ a distribution coefficient.

This equation, it turns out, is not convenient for calculation of energy consumed by liquid mixing for liquids with properties close to water, but it is possible to use the following simplified formula[27]

$$E \sim 0.9\left(\frac{n^3 D_m^{5.22}}{T^{0.4} q^{0.2}}\right)$$

We can express the dissipation energy ϵ for $\mathrm{Re} > 10^4$ in terms of the process parameters that characterize the turbulent regime in an apparatus with mixers as[44]

$$\epsilon = \frac{E}{Q_l} \sim \frac{n^3 D_m^5}{T^2 x_1 q^{0.2}} \tag{6.68}$$

Now, taking into account that

$$\epsilon = \frac{U^3}{l}$$

and that in this application

$$l \simeq D_m \tag{6.69}$$

equation (6.68) gives

$$U = \text{const.}\,\frac{n D_m^2}{(T^2 x_1 q^{0.2})^{1/3}}, \qquad \mathrm{Re} > 10^4 \tag{6.70}$$

Equations (6.69) and (6.70) conform well with experimental data, for systems of various geometries, over a broad range of hydrodynamic parameters.[28]

So, substituting values l and U from equations (6.69) and (6.70) in equation (6.62) gives us an expression for the relative velocity of microparticles in turbulent flow in a bubble apparatus with a mixer:

$$v_r = \text{const.}\left[1 - \frac{V + V_a}{V(\rho_p/\rho_f) + V_a}\right]^{1/2}\left(\frac{n^3 D_m^5 \nu_f}{T^2 x_1 q^{0.2}}\right)^{1/4}$$

6.10 Motion of a macroparticle suspended in a turbulent flow

Let us now consider the motion of a solid particle whose size is significantly greater than the size of small-scale eddies, noting that this will include taking into account the unsteady velocity field. The motion of a particle in this case is determined by the large-scale eddies, and the liquid acceleration depends on ϵ and λ in the form[36]

$$a_e \simeq (\epsilon^2/\lambda)^{1/3} \tag{6.71}$$

where a_e is now the acceleration of large-scale eddies and λ the scale of these eddies.

For $\rho_p \neq \rho_f$, the particle acceleration a_p will be less than the liquid flow acceleration, which we take to be equal to the acceleration a_e of large-scale eddies, so that we will have

$$a_p = a_e - a_r \tag{6.72}$$

where a_r is the relative acceleration. Now, the equation of relative motion of a particle is similar to equation (6.52), and we can determine the acceleration term a_p from this equation. The relative acceleration is given by equation (6.55) as

$$a_r = v_r/\tau \tag{6.73}$$

where the period of particle motion[36] is

$$\tau = r_p^2/\nu_f \tag{6.74}$$

so equation (6.72) becomes

$$a_p = \left(\frac{\epsilon^2}{\lambda}\right)^{1/3} - \frac{v_r \nu_f}{r_p^2} \tag{6.75}$$

From dimensional theory, we can determine the eddy period as[36]

$$\tau = (\lambda^2/\epsilon)^{1/3}$$

and taking this into account, with equation (6.74), we can now substitute into equation (6.53) the liquid acceleration a_e from equation (6.71) and particle acceleration a_p from equation (6.75), and write the relative velocity as

$$v_r = \left(\frac{\epsilon}{\nu_f}\right)^{1/2} r_p \left[\frac{V(\rho_p/\rho_f) - V}{V(\rho_p/\rho_f) + V_a}\right] \tag{6.76}$$

In reality, when there is a system of discrete particles in a turbulent flow, each particle is influenced by the particles that are in its immediate neighborhood, as well as those more distant, so the pattern of the discrete particles' motion is to a large measure determined by their concentration.[43] For determination of this effect of constrained motion we will use the *cell model.* Let us suppose that each particle is in the center of a sphere formed by adjacent particles. With this assumption, the problem of constrained motion may be reduced to the problem of liquid motion between two concentric spheres, the inner one being the particle, the outer one the surrounding particles.

Now, the influence of particles in the turbulent energy spectrum, $E(\kappa)$, has been investigated qualitatively,[43] and it seems that the addition of particles does not change the spectrum in the range of $\kappa \ll \kappa_d \sim 1/r_d$ (where κ_d and r_d are the wavenumber and eddy size, respectively, at which viscous effects first appear significant) but has the effect of extinguishing eddies for $\kappa \gg \kappa_d$. To simplify analysis of the problem, we can thus validly consider the short-wavelength range of $E(\kappa)$, starting from some characteristic wavenumber κ_0, to be eliminated and that, in general, for $\kappa < \kappa_0$ the spectrum $E(\kappa)$ is not changed by adding particles. In fact, if the particle concentration is sufficiently large, eddies smaller than the particles are completely extinguished.

As a first approximation, at a sufficiently high concentration of particles, when the radius of the outer sphere mentioned above becomes comparable to about twice the diameter of a particle, the motion of liquid inside the sphere may be considered as laminar. Further, under conditions of high particle concentration, and because the particle size is less than the eddy scale λ, the cluster of neighboring particles will be involved in the same motion as the central particle, being in the same eddy, and consequently the central particle and *spherical shell,* composed of neighboring particles, are moving with equal velocities. In other words, we may consider the outer shell to be stationary with respect to the central particle, and in this case we can write expressions for radial and tangential components of the liquid velocity, relative to the sphere center:[45]

$$v_R = \left(\frac{A}{r^3} + \frac{B}{r} + C + Dr^3\right)\cos\theta - v_r\cos\theta \tag{6.77}$$

$$v_\theta = \left(\frac{A}{2r^3} - \frac{B}{2r} - C - 2Dr^3\right)\sin\theta + v_r\sin\theta \tag{6.78}$$

where

$$A = \frac{-v_r r_m^3}{2 - 3\gamma + 3\gamma^5 - 2\gamma^6}$$

$$B = \frac{v_r r_m(2\gamma^5 + 3)}{2 - 3\gamma + 3\gamma^5 - 2\gamma^6}$$

$$C = \frac{-v_r(2\gamma^6 + 3\gamma)}{2 - 3\gamma + 3\gamma^5 - 2\gamma^6}$$

$$D = \frac{v_r\gamma^3}{(2 - 3\gamma + 3\gamma^5 - 2\gamma^6)r_m^2}$$

and $\gamma = r_m/r_p$, where r_m is the radius of the outer shell and r_p the particle radius. Here r and θ are the radial and angular coordinates, with velocity components v_R and v_θ respectively.

From physical considerations, for the case when the Schmidt number Sc $\gg 1$, equations (6.77) and (6.78) may be reduced to the form[28]

$$v_\theta = v_r\Psi(\gamma)\frac{\eta}{r_m}\sin\theta$$

$$v_R = -v_r\Psi(\gamma)(\eta/r_m)^2\cos\theta$$

where $\eta = r - r_m$. Thus the constraint conditions we have introduced for the motion may be represented by a separate factor $\Psi(\gamma)$, so that

$$v_r \sim v_{r_0}\Psi(\gamma)$$

and

$$\Psi(\gamma) = \frac{3 - \frac{9}{2}\gamma + \frac{9}{2}\gamma^3 - 3\gamma^6}{3 + 2\gamma^5}$$

The relative velocity of a particle in this constrained motion may then be obtained from equation (6.76) as

$$v_r = \left(\frac{\epsilon}{\nu_f}\right)^{1/2} r_p\left(\frac{\rho_p - \rho_f}{\rho_p + 0.5\rho_f}\right)\Psi(\gamma) \qquad (6.79)$$

It follows from equation (6.79) that the relative velocity of a particle in the constrained motion is determined by the particle size, the energy dissipation, and the volume content of the solid phase (the latter due to γ). Now, with the energy dissipation being $\epsilon \simeq U^3/l$, and particle relative velocity being $v_r \sim U^{3/2}$, using equations (6.70) and (6.65) (which determine the flow patterns), the relative velocity of a particle is obtained as follows:

1. For a tubular apparatus,

$$v_r = \text{const.}\left(\frac{\rho_p - \rho_f}{\rho_p + 0.5\rho_f}\right)\Psi(\gamma)\frac{U^{3/2}r_p}{(l\nu_f)^{1/2}}$$

2. For a bubbling apparatus,

$$v_r = \text{const.}\left(\frac{\rho_p - \rho_f}{\rho_p + 0.5\rho_f}\right)\Psi(\gamma)\,\Delta p_{\text{st}}^{3/4}\frac{r_p}{(x_1\nu_f)^{1/2}\rho_f}$$

3. For a bubbling apparatus with a mixer,

$$v_r = \text{const.}\left(\frac{\rho_p - \rho_f}{\rho_p + 0.5\rho_f}\right)\left(\frac{E^3D_m^5}{T^2x_1}\right)\frac{r_p}{(\nu_f\rho_f)^{1/2}}$$

References

1. Leva, M., *Fluidization,* McGraw-Hill, New York, 1959.
2. Zenz, F. A., and Othmer, D. F., *Fluidization and Fluid Particle Systems,* Reinhold, New York, 1960.
3. Orr, C., Jr., *Particulate Technology,* Macmillan, New York, 1966.
4. Soo, S. L., *Fluid Dynamics of Multiphase Systems,* Blaisdell, Waltham, Mass., 1967.
5. Brodkey, R. S., *The Phenomena of Fluid Motions,* Addison-Wesley, Reading, Mass., 1967.

6. Wallis, G. B., *One-Dimensional Two-Phase Flow*, McGraw-Hill, New York, 1969.
7. Zenz, F. A., *Pet. Refiner*, *36*(4), 173 (1957); *36*(5), 261 (1957); *36*(6), 133 (1957); *36*(7), 175 (1957); *36*(8), 147 (1957); *36*(9), 305 (1957); *36*(10), 162 (1957); *36*(11), 321 (1957).
8. Frantz, J. F., *Chem. Eng.*, *69*(Sept. 17), 161 (1962); *69*(Oct. 1), 89 (1962); *69*(Oct. 29), 103 (1962).
9. Julian, F. M., and Dukler, A. E., *AIChE J.*, *11*, 853 (1965).
10. Govier, G. W., and Aziz, K., *The Flow of Complex Mixtures in Pipes*, Van Nostrand Reinhold, New York, 1972.
11. Shook, C. A., and Daniel, S. M., *Can. J. Chem. Eng.*, *47*, 196 (1969).
12. Lamb, H., *Hydrodynamics*, Dover Publications, New York, 1945.
13. Oseen, C., *Neuere Methoden und Ergebnisse in der Hydrodynamik*, Leipzig, 1927.
14. Langhaar, H. L., *Dimensional Analysis and Theory of Models*, Wiley, New York, 1951.
15. Handl, *Exp. Phys.*, *4*(4), 13 (1932).
16. Klyachko, L., *Otoplenie Vent.*, no. 4 (1934).
17. Basset, A. B., *A Treatise on Hydrodynamics*, vol. 2, chap. 5, Dieghton, Bell and Co., Cambridge, 1888; Dover Reprints, 1961.
18. Boussinesq, J., *Thêorie analitique de chaleur*, vol. 2, Paris, 1903, p. 224.
19. Lumley, J., *Some Problems Connected with the Motion of Small Particles in Turbulent Fluid*, Ph.D. thesis, Johns Hopkins University, 1957.
20. Soo, S. L., *Chem. Eng. Sci. 5*, 57 (1956).
21. Friedlander, S. K., *AIChE J.*, *3*, 381 (1957).
22. Liu, V. C., *J. Meteorol. 13*, 399 (1956).
23. Tchen, C. M., Ph.D. thesis, Delft, 1947.
24. Corsin, S., and Lumley, J., *Appl. Sci. Res.*, *A6*, 114 (1956).
25. Hinze, J. O., *Turbulence*, McGraw-Hill, New York, 1959.
26. Hjelmfelt, A. T., Jr., and Mockros, L. F., *Appl. Sci. Res.*, *16*, 149–161.
27. Braginski, L. N., and Pavlushenko, L. S., *Sbornik progressy khimicheskoy tekhnologii*, Acad. Nauk, Moscow, 1965.
28. Schwartzberg, H. G., and Treyball, R. E., *Ind. Eng. Chem. Fund.*, *7*(1), 6 (1968).
29. Barenblatt, G. I., *Zh. Prikl. Mat. Mech.*, *17*, 3 (1953); *19*, 1 (1955).
30. Barenblatt, G. I., *Vestn. Mosk. Univ.*, *8* (1955).
31. Kolmogoroff, A. N., *Dokl. Acad. Nauk SSSR*, *30*, 229 (1941).
32. Kolmogoroff, A. N., *Dokl. Acad. Nauk SSSR*, *31*, 538 (1941).
33. Kolmogoroff, A. N., *Dokl. Acad. Nauk SSSR*, *32*, 19 (1941).
34. Batchelor, G. K., *The Theory of Homogeneous Turbulence*, Cambridge, University Press, New York, 1953.
35. Levich, V. G., *Physicochemical Hydrodynamics*, Prentice-Hall, Englewood Cliffs, N.J., 1962.
36. Landau, L., and Lifshitz, E., *Fluid Mechanics*, Pergamon Press, London, 1959.
37. Prandtl, L., and Tietjens, *Hydro- and Aerodynamics*, Dover Publications, New York, 1957.
38. Kochin, N. E., Kibel, A. I., and Roze, N. V., *Teoreticheskaya gidrodinamika* (Theoretical Hydrodynamics), Gostekhizdat, Moscow, 1955.
39. Smith, T. A., "Heat Transfer and Carryover of Low Pressure Water in a Heated Vertical Tube," M. A. thesis, Massachusetts Institute of Technology, 1976.
40. Ramm, V. M., *Absorption of gases*, Chimia, Moscow, 1966. (In Russian.)
41. Azbel, D. S., and Narozhenko, A. F., *Teor. Osn. Khim. Tekhnol.*, *3*, 508 (1969).
42. Kasatkin, A. G., *Osnovnye protressy i apparaty khimicheskoi tekhnologii*, Goskhimizdat, Moscow, 1966.
43. Ruckenstein, E., *Chem. Eng. Sci.*, *19*(2), 131 (1964).
44. Piterskikh, P. P., and Valashek, E. R., *Khim. Prom. (Moscow)*, no. 35, 1 (1956).
45. Ruckenstein, E., *Chem. Eng. Sci.*, *19*, 131 (1964).

PART II

Mass transfer in two-phase flows

7

Mass transfer at the phase boundary in bubble and droplet processes

7.1 Introduction

Many of the industrial processes in which mass transfer is important involve transport of a diffusing solute from the bulk fluid of one phase into the interface, and then from the interface to the bulk of the second phase. Consider, for example, mass flux, under turbulent conditions, from one phase to another when the first phase moves relative to the second. (This two-phase flow might be liquid–gas or liquid–liquid.) The diffusing solute transfers from phase P_1, where its concentration is higher than equilibrium, into phase P_2. Thus mass transfer takes place from the bulk fluid of phase P_1 to the interface [in doing so overcoming resistance to mass flux due to the interface (if this has a significant value)] and then from the interface to the bulk fluid of the other phase, P_2. In the interior of bulk fluids, mass flux is accomplished by mixing due to turbulent fluctuations, and the concentration of the diffusing solute in the interiors is thus practically constant, as shown in Figure 7.1. In the boundary layer near the interface, the turbulence diminishes, which causes a concentration gradient to develop as the interface is approached, until at the interface the mass flux encounters resistance due only to the mechanism of diffusion. The concentration gradient is steepest at the interface, and mass transfer ceases when equilibrium is established between the bulk fluid and the interface. Conditions in the immediate vicinity of the interface are difficult to explore experimentally, and hence the most appropriate approach to the problem is to develop a theoretical model for mass transfer in the interface region and see if the results of such an analysis are consistent with experimental overall mass-transfer results. Good agreement would then imply that the model describes the events that actually occur in the interface region.

Several different theoretical models have been developed, each representing a simplified mode of mass transfer, but most models are based on the following assumptions:

1. The total resistance to mass flux is the sum of the resistance of each phase and the resistance of the interface. However, in many cases the interface

resistance may be taken to be negligible, and in this case, taking into account that the mass flux within a single phase is independent of the other phase, the total resistance to mass transfer may be considered as the sum of phase resistances (this is known as the *additivity of resistances*).

2. Near the interface the two phases are in equilibrium, and this equilibrium is established more rapidly than the change of average concentration in either bulk fluid.

The two major models of the mechanism of mass (and also heat) transfer between two phases are the film theory and the penetration theory. The *film theory* assumes that there is a region in which steady-state molecular diffusion is the transfer mechanism; the *penetration theory* assumes that the interface is continuously being impinged upon by eddies, and that in these eddies mass transfer is controlled by unsteady molecular diffusion. Next, we consider each model in detail.

7.2 The film model

The film model was first proposed in 1904[1] and has been applied with some success to problems of both heat and mass transfer, but it was realized very early that the concept was a gross oversimplification of the actual conditions near a phase boundary. The weakest element of the theory is the positing of a stagnant film of a definite (but unknown) thickness, this thickness being evaluated in light of experimental results. Thus at best the theory is of a semiempirical nature. However, it does give quite reliable predictions of the rate of mass transfer where there is a simultaneous chemical reaction, compared with the rate under the same conditions without reaction, and it is also helpful in predicting the effect of large mass-transfer rates on heat transfer.

According to this model, steady-state mass transfer occurs by molecular diffusion across a stagnant, or laminar-flow film at the interface between the

Figure 7.1 Concentration distribution in phases during mass transfer.

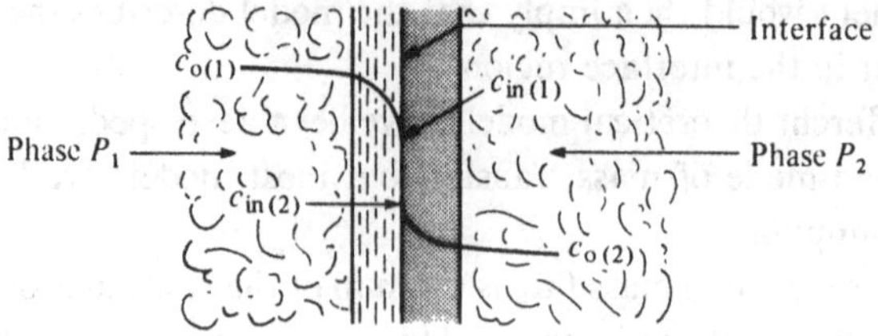

phases, in which the fluid is turbulent. All the resistance is assumed to be confined inside the film, and therefore mass concentration gradients arise only inside this film in bulk fluids, the concentrations being constant and equal to some average values. With this system it is logical to treat mass transfer at a phase boundary by use of the equations of molecular diffusion. The mass $\dot{m}'$ transferred across a unit area of interface, per unit time, is proportional to the concentration gradient between the bulk fluid and the interface, so that

$$\dot{m}' = \frac{-D}{\delta_{\text{eff}}}(c_0 - c_{\text{in}}) = -k(c_0 - c_{\text{in}}) \qquad (7.1)$$

where c_0 and c_{in} are the average concentrations in the bulk fluid and at the interface, respectively; D is the diffusivity; and δ_{eff} is the effective film thickness. For the phase on the other side of the interface, $\dot{m}'$ is similarly proportional to the concentration gradient between the interface and the bulk fluid.

In equation (7.1) $k = D/\delta_{\text{eff}}$ is a mass-transfer coefficient that characterizes the rate of mass flux, and the value δ_{eff} is, by definition, the thickness of some boundary layer in which molecular diffusion has the same effective flux resistance as that actually determined by the convective diffusion. The equation reveals a major flaw of the theory; that is, it indicates linearity between mass flux and the molecular diffusion coefficient D. In practice, turbulence in the bulk of the fluid diminishes only gradually as the film surface is approached, and consequently transition from eddy diffusion (where $\dot{m}' \propto D^0$, or mass flux is independent of molecular diffusion) to molecular diffusion (where $\dot{m}' \propto D^{1.0}$) is gradual, and a more reasonable value for the exponent of D in equation (7.1) would be some value between zero and unity. This is in fact found to be the case, and this discrepancy is a direct result of the main weakness of the theory: the introduction of a definite film thickness δ_{eff}.

Actual conditions near the interface are taken into account in the modified film theory known as the *border diffusion layer model.*[2] This model, in contrast to the simpler film model, considers:

1. The effect of liquid motion on mass transfer.
2. Molecular and convective diffusion in both the radial and tangential directions.
3. The absence of a clearly defined layer thickness.

Distribution of the diffusing matter in a turbulent stream has a four-layer structure, as shown in Figure 7.2. The constant concentration of matter in the bulk fluid, where there is a regime of fully developed turbulence, decreases slowly in the turbulent boundary layer, where both momentum and matter are transferred by means of turbulent eddies. Still closer to the interface (or wall in the case of fluid motion inside a solid structure), in the viscous sublayer, turbulent eddies become so small that momentum transfer by molecular viscosity exceeds that by turbulent eddies, and the concentration decreases appreciably faster. Here, under the action of viscous forces, the flow approaches a laminar state, and the effect of molecular diffusion is increased still more. However, more mass is still transferred by turbulent diffusion in the viscous sublayer, because the diffusivity D is typically three orders of magnitude smaller than the eddy coefficient. Only in the immediate vicinity of the interface, inside the diffusion sublayer, does the molecular diffusion mechanism predominate over the turbulence mechanism for mass transfer.

In the turbulent boundary layer the concentration is given by the familiar logarithmic relation

$$c \sim \ln y$$

where c is the concentration of the solute and y is the perpendicular distance from the interface. The viscous sublayer can be treated in one of two ways.[3] If molecular diffusion is the major transfer mechanism, it can be shown that

$$c \sim y$$

Conversely, if turbulent diffusion is considered to diminish gradually in the viscous sublayer rather than discretely at its upper boundary, we find that

$$c \sim 1/y^3$$

In the diffusion sublayer, we obtain a linear relation

$$c \sim y$$

which comes directly from equation (7.1) when the right-hand side is written as a local concentration gradient. Uncertainty as to the correct interpretation of the behavior in the turbulent boundary layer leads us to the

Figure 7.2 Distribution of the diffusing substance in a turbulent stream. (From V. G. Levich, *Physicochemical Hydrodynamics*, Prentice-Hall, Englewood Cliffs, N.J., 1962.)

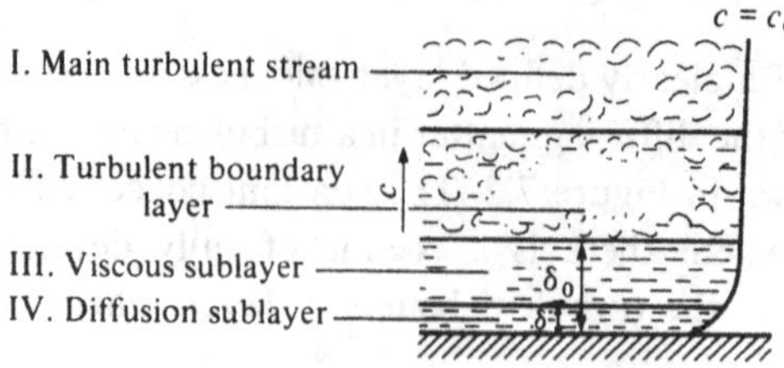

following relation between the thicknesses of the diffusion sublayer, δ, and the viscous sublayer, δ_0:

$$\delta = (D/\nu)^{1/n}\delta_0$$

where ν is the kinematic viscosity and n is a constant to be determined experimentally. Thus in this model, damping of turbulence is gradual and continuous and becomes identically zero only at the interface (or wall). For gas (vapor)–liquid and liquid–liquid systems it should be noted that surface-tension forces act in the same manner as viscous forces do at a solid surface. According to experimental data, it is found that $n \simeq 3$ for liquid–solid flows and that $n = 2$ for gas (vapor)–liquid systems. It can then be shown, using equation (7.1), that $k \sim D^{1/2}$.

7.3 The penetration model

In the penetration or surface-renewal model, mass transfer is observed as an unsteady time-dependent process. In the oldest model of this group, the Higbie penetration model (1935),[4] mass transfer is assumed to take place during brief, repeated contacts of matter (gas or solid) with the interface, this motion being generated by turbulent fluctuations in the bulk fluid. Fresh liquid elements continually replace those interacting with the interface, and consequently mass transfer is effected by the interface being systematically renewed.

The exposure time of such fluid elements to mass-transfer effects at the interface is so short that steady-state characteristics do not have time to develop, and any transfer that does take place is due to unsteady molecular diffusion. The theory originally conceived of gas bubbles rising in a liquid, and liquid elements being in contact with the bubble surface for a time equal to that for the bubble to rise one diameter. However, the model can be generalized to the case of a liquid moving in a turbulent manner, with its free surface contacting the gas. Mass transfer is effected during a cycle in which each eddy in the liquid comes to the surface, diffusion takes place and the eddy randomly returns to the bulk liquid. The eddies are all assumed to remain at the interface for the same time duration, τ. For this time duration, gas diffuses into the liquid eddy in a manner described by the diffusion equation:

$$\frac{\partial c}{\partial t} = D\frac{\partial^2 c}{dy^2} \tag{7.2}$$

Here c is the local gas concentration, and as before D is a diffusivity and y is the distance from the interface. For a small diffusion rate, and under the assumption that the time duration τ is small, the size of the eddy is effectively infinite, and we can apply the following boundary conditions on equation (7.2):

$$c = \begin{cases} c_0 & \text{at } t = 0, \quad y \geq 0 \\ c_0 & \text{at } y \to \infty, \quad t \geq 0 \\ c_{in} & \text{at } y = 0, \quad t > 0 \end{cases}$$

where c_0 is the average gas concentration in the liquid bulk and c_{in} is the concentration at the interface (which may be taken as the equilibrium solubility of the gas in the liquid).

The solution of equation (7.2) with these boundary conditions is

$$c = c_0 + (c_{in} - c_0)\left(1 - \frac{2}{\pi}\int_0^{y/2(Dt)^{1/2}} e^{-\eta^2}\, d\eta\right) \tag{7.3}$$

and the mass-transfer rate per unit area at the surface, at any time t, is then

$$\dot{m}' = -D\left(\frac{\partial c}{\partial y}\right)_{y=0} = (c_{in} - c_0)\left(\frac{D}{\pi\tau}\right)^{1/2} \tag{7.4}$$

on using equation (7.3). Averaging equation (7.4) over the time period τ leads to the average mass flux for each eddy:

$$\dot{m}'_{av} = \frac{\int_0^\tau (c_{in} - c_0)(D/\pi t)^{1/2}\, dt}{\tau} = 2(c_{in} - c_0)\left(\frac{D}{\pi\tau}\right)^{1/2}$$

and comparison of this result and equation (7.1) leads to a mass-transfer coefficient for the penetration model,

$$k = 2(D/\pi\tau)^{1/2} \tag{7.5}$$

Note that we have obtained the desired $\dot{m}' \sim D^{1/2}$ directly with this theory, as compared to the unrealistic $\dot{m}' \sim D$ for the film model.

7.4 Modified penetration models

More recent surface-renewal models have attempted to improve on the rather simple approach used in the penetration model. It will be remembered that the penetration model assumed each fluid element to be exposed to the free surface for a certain (constant) time period τ, and one of these more sophisticated models[5] eliminates this assumption, so that fluid elements can have a surface residence time anywhere in the range zero to infinity. This results from the new assumption that the probability of any given surface element being replaced by another is independent of how long it has been on the surface. If s is the fractional rate of renewal of elements of any age group, it can be shown that

$$\phi = se^{-st} \tag{7.6}$$

where ϕ is the probability density for any given element of area to be exposed to the surface for a time t before being replaced by a new element. The steady-state mass flux per unit interface area can then be obtained by combining equation (7.4) for the mass flux due to one fluid element's residence time t with equation (7.6) and integrating over time:

$$\dot{m}'_{av} = (c_{in} - c_0) \int_0^\infty \left(\frac{D}{\pi t}\right)^{1/2} se^{-st}\, dt = (c - c_0)(Ds)^{1/2}$$

Thus, by comparison with equation (7.1) we find the mass-transfer coefficient to be

$$k = (Ds)^{1/2} \tag{7.7}$$

Equation (7.7) reveals that this modified penetration theory still has an unknown quantity to be determined, in this case the fractional renewal s; in the earlier model it was the time duration τ. However, the modified theory is an improvement because s can be measured under controlled conditions (there is difficulty in obtaining a τ because in reality the residence times are not constant). Experiments have been conducted on the absorption of SO_2 in water[6] that is stirred at a known rate, and Figure 7.3 shows results obtained for a stirring rate of 230 rpm, as well as the closest fit of equation (7.6), to give $s = 2.81\ s^{-1}$. Different values of s were found for different stirring rates, and it appears from the data that this theory is a reasonably good description of actual flow conditions.

A slightly different approach[7] leaves the probability function as the unknown factor (instead of its functional argument s), and equation (7.7) then becomes

$$k = \frac{1}{(\pi)^{1/2}} \int_0^\infty \phi(t)\left(\frac{D}{t}\right)^{1/2} dt \tag{7.8}$$

When we average over all time durations τ, we can assume this probability function to take the form

$$\phi(\tau) = Ae^{-(\tau/T)^n} \tag{7.9}$$

where n, A, and T are constants, and we have the obvious condition on ϕ:

$$\int_0^\infty \phi(\tau)\, d\tau = 1 \tag{7.10}$$

Equations (7.8), (7.9), and (7.10) generate the following relations between the constants A and T, and hence between the mass-transfer coefficient k and T:

$$n = 1: \quad A = \frac{1}{T}, \qquad k = \left(\frac{D}{T}\right)^{1/2} \tag{7.11a}$$

$$n = 2: \quad A = \frac{2}{\pi^{1/2} T}, \qquad k = \frac{\Gamma(\frac{1}{4})}{\pi}\left(\frac{D}{T}\right)^{1/2} \tag{7.11b}$$

$$n = 3: \quad A = \frac{3}{T\Gamma(\frac{1}{3})}, \qquad k = \frac{\Gamma(\frac{1}{6})}{\Gamma(\frac{1}{3})}\left(\frac{D}{\pi T}\right)^{1/2} \tag{7.11c}$$

$$n = 4: \quad A = \frac{4}{T\Gamma(\frac{1}{4})}, \qquad k = \frac{\Gamma(\frac{1}{8})}{\Gamma(\frac{1}{4})}\left(\frac{D}{\pi T}\right)^{1/2} \tag{7.11d}$$

as well as equation (7.5) for the case $n = 0$. It turns out that when the constant T is computed from equations (7.11) using known values of the mass-transfer coefficient, the case $n = 0$ and $n = 1$ provide the best cor-

Figure 7.3 Surface-renewal model; experimental data on surface-age distribution function versus surface-element age for water stirred at 230 rpm. (From T. K. Sherwood, R. L. Digford, and C. R. Wilke, *Mass Transfer*, McGraw-Hill, New York, 1975.)

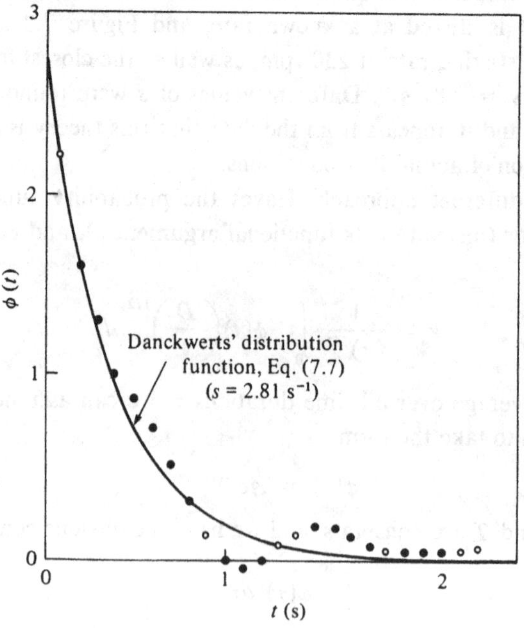

relation with measured concentration distributions. This indicates that of all the possible forms for the probability density function ϕ, the most realistic is, first, that where the residence time is constant, which is the Higbie model, and second, that given by equation (7.11a), which by comparison with (7.6) is simply the modified penetration model described above.

Yet another approach has been developed[8] which tries to unify the film and penetration models into one theory by noting that in the film theory we have $k \sim D$, as a result of the assumption that the residence time is long enough for a quasi-steady-state condition to exist, whereas in the penetration models we find $k \sim D^{1/2}$, with the assumption that the fluid elements or eddies have effectively infinite depth. Hence it is apparent that $k \sim D^n$, where n is dependent on the physical conditions, and that the two approaches might be linked by either allowing for a finite liquid element thickness, or for eddies of a limited time duration.

In the boundary conditions of equation (7.2), if we replace the second boundary condition with

$$c = c_0 \qquad \text{at } y = y_e$$

where y_e is the finite element thickness, it can then be shown that Higbie's model yields

$$k = \frac{D}{y_e}\left[1 - \frac{y_e^2}{3D\tau} - \frac{2y_e^2}{\pi^2 D\tau}\sum_{n=1}^{\infty}\frac{1}{n^2}\exp\left(\frac{-\pi^2 n^2 D\tau}{y_e^2}\right)\right] \qquad (7.12)$$

Equation (7.12) is shown in Figure 7.4, and it is clear that for $(D\tau)^{1/2}/y_e > 0.6$, we have k being linear with D, and hence the Higbie

Figure 7.4 Point transfer rate as a function of time. [From H. L. Toor and J. M. Marchello, *AIChE J.*, 4(1), 97 (1958).]

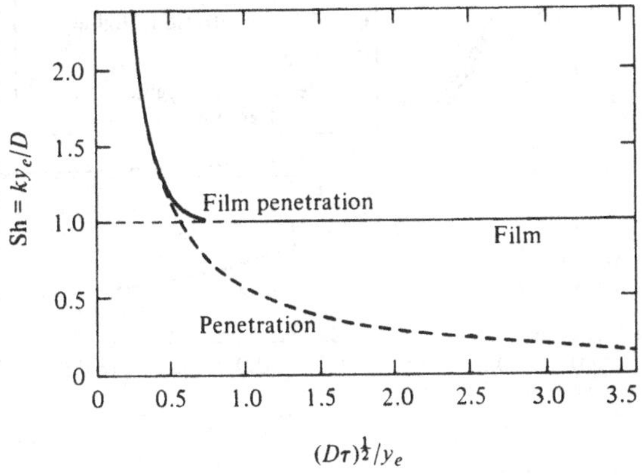

model decomposes to the film model for large values of D (rapid penetration), large values of τ (exposure or residence time), or small values of y_e (fluid element thickness). Conversely, the figure shows that for small D or τ, or for large y_e, this film-penetration model approaches the penetration model.

With the modified penetration model and the assumption of finite element thickness, it can further be shown that instead of equation (7.7), we obtain

$$k = (Ds)^{1/2} \coth\left(\frac{sy_e^2}{D}\right)^{1/2}$$

and this is illustrated in Figure 7.5. As the renewal rate becomes small or the penetration rate becomes high, we asymptotically approach the film theory limit of $k \sim D$. Under the converse conditions we approach the limit $k \sim D^n$, with n taking some value between 0.5 and 1.0, and this is the limit that describes many observed flows. Experiments have been conducted[9] on the absorption of gases in water which show n to be in the range 0.65 to 0.985, the smaller values coinciding with increasing turbulence (producing smaller values of s).

Other methods have been attempted to modify the regional penetration model to take it more realistic. For instance, one model[10] allows for a nonequilibrium interface condition. Another model introduces an eddy diffusivity into the penetration model which has a power function dependency on

Figure 7.5 Mean transfer rate as a function of time. [From H. L. Toor and J. M. Marchello, *AIChE J.*, *4*(1), 97 (1958).]

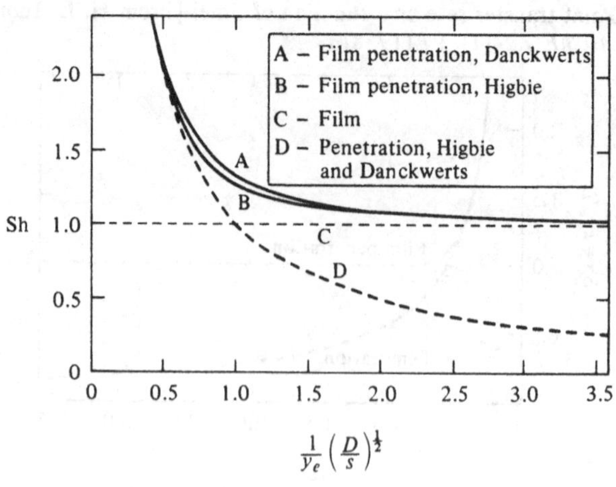

the distance from the surface (the exponent is found to have the value 4), and the time duration τ is left as a variable.

Yet another model attempts to combine the film and penetration models for the case of a wall–liquid interface[11] by considering periodically replaced developing boundary layers. This scheme results in a semipermanent thin fluid layer adjacent to the wall as well as the intermittently surfacing fluid elements, and mass transfer is considered to be affected by two mechanisms: the resistance to mass transfer due to the viscous sublayer at the wall, and the transfer due to eddies periodically penetrating the boundary layer (to varying depths). For the mass transfer due to fluid motion, we have the defining boundary-layer equations for the assumed two-dimensional flow,

$$\frac{\partial u}{\partial t} + u\frac{\partial u}{\partial x} + v\frac{\partial u}{\partial y} = \nu\frac{\partial^2 u}{\partial y^2}$$

$$\frac{\partial c}{\partial t} + u\frac{\partial c}{\partial x} + v\frac{\partial c}{\partial y} = D\frac{\partial^2 c}{\partial y^2}$$

$$\frac{\partial u}{\partial x} + \frac{\partial v}{\partial y} = 0$$

where u, v and x, y are the local velocities and directions parallel to and perpendicular to the wall, respectively, and c is the local concentration. We have boundary conditions

$$t = 0, \quad u = u_0, \quad c = c_0$$

$$x = 0, \quad u = u_0, \quad c = c_0$$

$$y = 0, \quad u = 0, \quad c = c_i$$

$$y \to \infty, \quad u \to u_0, \quad c \to c_0$$

where c_0 is the bulk concentration, c_i is the concentration at the wall, and u_0 is the velocity outside the boundary-layer region. Now, for small values of the Schmidt number ($\mathrm{Sc} = \nu/D$), the foregoing equations, describing the periodically developing viscous sublayer completely, determine the mass transfer. Solving these equations for flow in a pipe by the momentum integral method (with additional information available[12] on the cyclic nature of the viscous sublayer in a turbulent flow) yields

$$\mathrm{Sh} = 0.0097\mathrm{Re}^{9/10}\mathrm{Sc}^{1/2}(1.10 + 0.44\mathrm{Sc}^{-1/3} - 0.70\mathrm{Sc}^{-1/6})$$

$$(0.5 < \mathrm{Sc} < 10) \tag{7.13}$$

Here the Sherwood number is given by $\mathrm{Sh} = kd/D$, and the Reynolds number is defined as $\mathrm{Re} = u_0 d/\nu$, where d is the diameter of the pipe in which the flow is developing. If the entire developing boundary is considered to be unsteady, averaging the mass-transfer coefficient k over the entire flow length results in the following [instead of equation (7.13)]:

$$\mathrm{Sh} = 0.0107\mathrm{Re}^{9/10}\mathrm{Sc}^{1/2} \tag{7.14}$$

which is of the same form as equation (7.5). As we increase the Schmidt number, we have mass transfer by convection, as described above, as well as that due to the concentration gradient, given by equation (7.1), and a combination of equations (7.1) and (7.13) produces

$$\mathrm{Sh} = \frac{0.0097\mathrm{Re}^{9/10}\mathrm{Sc}^{1/2}(1.10 + 0.44\,\mathrm{Sc}^{-1/3} - 0.70\mathrm{Sc}^{-1/6})}{1 + 0.064\mathrm{Sc}^{1/2}(1.10 - 0.44\mathrm{Sc}^{-1/3} - 0.70\mathrm{Sc}^{-1/6})}$$

$$(10 < \mathrm{Sc} < 1000) \tag{7.15}$$

For yet higher values of the Schmidt number, mass transfer is affected by the renewal rate of the thin-wall layer, which is governed by the equation

$$u\frac{\partial c}{\partial x} = D\frac{\partial^2 c}{\partial y^2}$$

with boundary conditions

$$y = 0, \qquad c = c_i$$
$$y \to \infty, \qquad c \to c_0$$

The solution of this equation is

$$\mathrm{Sh} = 0.0102\mathrm{Re}^{9/10}\mathrm{Sc}^{1/3} \qquad (\mathrm{Sc} > 1000) \tag{7.16}$$

Figure 7.6 Comparison of the proposed model with experimental heat and mass-transfer data. [From W. V. Pinczewski and S. Sideman, *Chem. Eng. Sci., 29,* 1969 (1974).]

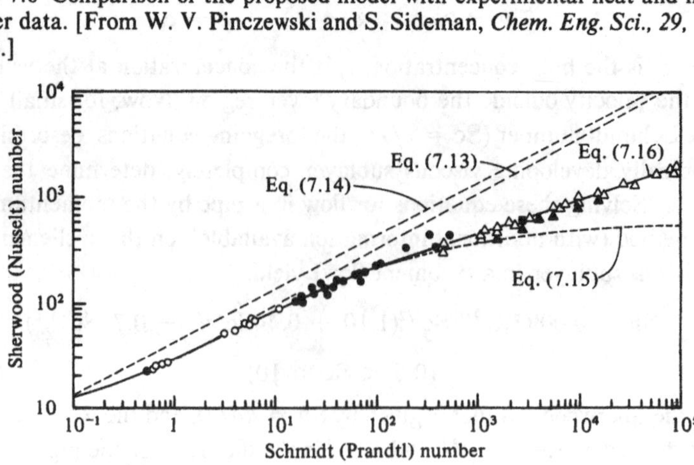

where use has again been made of results from turbulent flow studies. Equations (7.13), (7.14), (7.15), and (7.16) are shown compared with experimental data in Figure 7.6, indicating good agreement with data over their ranges of applicability. It is also noteworthy that this analysis only requires fluid mechanical data to be available, not specifically mass-transfer data.

We can, in view of recent developments in mass transfer as described in earlier sections, conclude that the film model has proved unsatisfactory, mainly because it takes no account of fluid-mechanical aspects. Nevertheless, the principle of additivity of resistances is still quite a useful concept for obtaining at least approximate results and has been subject to experimental and theoretical examination,[13–19] and the principle underlies modern methods of designing plate columns for absorption processes and other industrial applications.

7.5 Effects of mass-transfer coefficients and contact areas

The description of mass-transfer processes using mass-transfer coefficients is based on equation (7.1), which we can rewrite as

$$dm = k dA \Delta \, dt$$

This relates, by use of a mass-transfer coefficient k, the mass dm transferred from one phase into the other through the interface area dA, in time dt, by the driving force Δ. The mass m transferred from one phase into the other by a steady flow may be written in integral form as

$$m = kA\Delta$$

where k is now some averaged value of the local mass-transfer coefficient, and each term is a complicated function of many hydrodynamic and physical parameters. Great practical difficulties arise in the determination of the total surface contact area, and this is the reason many investigators have attempted definition of overall (fictitious) mass-transfer coefficients specified by per unit volume of the two-phase mixture (k_v) or by per unit effective surface of the diffusing surface (k_s) (e.g., a plate in the case of bubble plates), instead of a mass-transfer coefficient k per unit transfer, or contact, area. These coefficients are obviously related by the formula

$$kA = k_s S = k_v V \tag{7.17}$$

where A, S, and V are the interface contact area, active surface of the plate, and the working volume of an apparatus, respectively. There are similar relationships for mass-transfer coefficients k_f and k_g for each phase. The use of coefficients k_f and k_g implies that all the information concerning both the kinetics of the process and the surface contact area is concentrated in these coefficients, and such a phenomenological approach does not permit an evaluation of the factors that influence the mass transfer from one phase to another. Indeed, using these coefficients means that the mass-transfer area on which they are defined becomes a separate factor of investigation.

This approach, of decomposing the phenomenon of mass transfer into the effect of mass-transfer coefficients and of contact areas, does however allow us to investigate the influence of dynamic and physical factors on the coefficients and surface contact area separately, which is a very important separation in the solution of problems such as maximizing mass transfer in preexisting industrial processes, and in the design of new devices for gas–liquid systems.

7.6 Calculation of the mass-transfer coefficient in a liquid phase

Because measurement of the true liquid mass-transfer coefficient, k_f, per unit contact area is more difficult than measurement of the "fictitious" coefficients, k_{f_s} and k_{f_v}, based on system area or volume, most experimental investigations have concentrated on the measurement of k_{f_s} and k_{f_v}. Nevertheless, obtaining information on k_{f_s} meets with considerable difficulties, and relationships for this coefficient are usually obtained by using similarity theory; application of the similarity method to analysis of the convective diffusion equations, with appropriate boundary conditions, leads to the following general relationship:[20–22]

$$\mathrm{Sh} = f(\mathrm{Re}, \mathrm{Sc}, \mathrm{Fo}, \mathrm{Ga}, \Gamma_1, \Gamma_2, \ldots) \tag{7.18}$$

where $\mathrm{Sh} = k_f d/D_f$ is the Sherwood number, $\mathrm{Re} = v_{s_f} d/\nu_f$ the Reynolds number, $\mathrm{Sc} = \nu_f/D_f$ the Schmidt number, $\mathrm{Fo} = D_f\tau/d^2$ the Fourier diffusion number, $\mathrm{Ga} = gd^3/\nu_f^2$ the Galileo number, and Γ_1 and Γ_2 are geometrical factors. Here, v_{s_f} is the liquid superficial velocity, ν_f the liquid kinematic viscosity, d a representative length, D_f the liquid diffusivity, and τ is now any representative time. It is assumed in this equation that the surface contact area is known, and k_f is referred to this area.

We can develop formulas for the form of equation (7.18) for the coefficient k_{f_s}, but preliminary investigations of the process are required for determination of characteristic velocities and geometric parameters. Investigators have developed many such formulas, some of which are shown in Table 7.1 and described below.

The following relationship[23] for the coefficient k_{f_s} has been developed using the diffusion equations for a single rising bubble:[24]

$$k_{f_s} = \frac{C\,\Delta p^{0.8} D_f^{0.5} \sigma^{0.15}}{\rho_f^{0.05} \mu_f^{0.9} g^{0.25}}$$

where Δp is a differential pressure; D_f the diffusivity; σ the surface tension; ρ_f and μ_f are the liquid density and dynamic viscosity, respectively; and C is a nondimensional constant. In deriving this equation, the following assumptions were made:

1. The rise velocity of a bubble in a bubble swarm is equal to the rise velocity of a single bubble.
2. The average bubble diameter, d_b, is proportional to a "capillary constant," χ,

$$d_b \sim \chi = \left(\frac{\sigma}{\rho_f g}\right)^{0.5}$$

3. The total diffusion on the bubble surface for a bubble in a bubble swarm is the same as for a single bubble rising in an infinite volume of liquid.

Some researchers insert $h = \Delta p/\rho_f g$ (the liquid depth) as a basic parameter instead of d, and it is assumed in this case that the influence of the other hydrodynamic and geometric parameters on the mass-transfer coefficient k_{f_s} is taken into account by the parameter h.

According to experimental data[25] on mass transfer using distillation on grid trays, the following relation holds:

$$\mathrm{Sh} = \mathrm{We}^{0.15}\mathrm{Ga}^{0.7}\mathrm{Sc}^{0.5}$$

where the Sherwood number $\mathrm{Sh} = k_{f_s}h/D_f$, the Weber number $\mathrm{We} = \sigma/\rho_f g h^2$, the Galileo number $\mathrm{Ga} = gh^3/\nu_f^2$, and the Schmidt number $\mathrm{Sc} = \nu_f/D_f$.

Experimental data[26] on mass transfer during desorption of CO_2 have been found to be correlated by

$$k_{f_s} = \Delta p(1.95 v_s - 0.41)^{-1}$$

for sieve plates, where v_s (m/s) is the gas superficial velocity, and by

$$k_{f_s} = 2.15 \Delta p v_s^{0.73}$$

for bubble cap trays. It was assumed in the analysis of this experimental data that complete liquid mixing on the plate was achieved. The superficial gas velocity v_s, as seen in the equations above, has different influences on sieve plates and bubble cap trays, and this can most easily be explained as

Table 7.1. *Experimental and theoretical expressions for k_{fs}, the liquid mass-transfer coefficient based on area*

(1) Reference	(2) Process analyzed	(3) System conditions	(4) Equation obtained by investigators	(5) Symbols	(6) Transformed equation	(7) Defining size	(8) Comment
23	Obtained by theoretical analysis	Mass, bubbling Sieve plate	$\mathrm{Sh} = C\frac{\Delta P^{0.8}\mathrm{Re}_f^{0.5}\mathrm{Sc}^{0.5}}{\mu^{0.4}\gamma^{0.3}g^{0.2}\sigma^{0.1}}$ (γ = specific gravity)	$\Delta P = h$ $\chi \sim \sqrt{\frac{\sigma}{\gamma_f}}$ $\mathrm{Re}_f = \frac{\chi u}{\nu_f}$ $\mathrm{Sh} = k_{fs}\chi$	$k_{fs} = CD_f^{0.5}h^{0.8}\mu^{-0.5} \times g^{-0.2}\gamma^{-0.05}$	χ	u: rising velocity of a bubble
25	Rectification of binary mixtures	Column ϕ = 120 mm Grid tray d_0 = 40 mm ψ = 25%	$\mathrm{Sh} = 17\mathrm{We}^{0.15}\mathrm{Ga}^{0.7}\mathrm{Sc}^{0.5}$	$\mathrm{Sh} = \frac{k_{fs}h}{D_f}$ $\mathrm{Sc} = \frac{\nu}{D_f}$ $\mathrm{Ga} = \frac{gh^3}{\nu_f{}^2}$ $\mathrm{We} = \frac{\sigma}{\gamma_f h^2}$	$k_{fs} = 17D_f^{0.5}h^{0.8}\mu^{-0.9} \times \sigma^{0.15}g^{-0.2}\gamma^{-0.05}$	h	Obtained by transforming equation (7.59)
26	Desorption of CO_2 by air from water	(a) Column ϕ = 200 mm S_0 = 6.95% $h_{w.h.}$ = 32 mm u_f 0.174 m^3/h Sieve plate	$k_{fs} = \Delta P_f(1.95\upsilon_s - 0.41)^{-1}$	$\Delta P_f \sim \Delta P$	$k_{fs} = h\gamma(1.95\upsilon_s - 0.41)^{-1}$	—	$k_{fs} \sim h$
		(b) Column ϕ = 200 mm Bubble cap Tray S_0 = 4.9–9.8% $h_{w.h.}$ = 7–52 mm u_f = 3.02 m^3/h	$k_{fs} = 2.15\Delta P_f\upsilon_s^{0.73}$	$\Delta P_f \sim \Delta P$ $h = \frac{\Delta P}{\gamma_f}$	$k_{fs} = 2.15h\gamma\upsilon_s^{0.73}$	h	$k_{fs} \sim h$

93	Absorption of CO_2 by air from water	Column $\phi = 200$ mm Sieve plate $d_o = 2$ mm $S_o = 6.9\%$ $h_{w.h.} = 0$–250 mm $u_f = 0.35$ m^3/h	$Sh = 0.23Re^{1.19} \times Sc^{0.5}Ga^{0.4}K^{-1}$	$Sh = \frac{k_{fs}h}{D_f}$ $Sc = \frac{\nu}{D_f}$ $Re = \frac{v_s h}{\nu}$ $Ga = \frac{gh^3}{\nu^2}$	$k_{fs} = 13.7D_f^{0.5}g^{0.24} \times \nu^{-1.17}v_s^{1.19} \times h^{1.91}K^{-1}$	h ν_s	Obtained in supposing that Sh = f(Re, Sc, Ga)
28	Desorption of CO_2 by air from water	Column $\phi = 200$ mm Sieve plate $d_o = 2$ mm $S_o = 6.9\%$ $h_{w.h.} = 10$–60 mm $u_f = 87$–$700 L/h$	$Sh = 8.5Ga^{0.5}Sc^{0.5}$	$Sh = \frac{k_{fs}h}{D_f}$ $Sc = \frac{\nu f}{D_f}$ $Ga_f = \frac{gh^3}{\nu^2}$	$k_{fs} = 8.5D_f^{0.5}\mu^{-0.5} \times \gamma^{0.5}h^{0.5}$	h	Obtained in supposing that h = f(Re, Sc, Ga)
30	Chemi-sorption oxidation, sulfate into sulfite	Column $\phi = 300$ mm Sieve plate $d_o = 0.9$ mm $S_o = 1.5$–2.0% (no liquid overflow)	$k_{fs} = 1.76h^{1.2}\phi^{1.7}$	—	—	h	Obtained from experiments
104	Oxidation of sulfate into sulfite	Column 240 × 130 mm Sieve plate $d_o = 0.9$–0.5 mm $u_f = 0.25$–5.0 m^3/h (with and without liquid overflow)	$k_{fs} = 1.22h^{1.13}\left(\frac{\phi}{1-\phi}\right)^{0.19}$	—	—	h	Obtained from experiments

Table 7.1 (*cont.*)

(1) Reference	(2) Process analyzed	(3) System conditions	(4) Equation obtained by investigators	(5) Symbols	(6) Transformed equation	(7) Defining size	(8) Comment
32	Desorption of CO_2 by air from water	Column 310 × 110 mm Sieve plate $d_0 = 2.0$ mm $S_0 = 10\%$ $u_f = 0.1; 0.2; 0.6\ m^3/h$	$k_{fs} = 7.3h^{0.4}v_s^{0.14}\mu^{-0.85}K^{0.43}$	—	—	—	Obtained with account of liquid mixing
33	Rectification of binary mixtures	Column 310 × 110 mm Sieve plate $d_0 = 2.0$ mm $S_0 = 10\%$ $u_f = 0.1; 0.2; 0.6\ m^3/h$	$k_{fs} = 3 \times 10^3 D_f^{0.5}h^{0.94} \times v_s^{0.33}\mu^{-0.63}K^{0.43}$	—	—	—	Obtained on the basis of equation (7.59)
34	Desorption of CO_2 by air from water	Column $\phi = 1000$ mm Tray with two zones of contact $d_0 = 2.0$ mm $S_0 = 6.85$ and 12.2% $u_f = 11; 17; 21\ m^3/h$	$k_{fs} = 1.05C_0^{0.18}D_f^{0.5} \times h^{0.94}v_s^{0.35}\mu^{-0.65}K^{0.43}$	—	—	—	Obtained with account of mixing
35	Desorption of CO_2 by air from water	Columns $\phi = 120; 400; 80$ mm $d_0 = 6$–12 mm Grid tray	$\mathrm{Sh} = 1.26 \times 10^5 \mathrm{Pe}^{0.5}\phi(H)$	$\mathrm{Sh} = \frac{k_{fs}H}{D_f}$ $\mathrm{Pe}_f = \frac{v_s H}{D_f}$	$k_{fs} = 1.26 \times 10^5 \times D_f^{0.5}v_s^{0.5}H^{1.9}$	H v_s	Obtained on the basis of Higbie model

36	Desorption of and rectification of binary mixtures	Columns $\phi = 80; 120;$ 400 mm and 250; 800 mm Grid tray $d_0 = 6.0$ mm $S_0 = 11–40\%$ $u_f = 5–30\ m^3/h$	$\mathrm{Sh} = 6.24 \times 10^5 \Gamma_C \mathrm{Pe}_f^{0.5} \times \left(\frac{\mu_g}{\mu_f + \mu_g}\right)^{1/2}$	$\mathrm{Sh} = \frac{k_f d}{D_f}$ $\mathrm{Pe} = \frac{v_s d}{(1-\phi) D_f}$		—	Obtained on the basis of Levich equation for a single bubble
27	Rectification of binary mixtures	Column $\phi = 170$ mm Sieve plate $s_0 = 9.0\%$ $d_0 = 2.0$ mm $h_{w.h.} = 25.0$ mm	$\mathrm{Sh} = 540 \mathrm{Re}^{0.29} \mathrm{Sc}^{0.45}$	$\mathrm{Sh} = \frac{k_{fs} h}{D_f}$ $\mathrm{Re} = \frac{v_s h}{\nu}$ $\mathrm{Sc} = \frac{\nu_f}{D_f}$	$k_{fs} = 540 D_f^{0.5} W_K^{0.29} \times h^{-0.71} \nu^{0.16}$	—	—
29	Desorption of O_2 from water	Columns $\phi = 100$ and 150 mm with one bubble cap $d_{bc} = 7.5$ and 10.0 mm $S_0 = 2.74\%$ and 5.4%	$k_f a = 6.97 v_s^{0.75} h^{0.5} \times D_f^{0.5}\ (s^{-1})$	a: specific surface contact area	$k_{fs} = 6.97 D_f^{0.5} \times v_s^{0.75} h^{0.33} K^{-1} a$	—	a is unknown
37	Desorption of gases of small solubility	—	$\mathrm{Sh} = 3.01 \times 10^{-5} \mathrm{Re}_f^{0.5} \times \left(\frac{D_K}{d_{eq}}\right) \frac{V_g}{V_f} \left(\frac{v_{sf}}{v_s}\right)$	—	V_g, V_f: specific molar volumes of gas and liquid D_K: diameter of a column; d_{eq}: equivalent size	—	—
38	Adsorption of CO_2 and rectification of binary mixtures	—	$N_f \mathrm{Sc}^{0.5} = 5 + 10 t_f \times [1 + 0.17 \times (1.22 v_s \gamma_g^{0.5} + 1)] \times (2 + 0.039h)$	—	—	—	This equation cannot be transformed

being due to different hydrodynamic mixing conditions occurring in the plates and trays.

Studies of the absorption of CO_2 in a tank of water[27] have also determined a correlation of k_f, with the superficial gas velocity v_s, gas content ϕ (or void fraction, defined as the ratio of the volume of gas to the total volume), and liquid static head h. Here we find that

$$Sh = 0.23Re^{1.13}Ga^{0.24}Sc^{0.5}\phi^{-1.0}$$

where the Reynolds number $Re = v_{s_f}h/\nu_f$ and v_{s_f} is the superficial liquid velocity.

In this case also a complete mixing of liquid at the plate was assumed, as well as that its concentration along the plate was equal to its final concentration when leaving the plate, and that the gas goes through the liquid without internal mixing. Additional experiments[28] on desorbtion of CO_2 from liquid solutions on sieve plates result in

$$Sh = 8.5Ga^{0.5}Sc^{0.5}$$

Experiments[27,29] have been carried out under similar conditions, but differing values for k_f, were obtained. Such disagreement may be explained by differences in the methods of analysis of the experimental data, and the use of different parameters. For example, as a linear dimension, instead of an actual characteristic height $h = d$ being used, some value conditional on the superficial velocity becoming zero was used in some studies.

Further experiments using an air–sulfite water solution, with the addition of a copper catalyst, were conducted in which the influence of mixing was eliminated by arrangement of the experiments such that the liquid flow rate was zero, and on the basis of these experiments the following correlation was proposed:

$$k_f = 1.76h^{1.2}\phi^{1.7}$$

Because variation of the empirical parameters was limited,[30] additional measurements were made in order to check the preceding equation, and as a result it was modified to the form

$$k_f = 1.22h^{1.13}\left(\frac{\phi}{1-\phi}\right)^{0.19}$$

A study of the desorbtion of CO_2 from a water solution on a sieve plate was made, this time taking into account the actual concentration distribution on the plate,[31,32] and this resulted in

$$k_f = 7.3h^{0.94}v_{s_f}^{0.14}\mu_f^{-0.85}\psi^{0.43}$$

where ψ is the "specific gravity" of the foam given, for example, in Section 4.6, and additional experimental data from studies of the rectification of binary mixtures[33] resulted in the preceding equation being made more precise:

$$k_f = 3.103 v_{sf}^{0.5} h^{0.94} \mu_f^{-0.63} \psi^{0.43}$$

By analyzing the experimental data on CO_2 desorption from a water solution by air, with a two-zone phase contact area, the following equation was obtained:[34]

$$k_f = 1.05 b^{0.18} D_f^{0.5} h^{0.94} v_{sf}^{0.35} \mu_f^{-0.65} \psi^{0.43}$$

This equation is similar to the preceding equation except that the values of D_f and μ_f were obtained more precisely, and an additional term b was added to take into account the rise of the disk.

It is clear from all of the equations above that the major hydrodynamic parameters influencing the mass-transfer coefficient k_f are h and ϕ (or ψ). In some studies the height of the gas–liquid mixture, x_1, obtained in Chapter 4, is taken as a determining hydrodynamic parameter. In addition, one study[35] made the assumption that characteristic surface area can be taken to be constant and independent of the turbulence rate, and an equation for calculating k_f then follows:

$$\text{Sh} = \text{Pe}^{1/2} f(x_1)$$

where $\text{Pe} = v_{sf} x_1 / D_f$ is the Peclet number. The function $f(x_1)$ is then obtained by analyzing experimental data. In another study[36] the following equation is given:

$$\text{Sh} = 1.84 \text{Pe}^{1/2} \frac{\Delta p}{\rho_f g d_b} \left(\frac{\mu_g}{\mu_g + \mu_f} \right)^{1/2} \times 10^5$$

where μ_g is the gas dynamic viscosity.

Further, in some studies the superficial gas velocity v_s is taken as a determining parameter. By investigating mass transfer in liquid through rectification of binary mixtures on sieve plates, the following correlation[37] was obtained:

$$\text{Sh} = 540 \text{Re}^{0.29} \text{Sc}^{0.45}$$

where, as a characteristic velocity in the Reynolds number, the superficial gas velocity v_s was used. The dependency of k_f on h in this work was different from ones previously considered and has not been confirmed by the experimental data of other investigators.

By analyzing the influence of the superficial gas velocity, v_s; the height of the stagnant liquid, h; the diffusivity, D_f; and physical properties of the liquid on the mass-transfer coefficient, the following correlation, describing experimental data obtained from studies using bubble cap trays and sieve plates, was found:[29]

$$k_f a = 6.97 v_s^{0.75} h^{-0.67} D_f^{0.5}$$

where a (cm^{-1}) is the specific surface contact area given in Section 3.4. The experiments were conducted using columns of small diameter (0.1 and 0.155 m, 3.9 and 6.1 in.), having in the center one bubble cap (of diameter 7.5 or 10 cm, 3 or 3.9 in.), the diameter of the sieve plate being 3.6 cm (1.4 in.) with orifices of 2.0 mm (0.79 in.) and free areas of 2.78 and 5.4 percent.

Finally, it has been proposed[38] that correlating experimental data on mass transfer in a liquid phase according to the equation

$$\mathrm{Sh} = 3.01 \mathrm{Re} \mathrm{Sc}^{0.5} \left(\frac{D_f}{d} \right) \frac{V_g}{V_f} \left(\frac{v_s}{v_{sf}} \right)^{0.2} \times 10^{-5}$$

where V_g and V_f are the molar specific volumes of gas and liquid, gives a reasonable correlation with experimental data.

After considering in some detail results obtained by correlating the mass-transfer coefficient with several different parameters, we now consider the effect of the individual independent variables on the mass-transfer coefficient.

7.7 Effect of the superficial gas velocity on the mass-transfer coefficient

The influence of the superficial gas velocity v_s on the mass-transfer coefficient k_f is indicated in studies[39–41] performed using sieve plates and bubble cap trays.[42,43] The complicated relationship of k_f with v_s was investigated in desorption of CO_2 from water,[39–42] and a maximum value for k_f was observed for v_s in the range 0.03 to 0.05 m/s (1.2 to 2 in./s). A similar result was observed by other investigators; for example, it was shown[44] that the coefficient k_f increases linearly with v_s up to $v_s \simeq 0.4$ m/s (1.3 ft/s), that for $v_s \simeq 0.5$ m/s (1.6 ft/s) to 0.6 m/s (2 ft/s) k_f was at a maximum, and then decreased for higher values of v_s until for $v_s > 0.8$ m/s (2.6 ft/s) the coefficient became almost constant. Other studies give[31,45] the same results except that the maximum was achieved in a different range of velocities, but this discrepancy is mostly a result of having different liquid levels in the experiments.

It has also been shown experimentally that the change in the height x_1 of the gas–liquid mixture with superficial velocity v_s is similar to the change of k_f with v_s, but that the k_f maximum is displaced in the direction of the higher values of v_s, and this difference between the maxima is associated, by some investigators, with rupture of the dynamic "froth." Analysis of the influence of the superficial gas velocity on the mass-transfer coefficient in a liquid phase reveals a close relationship between the hydrodynamic conditions in the system and the mass-transfer process. In a bubble-by-bubble regime, k_f increases almost proportionately to v_s; in the regime of frothing, k_f is only slightly dependent on v_s, and this explains why, in the relationships described above, the values of v_s giving maxima differed according to the hydrodynamic regime in the experiments. Hence, instead of the dependence on superficial velocity, we can study dependence on the hydrodynamic condition created by the gas velocity, specifically the gas void fraction.

7.8 Effect of the void fraction on the mass-transfer coefficient

The height of the gas–liquid layer x_1 and the specific surface contact area a are substantially dependent on the gas void fraction ϕ, as discussed in Chapters 3 and 4. Because the coefficient k_f is dependent on x_1 and a, it must also depend on the gas void fraction, but this dependence is quite complicated, and as a result investigators have tried several different approaches to the problem.

Experimental data on mass-transfer studies in bubble cap trays[46] and sieve plates[47,48] led researchers to the conclusion that the intensity of mass transfer is determined by the height of the gas–liquid mixture and by the void fraction. Additional support for such a conclusion was obtained from a study[35] in which equations were given for sieve plate, bubble cap tray, and grid tray applications. In a further experimental study[49] dependency of mass transfer on the height of the mixture or froth could not be found, this being explained by the changing structure of the gas–liquid mixture as the height was increased. These changes in structure were indicated by changing values of the relative specific gravity ψ of the froth.

The influence of hydrodynamic conditions on the coefficient k_f is, therefore, affected by the liquid static head, the relative specific gravity of the froth, or some combination of the two. (We have already noted that the influence of the relative specific gravity is in fact a result of the influence of the gas superficial velocity and the physical properties of the liquid.) Unfortunately, a comparison of experimental results[31,49] indicates considerable uncertainty as to the influence of the relative specific gravity of the froth, and this is at least partly due to the fact that the range of ψ (or ϕ) usually studied is quite small, and it is very difficult to evaluate experimentally the influence of ϕ or ψ on the mass-transfer coefficient.

7.9 Mass transfer from a single bubble

A second approach to the study of mass transfer, alluded to in Section 7.5, assumes a separate analysis of mass-transfer kinetics and of the surface contact area, and it is expedient to start this kind of investigation from the simplest case – mass transfer from a single bubble into liquid. The dissolution of gas from a single bubble for low-solubility gases is, of course, a gross simplification of real conditions, in which the process is developed in a statistical system composed of many elements, each bubble being such an element. However, to build a physical model for a real process, we must divide the complicated process into its elementary parts, which are not complicated by secondary and random effects, and then study these elementary parts thoroughly.

The rate of mass transfer from small bubbles into liquid can be evaluated with reasonable accuracy, the bubbles in this case being spherical and rising rectilinearly, and theoretical solutions have been derived using various methods. A rigorous theoretical treatment was made[24] in which the total diffusion flux on the bubble surface was written as

$$\dot{m} = 4\left(\frac{2\pi D v_r}{r_b}\right)^{1/2} r_b^2 c_0 \tag{7.19}$$

where D is the diffusion coefficient, v_r the relative velocity (between phases), r_b the bubble diameter, and c_0 concentration in the bulk, or, in dimensionless form,

$$\mathrm{Sh} = \left(\frac{2}{\pi}\,\mathrm{ReSc}\right)^{1/2} \tag{7.20}$$

This system has been investigated[50–52] for the case of $\mathrm{Re} < 1$ and results were obtained that practically coincide with equation (7.20). On the other hand, the irregular motion of large bubbles does not allow quite the same rigor of theoretical treatment; however if we assume that the change of shape and trajectory of large bubbles has only a slight influence on the mass-transfer rate, then it is possible to obtain approximate values for the rate for large bubbles by considering a model in which the shapes of the bubbles are strictly fixed and in which they move in a straight line. The bubbles, as they deform, acquire the shape of ellipsoids or spherical caps, and it is convenient to assume that all the bubbles are symmetrical about an axis parallel to the direction of bubble rise.

In cases where the thickness of the hydrodynamic boundary layer around the surface of the bubble is much less than an equivalent radius of the dissolving bubble, the equations of continuity and convective diffusion may be written[53,54] in the dimensionless coordinates

$$u' = \frac{u}{U}; \qquad v' = \frac{v}{U}; \qquad r' = \frac{r}{r_e}; \qquad x' = \frac{x}{r_e}; \qquad y' = \frac{y}{r_e}$$

where r_e is the radius of a sphere having the same volume as the axisymmetrical bubble; U is a characteristic velocity (e.g., the relative velocity or the superficial velocity); u and v are local velocities; x and y are the corresponding linear dimensions parallel and perpendicular to the bubble surface, respectively; and r is the radial dimension.

Using these terms, the continuity equation may be written as

$$\frac{\partial(u'r')}{\partial x'} + \frac{\partial(v'r')}{\partial y'} = 0 \tag{7.21}$$

Similarly, the equation of convective diffusion may be written in the following boundary-layer form:

$$u'\frac{\partial c'}{\partial x'} + v'\frac{\partial c'}{\partial y'} = \frac{2}{\text{Pe}}\frac{\partial^2 c'}{\partial y'^2} \tag{7.22}$$

with $c' = c/c_{eq}$, c_{eq} being the equilibrium concentration, c the local concentration, and Pe the Peclet number (assumed large).

If the thickness of the diffusion boundary layer is less than the thickness of the hydrodynamic boundary layer, mass transfer will occur in a region where the velocity tangential to the bubble surface may be presented as

$$u' = u_o' + u_o''y' \tag{7.23}$$

where u_o' is a measure of the relative motion in the interior of the bubble and u_o'' is a measure of the boundary-layer velocity gradient. Clearly, as the Schmidt number Sc is a measure of the ratio of the thickness of the hydrodynamic layer to the diffusional one, this simplification is possible only when Sc $\gg$ 1. For neatness, in the remainder of this section we will ignore the primes on the symbols. This should cause no confusion for the reader.

Using equations (7.21) and (7.23) in equation (7.22), we find that

$$(u_o + u_o'y)\frac{\partial c}{\partial x} - \frac{1}{r}\frac{\partial}{\partial x}\left[\left(u_o y + \frac{1}{2}u_o'y^2\right)r\right]\frac{\partial c}{\partial y} = \frac{2}{\text{Pe}}\frac{\partial^2 c}{\partial y^2} \tag{7.24}$$

To this equation, describing the concentration distribution for the dissolving, axisymmetrical, gas bubble in liquid (for large Pe and Sc values), must be added the boundary conditions

$$c = 1; \qquad y = 0$$

$$c \to 0; \qquad y \to \infty$$

and we can then consider solutions of equation (7.24) for two limiting cases:

1. Stationary contact surface, $u_o = 0$.
2. Moving contact surface, $u_o > u'_o y$, where movement of the contact surface is relative to the bubble center.

For the first case the equation may be written

$$u'_o y \frac{\partial c}{\partial x} - \frac{1}{r}\frac{\partial}{\partial x}\left(\frac{u'_o y^2 r}{2}\right)\frac{\partial c}{\partial y} = \frac{2}{\text{Pe}}\frac{\partial^2 c}{\partial y^2}$$

This equation has been solved[55] in the form

$$\text{Sh} = 0.641\left(\frac{4\pi r_e^2}{S}\right)\left[\int_0^x (u'_o r)^{1/2} r\,dx\right]^{2/3} \text{Pe}^{1/3} \qquad (7.25)$$

where S is now the surface area of the bubble. Specific information on the shape of the bubble and character of the motion permits simplification of equation (7.25) for each application.

For the second case, equation (7.24) may be written

$$u_0 \frac{\partial c}{\partial x} - \frac{1}{r}\frac{\partial}{\partial x}(u_0 y r)\frac{\partial c}{\partial y} = \frac{2}{\text{Pe}}\frac{\partial^2 c}{\partial y^2}$$

and the solution of this equation with the same boundary conditions is

$$\text{Sh} = \left(\frac{2}{\pi}\right)^{1/2}\frac{4\pi r_e^2}{S}\left(\int_0^x u_o r^2\,dx\right)^{1/2} \text{Pe}^{1/2} \qquad (7.26)$$

Let us now consider some particular solutions of the mass-transfer equations (7.25) and (7.26) for relatively immobile and mobile bubbles, respectively. First, in the case of solid spheres, where $r = \sin\theta$, $dx = d\theta$, and $S = 4\pi r_e^2$, equation (7.25) is appropriate, and it may be rewritten as

$$\text{Sh} = 0.641\left[\int_0^\pi (u'_0 \sin\theta)^{1/2} \sin\theta\,d\theta\right]^{2/3} \text{Pe}^{1/3} \qquad (7.27)$$

and we can apply equation (7.27) to different Reynolds number ranges, as follows.

Case (a): Re < 1. With the inertia terms in the Navier–Stokes equations omitted, the classical solutions of Stokes and Lamb[55] for the stream function around a solid sphere can be utilized, and we obtain the result

$$u_o' = \tfrac{3}{2} \sin \theta$$

Substituting this expression into equation (7.27) and performing the integration gives

$$\mathrm{Sh} = 0.99\mathrm{Pe}^{1/3} \tag{7.28}$$

This result has been obtained elsewhere.[56] Equation (7.28) is experimentally verified for the dissolution of small bubbles,[57] liquid drops, and solid spherical particles.[58]

Case (b): Re > 1. For this regime the inertia terms of the Navier–Stokes equation cannot be ignored. The appropriate equation for the boundary layer around the moving sphere[53] is

$$u \frac{\partial u}{\partial x} + v \frac{\partial u}{\partial y} = U_\infty \frac{\partial U_\infty}{\partial x} + \nu \frac{\partial^2 u}{\partial y^2} \tag{7.29}$$

where U_∞ is the "free stream" velocity and ν the kinematic viscosity.

Assuming that the velocity profile is described by

$$\frac{u}{U_\infty} = 4\eta - 6\eta^2 + 4\eta^3 - \eta^4 \tag{7.30}$$

where $\eta = y/\delta$ and δ is the thickness of the hydrodynamic layer, equation (7.29) can then be used for obtaining δ, giving a value of the velocity gradient u_0', hence yielding, from equation (7.27),

$$\mathrm{Sh} = 0.62\mathrm{Re}^{1/2}\mathrm{Sc}^{1/3}$$

This equation was derived for a sphere, assuming no flow separation. Actually, for the Reynolds number range considered, separation occurs at $\theta \simeq 108^\circ$, so only the mass transfer occurring over the surface of the sphere upstream of the separation point will be given by equation (7.30) after area compensation, so that we find

$$\mathrm{Sh} = 0.56\mathrm{Re}^{1/2}\mathrm{Sc}^{1/3} \tag{7.31}$$

A similar equation was obtained in another study,[59] but with the constant equal to 0.59. It has been shown[57] that the dissolution of small bubbles, in flows where Re > 1, is described by equation (7.31).

Now, in the case of a mobile surface boundary, we must use equation (7.26), which may be rewritten as

$$\mathrm{Sh} = \left(\frac{2}{\pi}\right)^{1/2} \left(\int_0^\pi u_o \sin^2 \theta \, d\theta \right)^{1/2} \mathrm{Pe}^{1/2} \tag{7.32}$$

and it is then possible to obtain a value of the Sherwood number when the interface velocity is known. As before, we consider two cases.

Case (a): Re < 1. From the Hadamard flow function[60] for the streamlines of a liquid sphere, we obtain

$$u_o = \frac{1}{2} \sin \theta \left(\frac{\mu_f}{\mu_f + \mu_g} \right)$$

and substituting this value of u_o into equation (7.32), we find

$$\mathrm{Sh} = 0.65 \left(\frac{\mu_f}{\mu_f + \mu_g} \right)^{1/2} \mathrm{Pe}^{1/2} \tag{7.33}$$

This equation agrees fairly well with the experimental data[61] on bubble dissolution in glycerol.

Equation (7.33) has also been obtained by[58] an approximate integral method, but the value of the constant in this case is equal to 0.61, the equation with this value agreeing fairly well with experimental data obtained for dissolution of water drops in cyclohexane.

Case (b): Re ≫ 1. By analyzing the boundary layer around spheres moving with large Reynolds numbers, the velocity distributions inside and outside the sphere can be obtained[62] as a perturbation of the potential flow, and an equation for u_o is obtained as follows:

$$u_o = \frac{3}{2} \sin \theta - \frac{2(3)^{1/2}[2 + 3(\mu_g/\mu_f)] \int_0^\theta \sin^3 \theta \, d\theta}{\mathrm{Re}^{1/2}[1 + (\rho_g \mu_g/\rho_f \mu_f)^{1/2}]\pi^{1/2} \sin \theta}$$

and using this equation in equation (7.32), we obtain

$$\mathrm{Sh} = 1.13 \left[1 - \frac{2 + 3(\mu_g/\mu_f)}{1 + (\rho_g \mu_g/\rho_f \mu_f)^{1/2}} \frac{1.48}{\mathrm{Re}^{1/2}} \right]^{1/2} \mathrm{Pe}^{1/2} \tag{7.34}$$

For gas bubbles the ratios μ_g/μ_f and $\rho_g \mu_g/\rho_f \mu_f$ are usually much less than unity, and equation (7.34) may then be rewritten

$$\mathrm{Sh} = 1.13 \left(1 - \frac{2.96}{\mathrm{Re}^{1/2}} \right)^{1/2} \mathrm{Pe}^{1/2}$$

and for larger values of Re it will be

$$\mathrm{Sh} = 1.13 \mathrm{Pe}^{1/2} \tag{7.35}$$

This equation has been used successfully[61,63] to correlate experimental data on the dissolution of large bubbles.

The boundary-layer approach[62] has been further developed[64] in order to calculate mass-transfer coefficients for single spherical bubbles (or drops) rising steadily through a liquid at large Reynolds numbers. In this approach, for dilute solutions of constant density and diffusivity, the Sherwood number is found to be

$$\mathrm{Sh} = \left(\frac{3}{4\pi}\right)^{1/2} f(\mathrm{Re}, \mu, \rho)\mathrm{Pe}^{1/2} \tag{7.36}$$

where $f(\mathrm{Re}, \mu, \rho)$ is a given function of the Reynolds number and the viscosity and density of the fluid. This boundary-layer approach predicts results 8 percent too high when we compare predictions with experimental data obtained from measurements on mass transfer to carbon dioxide bubbles in water.[65]

Gas bubbles, with $r_e \simeq 0.9$ cm (0.35 in.), approximate oblate spheroids, and for the Reynolds number of these bubbles being large, the streamline can be considered (outside the boundary layer) as those of potential flow, and for this case a proposal[66] for velocity u_o is

$$u_o = (1 + k) \sin \theta$$

where k is a complicated function of the spheroidal eccentricity. Use of this term in either equation (7.32) or (7.36) gives, after manipulation, the following expression for the Sherwood number:

$$\mathrm{Sh} = [\tfrac{2}{3}(1 + k)]^{1/2}\mathrm{Pe}^{1/2}\left\{\frac{2.26 l_c^{1/3}(l_c^2 - 1)^{1/2}}{l_c(l_c^2 - 1)^{1/2} + \ln[l_c + (l_c^2 - 1)^{1/2}]}\right\} \tag{7.37}$$

where l_c is the axis ratio.

It follows from equation (7.37) that "flattening" of the bubble does not greatly affect the mass-transfer coefficient, and therefore equation (7.35) describes very well the mass-transfer process for large oblate spheroids.[61,63]

Gas bubbles with $r_e > 0.9$ cm assume toroidal and cap shapes and have a potential flow with the interface velocity given by

$$u_o = \tfrac{3}{2} \sin \theta$$

If we neglect mass transfer from the wake, the preceding equation yields

$$\mathrm{Sh} = 1.79\left(\frac{3l_c^2 + 4}{l_c^2 + 4}\right)\mathrm{Pe}^{1/2} \tag{7.38}$$

Various authors[67,68] have assumed that for toroidal bubbles, $l_c = 3.5$, and in this case equation (7.38) may be rewritten as

$$\mathrm{Sh} = 1.28\mathrm{Pe}^{1/2} \tag{7.39}$$

which also does not greatly differ from equation (7.35). Comparing the mass-transfer equation (7.28) with equations (7.35) or (7.39) for a single rising bubble in a pure liquid phase, we see that the mass-transfer coefficient for large bubbles is somewhat greater than for small ones.

Equations (7.35) and (7.38) have been evaluated in experiments that permitted the measurement of local mass-transfer coefficients.[69,70] Also, by measuring bubble sizes and rise velocities in low-viscosity liquids, average mass-transfer coefficients were evaluated [on a bubble-surface-area basis, as in equation (7.35)] leading to the formula

$$k_f = 1.13\left(\frac{D_f v_b}{r_e}\right)^{1/2} \tag{7.40}$$

where D_f is the liquid diffusivity and v_b and r_e are the bubble rise velocity and equivalent bubble radius, respectively.

For more viscous liquids (e.g., a water solution of glycerol) the values of the mass-transfer coefficients k_f obtained are higher than predicted by equation (7.39), and this reduction of k_f values with increasing viscosity may be explained by a decrease of the bubble rise velocity and an increase in contact time of liquid elements on the surface of the bubble, allowing "saturation" of these elements. This phenomenon is indicated by a theoretical study[71] that analyzed the problem of unsteady diffusion from a moving drop in which the resistances to mass transfer of continuous, and dispersed, phases are comparable, the velocity distribution in these phases being defined by the Rybczynski–Hadamard formula.[60] The following equations for the diffusion mass transfer per unit area and time were obtained:

$$\frac{\dot{m}}{S} = \frac{c_{\text{in}} - \alpha c_0}{\alpha + (D_g/D_f)^{1/2}}\left(\frac{4D_f}{\pi t}\right)^{1/2} \tag{7.41}$$

where c_{in} and c_0 refer to concentrations inside the bubble and in the bulk liquid, respectively, and t is the time, and here $t \ll \tau_r$; and

$$\frac{\dot{m}}{S} = \frac{c_{\text{in}} + \alpha c_0}{\alpha + (D_g/D_f)^{1/2}}\frac{D_f}{r_e}\left[\frac{2\text{Pe}_f}{3\pi(1 + \mu_g/\mu_f)}\right]^{1/2}$$

where τ_r is the relaxation time of the diffusion boundary layer and α is a distribution coefficient measuring the geometric relation among the bubbles (or droplets). If the mass-transfer rate is determined mostly by the external medium (liquid in the case of gas bubbles), equation (7.41), as a result of the substitution of $D_g \to \infty$ and $\alpha = 1$, is reduced to equation (7.40). Note however, that the use of equation (7.40) when $t \sim r_e/v_b$ does not lead to compatible results.

The distinctive features of these equations are that the diffusion mass transfer is proportional to the square root of the diffusivity and the rise velocity of the bubbles and is inversely proportional to the square root of the bubble diameter. The dependence of the mass-transfer coefficient on bubble rise velocity and bubble size becomes of critical importance when we move from a study of a single bubble to the study of mass-bubbling processes, but unfortunately, the literature available reveals no common approach to this problem.

For large deformed bubbles, the total diffusion flow per unit area of bubble surface is only weakly dependent on bubble size, as indicated experimentally.[72] By investigating absorption from large single bubbles (r_e = 0.4 to 2.1 cm) (0.16 to 0.83 in.) with surface-active contaminants, in water the following correlation was obtained:[73]

$$k_f = \text{const.}(r_e D_f^{\ 1/2} g^{1/4})$$

with the (dimensional) constant lying in the range 0.475 to 0.495 $m^{9/4}/s$.

A study[74] of CO_2 absorption in water was made to evaluate the mass-transfer coefficient k_f, and a relation was found between instantaneous values of the mass-transfer coefficient and bubble diameters by measuring their rise times. As bubble diameters were increased to about 3 mm (0.12 in.), the coefficient k_f increased, but for any further increase in the bubble diameter, k_f was independent of bubble size, thus making a one-to-one correspondence of k_f and rise time feasible. The variation of bubble diameters in liquid phase with the mass-transfer coefficient is indicated in Table 7.2, obtained from several sources,[65,73,74–80] and it is clear that, despite a wide range of bubble size, the mass-transfer coefficient does not vary greatly.

There has been some disagreement on the significance of mass transfer during bubble formation at the orifice. Some authors[76,81] have assumed that

Table 7.2. *Variation of mass-transfer coefficient k_f with bubble diameter d_b*

System	d_b (mm)	k_f (m/h)	T (°C)	Reference
H_2O–O_2	3.2	1.08	20	74
H_2O–CO_2	3.6	1.11	20	74
H_2O–CO_2	5.0	1.08	20	73
H_2O–CO_2	4.55	1.25	18	75
H_2O–air	0.3	0.45	17	61
H_2O–air	1.6	1.15	20	77
H_2O–air	1.5–1.3	1.4	20	78
H_2O–O_2	2.5	1.15	17	61
H_2O–CO_2	2.5	1.15	17	61
H_2O–CO_2	4.6	1.08	20	79
H_2O–CO_2	5.0	1.05	20	80
H_2O–O_2	3.2	0.9	20	80
H_2O–CO_2	3.6	1.15	20	61

most of the mass transfer from a bubble into a liquid occurs during bubble generation. However, it has been shown[82] that the average mass-transfer coefficient estimated during the period of generation at the submerged orifice is less than that during the ensuing rise of the bubble. Other studies, using contact plates, confirm that the role of mass transfer during bubble generation is not significant.

7.10 Decay of the mass-transfer coefficient k_f with bubble age

Although a considerable amount of work has been done on mass transfer from a single bubble, the decay of mass-transfer coefficient k_f with bubble age has not been considered in many studies, and therefore results of these studies might not be applicable for bubble–liquid systems where solubility of the gas is high or the bubble diameter is small, both conditions being ones in which bubbles are quickly absorbed. This phenomenon of a decrease of mass-transfer coefficient k_f with age of a single bubble has been observed in some studies[65,80,83] and not in others.[68–80,84–86] One possible explanation[85] for the effect due to bubble aging is that surface-active substances, accumulated over time at the gas–liquid interface, retard both the surface motion and mass transfer. Bubble rise velocity may be reduced by the surface-tension gradient, which is produced by the concentration gradient of these surface-active substances along the gas–liquid interface. Because the net surface-tension force exerted at the surface of a bubble is proportional to $\partial\sigma/\partial\Gamma$, where σ is the surface tension and Γ the local concentration of surface-active substances, bubble-rise-velocity reduction is greatest in a liquid highly contaminated with surface-active substances. Data[68] on such surface-active effects agree well with results obtained from numerical solutions[87] of the Navier–Stokes equations for the flow around spheres (with internal circulation) for Reynolds numbers up to 200.

However, even in clean liquid, a decrease of k_f with bubble age has been observed,[80] this effect apparently being due to a reduction of the concentration gradient at the interface, with consequent radial convection of liquid to the interface. This explanation of decay of the mass-transfer coefficient with bubble age has been examined in a study[88] which postulated that, for a liquid where the solubility and the diffusivity (of carbon dioxide in water, for example) is large, the volume of the interface film always remained constant, and for the case of a bubble with constant rise velocity, an empirical equation for the mass-transfer coefficient was then proposed taking into account the effect of bubble age.

7.11 Real mass-transfer coefficients and surface contact areas

Surface-contact-area measurements have been made in various studies,[89–92] and at the same time mass-transfer coefficients defined in terms of either plate surface area or the total gas–liquid volume have been experimentally obtained. From such values of the specific phase contact area and "fictitious" mass-transfer coefficients, the real coefficients (those defined by the surface contact area) can be obtained, the relationship between the various mass-transfer coefficients being given by equation (7.17). Correlating experimental data on mass transfer, investigators have[57] obtained two relations for mass-transfer coefficients for mass bubbling in a liquid phase, one for bubbles with an average diameter of less than 2.5 mm (0.98 in.),

$$k_f \mathrm{Sc}^{2/3} = 0.3\left[\frac{g\mu_f(\rho_f - \rho_g)}{\rho_f^2}\right]^{1/3} \tag{7.42}$$

and the other for bubbles with a diameter greater than 2.5 mm,

$$k_f \mathrm{Sc}^{1/2} = 0.42\left[\frac{g\mu_f(\rho_f - \rho_g)}{\rho_f^2}\right]^{1/3} \tag{7.43}$$

As a result of these studies, it appears to be a reasonable conclusion that the coefficient of mass transfer in a liquid phase is determined only by the physical properties of the system. However, the fact that we have two correlations, as shown above, indicates that a more appropriate analysis of the experimental data would include hydrodynamic parameters, so equations (7.42) and (7.43) must be considered as correlations describing only particular sets of experimental data.

Similar results were obtained in another study,[89] in which an equation for calculating the phase contact surface corresponding to 1 m^2 of plate area, was given:

$$A = ax_1 = c\left(\frac{\Delta p_f}{\rho_f g d_b}\right)^n \tag{7.44}$$

where a is the specific surface area, x_1 is the two-phase mixture height, $\Delta p_f = \Delta p + \Delta p_\sigma$ and Δp_σ is the pressure difference due to surface tension, and c and n are constants, and the bubble diameter was calculated using the following equation:

$$d_b = 1.2\left(\frac{\sigma^{0.625}}{\Delta p_f^{0.25}(\rho_f g)^{0.375}}\right)\left(\frac{\nu_g}{\nu_f}\right)^{0.375}$$

where ν_g and ν_f are the gas and liquid kinematic viscosities, respectively.

The values of factors c and n in equation (7.44) depend on the bubbling regime and the type of plate used; for foams in overflow plates $c = 28$ and $n = 0.25$, and for froth and grid trays, $c = 17$ and $n = 0.8$. These experimental data were correlated in the form

$$\mathrm{Sh} = 0.7\mathrm{Re}^{2/3}\mathrm{Sc}^{1/2}$$

where

$$\mathrm{Sh} = \frac{kd_e}{D}, \qquad \mathrm{Re} = \frac{Q_f d_e}{(1-\phi)\nu \cdot 3600}, \qquad \text{and} \qquad d_e = \frac{4(1-\phi)}{d}$$

and Q_f is the volumetric liquid flow rate.

It is therefore clear that it is impossible to study mass-transfer processes independently of the hydrodynamics of the flow system, no matter which mass-transfer coefficients k_f, k_{f_s}, or k_{f_v} are considered. In fact, most mass-transfer studies are concerned with experimental measurement of the "fictitious" coefficients k_{f_s} and k_{f_v}, but, unfortunately, numerous studies correlating these mass-transfer coefficients with hydrodynamic parameters do not clearly indicate which parameters should be considered as the determining ones, or even how these parameters influence mass transfer in a liquid phase.

For example, as shown in Table 7.1, equations for calculation of k_{f_s} and k_{f_v} differ, depending on whether local or mean values of the various factors are used, raising questions as to the practical application of the results. Theoretical analysis of mass transfer in mass-bubbling processes is, of course, complicated by the fact that the hydrodynamic equations of one-phase flow cannot easily be applied to a two-phase flow divided into numerous discrete elements. Solution of the equations of motion, together with the convective diffusion equation and the continuity equation for such flows is associated with serious mathematical difficulties, and therefore the basic approach to the study of mass transfer in bubbling processes must, as we have indicated throughout this chapter, be experimental.

However, a difficulty in most experimental studies is that of trying to obtain suitable equations for correlating data in a meaningful way, and, as Table 7.1 indicates, despite many proposed equations, no consistency has been obtained. Clearly, the method of analysis of the experimental data is very important. For example, in two independent studies with similar experimental conditions, the correlations obtained were different,[93] and this can only be explained by the functional relationships used for analyzing the data. We can conclude that great caution must be used in setting up formulas to correlate experimental data. Several approaches to this problem will now be discussed.

7.12 Mass transfer from a single bubble suspended in a turbulent stream

In a theoretical study of transfer of a substance from one phase to another, mathematical difficulties arise due in large measure to the complex effects of hydrodynamic factors on the process of mass transfer. We have considered several approximate methods for dealing with the study of mass transfer in a liquid phase, the approximations consisting of selecting geometrical and flow conditions to give us the simplest model possible. Flow regimes were considered as known and similar in all regions of the system, and random changes of regime in time and space were not considered. However, in contrast to these analyses, in industrial practice it is usually desired to achieve the maximum possible mixing to give a high rate of transfer, and in these situations, our earlier results are inapplicable.

Mass transfer in these applications takes place in a system of bubble swarms, and allowance must be made for such mass-bubbling phenomena as fluid-mechanical interactions between bubbles and nonuniformity of the bubble volume distribution. We consider a model of mass transfer from a single bubble suspended in a turbulent liquid flow, this flow being created by mixing of the liquid by a bubble swarm.[94] Clearly, this is the simplest model we can develop and still hope to allow for mass-bubbling hydrodynamic effects, and in fact the surface-renewal concept[4,5] discussed in Section 7.3 has been incorporated into more elaborate turbulent models in a number of recent studies.[70,95,96] Also, turbulent surface renewal in terms of eddy diffusivity (which can only be related empirically to the turbulence due to mass-transfer behavior) has been proposed[97,98] as a model.

The turbulent transfer at the gas–liquid interface can only be understood when an appropriate physical model of the velocity field near the interface is available, and some aspects of this problem have been considered,[97–103] but no satisfactory quantitative link between mass-transfer rates and the turbulent motion has emerged, even though one model has been developed[95] which does, to some degree, provide a link between the observed mass-transfer behavior and the state of the turbulent field. In this latter study, the overall effect of turbulence on the mass-transfer rate was determined from a consideration of the eddy and energy spectra, according to a postulated structure of turbulent flow.

As we have described in Section 7.2 and Figure 7.2, the turbulent boundary layer is a layer in which matter is transferred by turbulent fluctuations and where molecular diffusion does not play a noticeable part. Closer to the interface is the viscous sublayer, in which matter is transferred both by molecular diffusion and turbulent fluctuations, but since the scale of the fluctuations is here so small, the momentum and mass transferred by molecular viscosity exceeds that transferred by turbulent pulsations. In the innermost portion of the viscous sublayer, in the diffusion boundary layer, the molecular diffusion mechanism completely predominates over the turbulence mechanism.

This description of the diffusional boundary layer permits us, using physically well-founded simplifications, to overcome the difficulties associated with the solution of the equation of convective diffusion. At moderate Reynolds numbers, the diffusion of matter can be calculated over the entire bubble surface, except for the separated region at the rear, and the total diffusion mass flow over the bubble surface is given by equation (7.19):

$$\dot{m} = 4\left(\frac{2\pi D v_b}{d_b}\right)^{1/2} r_b^2 c_0$$

where D is the diffusivity, v_b the bubble rise velocity, r_b the bubble radius, and c_0 the bulk concentration (indicative of the mass-transfer driving force).

The diffusion flux over unit area of bubble surface is then

$$\dot{m}' = \left(\frac{D v_b}{\pi r_b}\right)^{1/2} c_0$$

For bubble motion in a swarm of bubbles, the velocity will differ from that of free rise, because the bubble is obviously affected by the flow field created by the other bubbles in motion. Based on the theory of isotropic homogeneous turbulence, we have obtained in equation (3.14) a formula for calculation of the velocity of a suspended bubble in a turbulent flow. Substituting the value of this velocity from equation (3.14) into equation (7.19), we get

$$\dot{m} = \text{const.}\ \frac{D^{1/2} U^{3/4} r_b^2 c_0 (1-\phi)^{1/4}}{(l v_f)^{1/4} (1-\phi^{5/3})^{1/4}}$$

where U is some flow system characteristic velocity, l the size of the largest eddies (approximately the flow system typical linear dimension), and ϕ the gas void fraction. In dimensionless form

$$\text{Sh} = \text{const.}\left(\frac{d_b}{l}\right)\frac{(1-\phi)^{1/4}}{(1-\phi^{5/3})^{1/4}}\,\text{Re}^{3/4}\text{Sc}^{1/2} \qquad (7.45)$$

where

$$\text{Sh} = \frac{k_f d_b}{D}; \qquad \text{Sc} = \frac{v_f}{D}; \qquad \text{and} \qquad \text{Re} = \frac{Ul}{v_f}$$

Here the Reynolds number can be calculated using the liquid overall relative velocity [because for all practical purposes this is the same as the velocity of large-scale fluctuations as determined by equation (3.2)], and the linear dimension l is comparable to the height of the gas–liquid mixture x_1. Most of the kinetic energy of the turbulent flow is confined in such large-scale fluctuations, and this energy is continually being transferred to ever-smaller turbulent eddies, until dissipated by molecular viscosity. Note that, because the turbulent flow is created by bubble motion, the energy of dissipation is a suitable measure of the total effect of this motion, and consequently the diffusion flow over the bubble surface is a function of the gas energy supply, which is ultimately dissipated in the liquid.

7.13 Mass transfer for a bubble swarm

We now attempt to determine the total diffusion from a group of bubbles, allowing for their distribution with respect to size in the flow, with the simplifying assumption that the bubbles behave independently in respect to diffusion. Application of this assumption leads to a law of distribution for bubble diameter d_b which agrees well with experimental data for liquid–liquid and gas-liquid systems.[81,82] This distribution, also given in Section 3.5, is

$$f(d_b, \alpha) = 4\left(\frac{\alpha^3}{\pi}\right)^{1/2} d_b^2 \exp\left(-\alpha d_b^2\right)$$

where

$$\alpha = \left[\frac{16(\pi N)^{1/2}}{3\phi}\right]^{2/3} > 0 \tag{7.46}$$

We have the constraint on $f(d_b, \alpha)$ that

$$\int_0^\infty f(d_b, \alpha)\, dd_b = 1$$

If the number of bubbles in the dispersed phase (per unit mixture volume) is equal to

$$N = \phi/\tfrac{4}{3}\pi \bar{d}_b^3 \tag{7.47}$$

where $\bar{d}_b$ is the mean volume diameter, defined as

$$\bar{d}_b = \left(\frac{\Sigma n_i d_{b_i}^3}{\Sigma\, n_i}\right)^{1/3}$$

and n_i is the number of bubbles with diameter d_{b_i}, then, substituting N from equation (7.47) into equation (7.46), we obtain, after manipulation,

$$\alpha = (4/\pi^{1/2}\overline{d}_b^3)^{2/3}$$

Thus

$$f(d_b, \alpha) = F(d_b, \overline{d}_b) = \frac{4}{\pi \overline{d}_b^3} d_b^2 \exp\left[-\left(\frac{4}{\pi^{1/2}\overline{d}_b^3}\right)^{2/3} d_b^2\right] \tag{7.48}$$

From equation (7.48), it follows that the distribution function can be completely determined if the parameter $\overline{d}_b$ is known as a function of the physical properties of the system and the process variables. The dispersed phase (gas) volume, consequently, is expressed by the relation

$$V_t d\phi = \tfrac{4}{3}\pi d_b^3 N V_t F(d_b, \overline{d}_b)\, dd_b \tag{7.49}$$

where V_t is the two-phase mixture volume, or

$$\phi = \tfrac{4}{3}\pi N \int_0^\infty d_b^3 F(d_b, \overline{d}_b)\, dd_b$$

To determine the overall mass diffusion M per unit time, it is necessary to integrate over the total surface of the bubbles, with allowance for the distribution F as a function of d_b and $\overline{d}_b$, so that

$$M = \int_0^A \dot{m}'\, dA \tag{7.50}$$

where A denotes the total bubble–liquid interface area and, from equation (7.45), the mass diffusion $\dot{m}'$ per unit bubble surface area is

$$\dot{m}' = \frac{D^{1/2}U^{3/4}c_0(1-\phi)^{1/2}}{4\pi(l\nu_f)^{1/4}(1-\phi^{5/3})^{1/4}}$$

Now, dA can be expressed as a function of bubble size, and we find that

$$dA = \frac{4\pi d_b^2 V_t}{4\pi d_b^3/3}\, d\phi = \frac{3V_t}{d_b}\, d\phi$$

and using equations (7.47) and (7.49), this equation becomes

$$dA = \frac{3\phi V_t}{\overline{d}_b^3} d_b^2 F(d_b, \overline{d}_b)\, dd_b \tag{7.51}$$

Thus, using equation (7.51) in equation (7.50), we find

$$M = \frac{3\phi V_t}{\overline{d}_b^3} \int_0^\infty \dot{m}'\, d_b^2 F(d_b, \overline{d}_b)\, dd_b$$

and after substituting the values of $\dot{m}'$ and $F(d_b, \bar{d}_b)$ into this equation and integrating, we obtain

$$M \simeq \frac{6\phi V_t}{\pi^2 \bar{d}_b^6} \frac{(1-\phi)^{1/2}}{(1-\phi^{5/3})^{1/4}} \frac{D^{1/2}U^{3/4}c_0}{(l\nu_f)^{1/4}} \frac{\Gamma(2.5)}{\alpha^{5/2}}$$

where Γ is the gamma function, and, finally,

$$M = \frac{2.62}{4\pi} \frac{\phi V_t}{\bar{d}_b} \frac{(1-\phi)^{1/2}}{(1-\phi^{5/3})^{1/4}} \frac{D^{1/2}U^{3/4}c_0}{(l\nu_f)^{1/4}} \tag{7.52}$$

Equation (7.52) permits us to calculate the total diffusion from the bubbles, and we can express this equation in the form

$$k_f a = \frac{2.62}{4\pi} \frac{\phi}{\bar{d}_b} \frac{(1-\phi)^{1/2}}{(1-\phi^{5/3})^{1/4}} \frac{D^{1/2}U^{3/4}}{(l\nu_f)^{1/4}} \tag{7.53}$$

Here k_f is liquid mass-transfer coefficient and a is the specific surface area (the ratio of total bubble surface area to total mixture volume). Now, using equations (7.48) and (7.51), the specific phase contact surface may be written as

$$a = \frac{\int_0^A dA}{V_t} = \frac{3\phi}{\bar{d}_b^3} \int_0^\infty \frac{16}{\pi \bar{d}_b^3} d_b^4 \exp\left[-\left(\frac{4}{\pi^{1/2}\bar{d}_b^3}\right)^{2/3} d_b^2\right] dd_b$$

and upon integrating,

$$a = \frac{2.619\phi}{\bar{d}_b} \tag{7.54}$$

Then equation (7.53) becomes

$$k_f \simeq \frac{D^{1/2}}{4\pi} \frac{U^{3/4}}{(l\nu_f)^{1/4}} \frac{(1-\phi)^{1/2}}{(1-\phi^{5/3})^{1/4}}$$

and, substituting the value of velocity U from equation (3.4) into this equation, we obtain

$$k_f = \text{const.} \frac{D^{1/2}}{(l\nu_f)^{1/4}} \left(\frac{\Delta P}{\rho_f}\right)^{3/8} \frac{(1-\phi)^{1/2}}{(1-\phi^{5/3})^{1/4}} \tag{7.55}$$

Recall that D is the diffusivity, ν_f the kinematic viscosity, ΔP the system pressure drop, l some characteristic length (say the height x_1 of the gas–liquid mixture), and ρ_f the liquid density.

Equation (7.55) offers the possibility of calculating the true coefficient of mass transfer in the liquid phase, for the system, and it follows from the equation that this coefficient depends on the molecular diffusion coefficient, the kinematic viscosity of the liquid, and the height and pressure drop in the two-phase mixture, as well as on the gas void fraction. Note that in deriving equation (7.55) we used quantities obtained from dimensional considerations, and therefore the equation is only correct to a constant factor.

7.14 Experimental data on mass transfer into a liquid phase

For verification of equation (7.55) we need experimental data on mass-transfer coefficients defined in relation to the total interface surface, and obviously these coefficients cannot be directly measured. In studying mass transfer into a liquid phase, one or more of the following simplifying criteria for selecting experimental systems is usually employed:

1. Systems in which there is no resistance in the gas phase.
2. Systems in which the dispersed component is a slowly soluble gas.
3. Systems with a chemically active liquid phase.

For systems of the first kind, the overall coefficient of mass transfer is equal to the mass-transfer coefficient in the liquid phase, k_f, which can then be defined as the experimentally measured overall mass-transfer coefficient. Gas–liquid systems of the second kind have attracted the most attention in the measurement of mass-transfer coefficients in a liquid,[16-18,22,35,36,104] and usually the dispersed component in these systems is CO_2, O_2, or H_2. The coefficients of mass transfer are calculated from the equation

$$\dot{m} = k_f A \Delta \tag{7.56}$$

where A denotes the area and Δ the driving force.

If the interface area cannot be determined, the mass-transfer coefficients are specified either by unit plate area or by unit volume of the gas–liquid mixture, and as we have noted in equation (7.17), these coefficients are related to k_f by

$$k_{f_s} = k_f a x_1 \tag{7.57}$$

and

$$k_{f_v} = k_f a \tag{7.58}$$

where a is the specific surface area and x_1 the mixture total height.

Using equations (7.57) and (7.58), experimental verification of equation (7.55) can be attempted. The precision of measurement of these coefficients is most significantly affected by the method of estimation of the process driving force Δ and, because the gas distribution in liquid is dependent on the gas and liquid flow rates, making an assumption, for example, of complete mixing of liquid[6,21,22] may lead to significant errors in calculation of mass-transfer coefficients by equation (7.56). It also follows that, for example, experimental mass-transfer coefficients k_{f_s} obtained without taking into account the actual concentration distributions imposed on the system cannot be used for determination of quantitative relationships between these coefficients and the hydrodynamic and physical parameters of the process.

Using systems with a chemically active liquid phase permits us to avoid the difficulty associated with the measurement of the process driving force. Experiments,[61] conducted to evaluate k_f in columns of heights 7.7, 15, and 30 cm (3, 5.9, and 11.8 in.) with porous, perforated and single-orifice sources indicate that the liquid diffusivity D, liquid density ρ_f, kinematic viscosity ν_f, average bubble diameter $\bar{d}$, and rise velocity of the bubbles U, all have some effect on the mass-transfer coefficient for the liquid phase k_f. From Figure 7.7 it is clear from data on absorption of oxygen into various liquids, with different kinds of perforated plates that k_f is affected by the average bubble diameter $\bar{d}_b$. The coefficient k_f approximately increases with $\bar{d}_b^{1/2}$. In Figure 7.8, k_f, $\bar{d}_b$, and surface tension σ are plotted against the gas void fraction ϕ, and we see that k_f is, at most, weakly dependent on ϕ (and also σ). Figure 7.9 shows the Sherwood number Sh plotted against the Gal-

Figure 7.7 Effect of average bubble diameter on liquid-phase mass-transfer coefficient. [From D. Hammerton and F. H. Garner, *Trans. Inst. Chem. Eng.*, *32*, 518 (1954).]

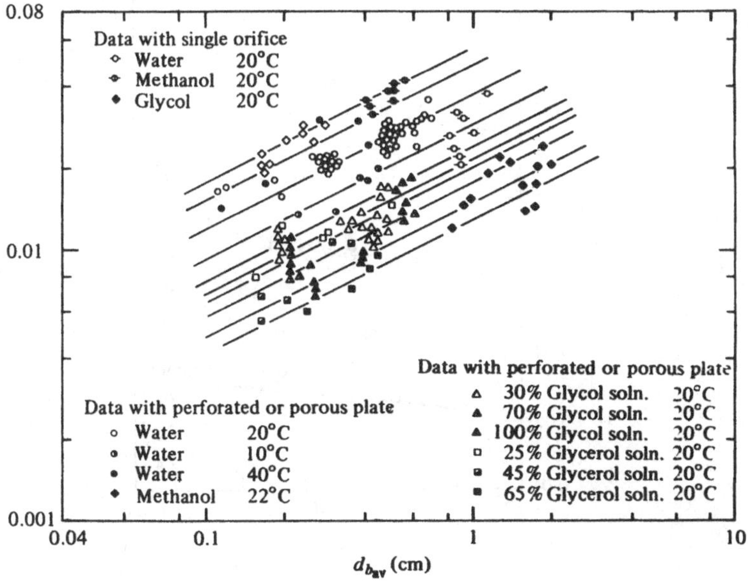

ileo number Ga for various liquids, and the slope of the curves is $\frac{1}{2}$, as we might have expected from the effect of the bubble diameter. The straight line in Figure 7.10, drawn through data points from 10 different systems, can be expressed by the following correlation:

$$\mathrm{Sh} = 0.5\mathrm{Sc}^{1/2}\mathrm{Ga}^{1/4}\mathrm{Bo}^{3/8} \tag{7.59}$$

where Sh is the Sherwood number $= k_f\overline{d}_b/D_f$, Sc is the Schmidt number $= \nu_f/D_f$, Ga is the Galileo number $= g\overline{d}_b^3/\nu_f^2$, and Bo is the Bond number $= g\overline{d}_b^2\rho_f/\sigma$. Equation (7.59) can be rewritten as

$$k_f = 0.5g^{5/8}D_f^{1/2}\rho_f^{3/8}\sigma^{-3/8}\overline{d}_b^{1/2}$$

and according to this equation k_f is not affected directly by ν_f, although we note that D_f and $\overline{d}_b$ are functions of ν_f.

Experiments have been conducted using an aqueous solution of sodium sulfite and atmospheric oxygen (oxidation of sodium sulfite into sodium sulfate), and because of this chemical reaction, the mass-transfer process in the liquid phase is accelerated to some extent, as compared with the purely

Figure 7.8 Standard deviation, average bubble diameter, and liquid mass-transfer coefficient versus void fraction. [From D. Hammerton and F. H. Garner, *Trans. Inst. Chem. Eng.*, *32*, 518 (1954).]

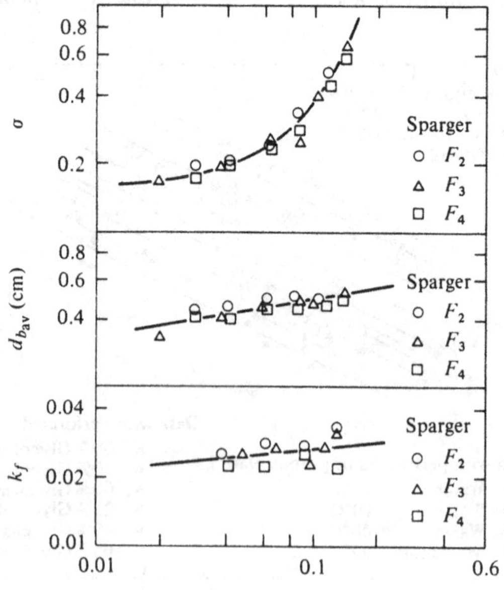

physical absorption process.[22] The mass-transfer coefficient in a liquid phase when a chemical reaction is taking place is given by

$$k_f' = \alpha k_f$$

where α is an "acceleration coefficient" for mass transfer. Under certain conditions, the chemical reaction for this system has no significant influence on the mass transfer;[30,83,104] in other words, $k_f' \simeq k_f$. However, when the sulfite concentration in the aqueous solution is higher than 0.03 kg mol/m^3 (1.8×10^{-3} lb mol/ft^3) and the concentration of Cu^{2+} ions is higher than 10^{-4} kg mol/m^3 (6.2×10^{-6} lb mol/ft^3), the absorbed oxygen in the bulk of liquid is all chemically combined, so its concentration in the solution is essentially equal to zero,[83] and in this case, mixing of the liquid phase need not be taken into account for calculation of the driving force to obtain mass-transfer coefficients. In the experiment, the concentration of sodium sulfite ranged from 20 to 30 kg/m^3 (1.2 to 1.9 lb/ft^3) down to 4 kg/m^3 (0.25 lb/ft^3) and, over this range, the oxidation rate is found to have a linear time dependence as shown in Figure 7.11. When the sodium sulfite concentration

Figure 7.9 Sherwood number versus Galileo number for various liquids. [From D. Hammerton and F. H. Garner, *Trans. Inst. Chem. Eng.*, *32*, 518 (1954).]

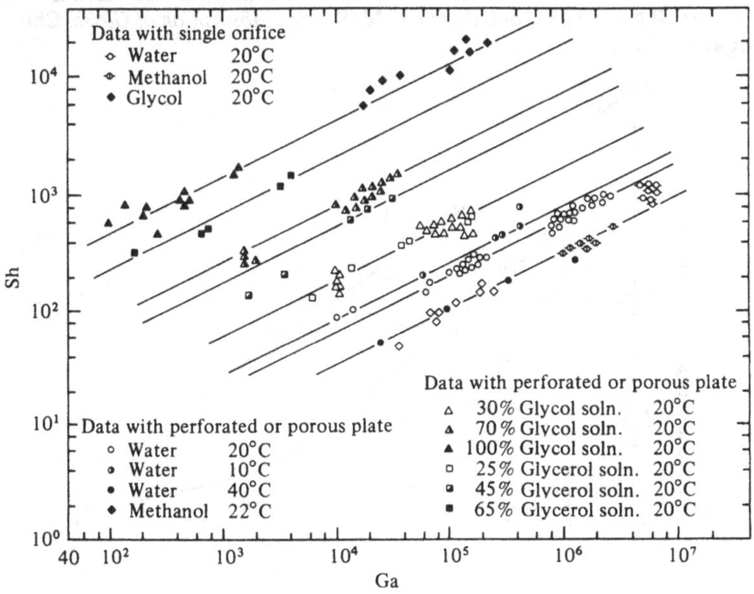

Figure 7.10 General correlation for liquid-phase mass-transfer coefficient. [From D. Hammerton and F. H. Garner, *Trans. Inst. Chem. Eng.*, *32*, 518 (1954).]

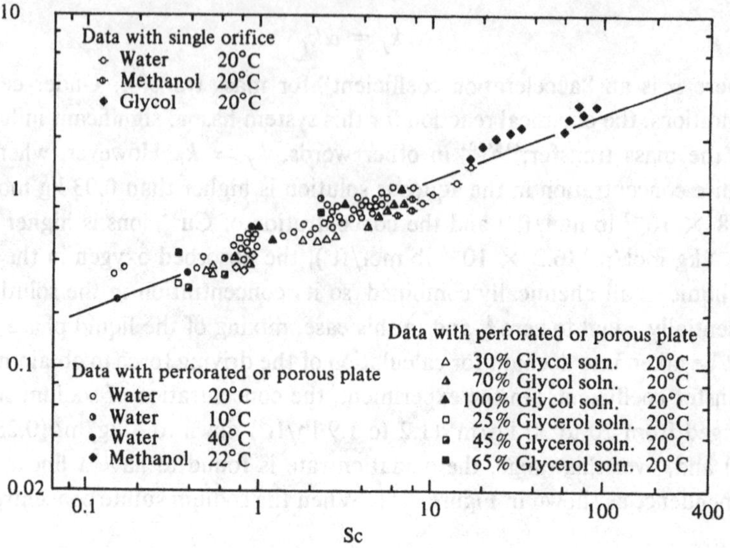

Figure 7.11 Sodium sulfite oxidation as a function of time. Liquid column, *h*: (1) 80 mm, (2) 50 mm, (3) 30 mm. (From V. M. Ramm, *Absorption of Gases*, Chimia, Moscow, 1966.)

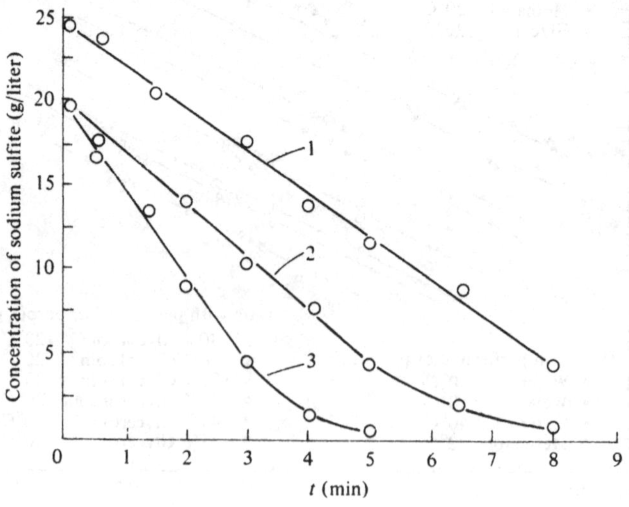

is about 4 kg/m^3 (0.25 lb/ft^3), the reaction rate is dependent on the sodium sulfite concentration in solution. Oxidation of sodium sulfite into sodium sulfate in the presence of atmospheric oxygen is described by the formula

$$Na_2SO_3 + \tfrac{1}{2}O_2 = Na_2SO_4$$

and the amount of absorbed oxygen is determined by the amount of reacted sodium sulfite in the ratio 8:63. The volumetric mass-transfer coefficient is then defined as

$$k_{f_v} = \frac{8(c_1 - c_2)}{63\tau\Delta} \tag{7.60}$$

where τ is the time duration of the experiment (given by the flow rate of the solution and the total volume of solution, for flowing systems), Δ is the driving force of the process, and c_1 is the initial and c_2 the final concentration of sodium sulfite. Also, using equations (7.57) and (7.58) in equation (7.60), we find that

$$k_{f_s} = \frac{8(c_1 - c_2)x_1}{63\tau\Delta}$$

The driving force can be calculated from the final equilibrium concentration:[83]

$$\Delta = \frac{c_e p_e}{m_{\text{equ}}} \tag{7.61}$$

where p_e is the pressure in the exhaust gas, m_{equ} the phase equilibrium constant, and c_e the oxygen concentration in exhaust gas at equilibrium, given by

$$c_e = \frac{cQ - (32M/\rho_g V_f \tau)}{Q - (32M/\rho_g V_f \tau)} \tag{7.62}$$

where Q is the air flow rate, c the oxygen concentration (by volume) in the air feed, M the mass of reacted sodium sulfite, and V_f the liquid volume. For high gas flow rates, c_e is practically equal to c.

With this approach the measured mass-transfer coefficients in the liquid phase are not subject to errors associated with measurement of the process driving force, and the data so obtained, in conjunction with equations (7.61) and (7.62), make it possible to calculate mass-transfer coefficients in relation to specific surface contact area, using equations (7.57) and (7.58). Ultimately, we can then check the validity of theoretical equation (7.55). As a result of such studies, relationships have been obtained between k_{f_s} and the irrigation intensity Q_f/b (where b is the weir length), liquid column height h, and plate (orifice) geometry, and analysis of these relationships

Figure 7.12 Mass-transfer coefficient, k_{f_s}, gas content, ϕ, and liquid column, h, plotted versus superficial gas velocity, v_s, h_{wh} = 100.0 mm; Q_f/b = 2.5 m^3/m·h. [From G. P. Solomakha, G. A. Rudevich, and P. I. Nikolaev, *Theor. Found. Chem. Eng. (USSR)*, *5* (1968).]

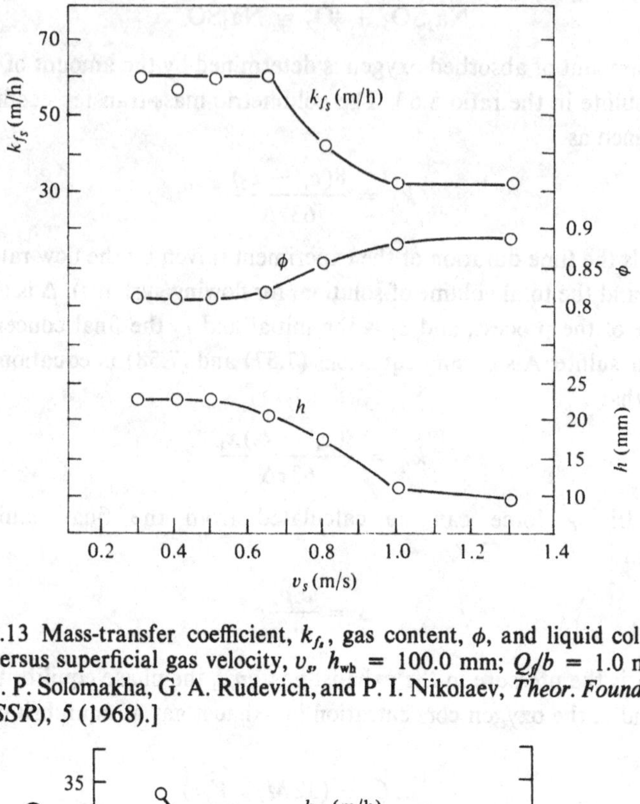

Figure 7.13 Mass-transfer coefficient, k_{f_s}, gas content, ϕ, and liquid column, h, plotted versus superficial gas velocity, v_s, h_{wh} = 100.0 mm; Q_f/b = 1.0 m^3/m·h. [From G. P. Solomakha, G. A. Rudevich, and P. I. Nikolaev, *Theor. Found. Chem. Eng.* (*USSR*), *5* (1968).]

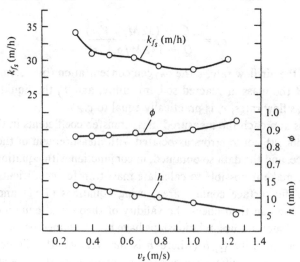

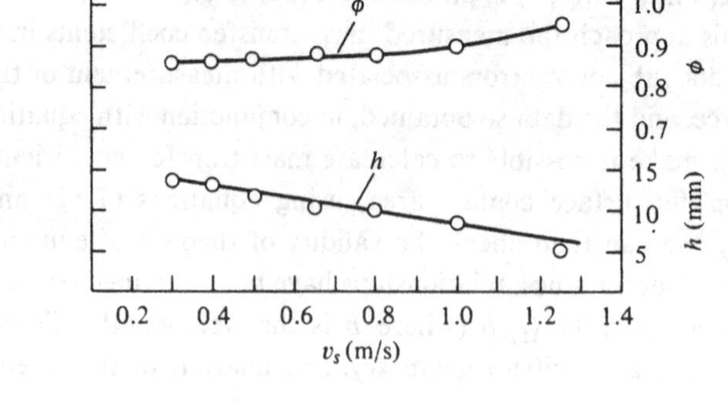

leads to the kind of data shown in Figures 7.12, 7.13, and 7.14. Here, the mass-transfer coefficient k_{f_s} is shown plotted against the superficial gas velocity v_s for weir height h_{wh} = 100 mm and for different Q_f/b values.[30] It appears from these figures that the relation of k_{f_s} to v_s has an extremum, determined[30] by changes of hydrodynamic regimes.

For a system consisting of an air–sodium sulfite aqueous solution, $\sigma = 7.3 \times 10^{-2}$ N/m (4.2×10^{-4} lbf/in.) and $\mu_g/\mu_f = 0.018$ and equation (3.26) for the average bubble diameter may then be rewritten as

$$\bar{d}_b = 0.61 h^{-0.2} \frac{\phi^{0.65}}{(1 - \phi)^{0.4}} \quad \text{cm} \tag{7.63}$$

where h is the height of liquid column (mm), and equation (3.44) for this system becomes

$$a = 98.0 h^{0.2} (1 - \phi)^{0.4} \phi^{0.35} \tag{7.64}$$

So the superficial gas velocity appears to have no significant effect on bubble size, at least over a limited range of gas velocities, as shown in Figure 7.15. The influence of liquid column height on bubble size, as indicated by equation (7.63), is shown in Figure 7.16, where, for shallow pools, the bubble sizes are significantly greater than for deep pools. The independence of specific surface contact area from the superficial gas velocity, as indicated in equation (7.64), is shown in Figures 7.17a and b; this result is similar to that obtained by other investigators.[82,89–91]

Figure 7.14 Mass-transfer coefficient, k_{f_s}, gas content, ϕ, and liquid column, h, plotted versus superficial gas velocity, v_s. h_{wh} = 100.0 mm; Q_f/b = 0.25 m³/m·h. [From G. P. Solomakha, G. A. Rudevich, and P. I. Nikolaev, *Theor. Found. Chem. Eng. (USSR)*, 5 (1968).]

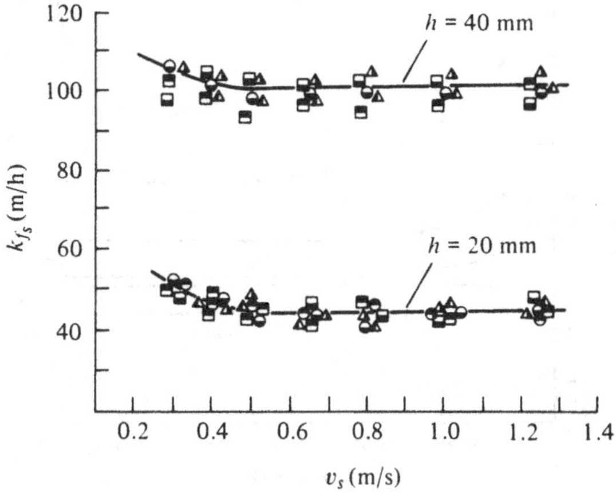

Figure 7.15 Relationship between an average bubble diameter and superficial gas velocity by different weir heights. [From G. P. Solomakha, G. A. Rudevich, and P. I. Nikolaev, *Theor. Found. Chem. Eng. (USSR)*, 5 (1968).]

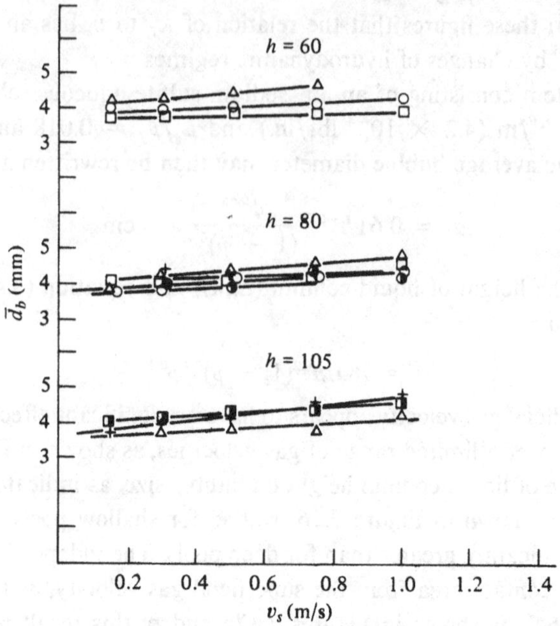

Figure 7.16 Relationship between an average diameter and superficial gas velocity for nonoverflow system. Liquid column, h: (1) 10.0 mm, (2) 20.0 mm, (3) 40 mm, (4) 50 mm, (5) 80 mm. Superficial gas velocity, v_s: (1) 0.3 m/s, (2) 1.0 m/s. (From L. F. Filatov, Candidate's dissertation, Moscow Institute of Chemical Engineers, 1969.)

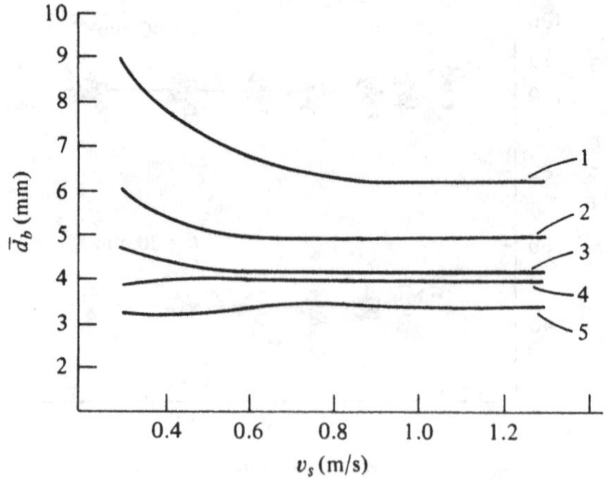

Figure 7.17 (a) Relationship between specific surface contact area and superficial gas velocity for different liquid flow rates and weir heights. (b) Relationship between specific surface contact area and superficial gas velocity for nonoverflow systems. Weir height: (1) 10.0 mm, (2) 20.0 mm, (3) 40 mm, (4) 50.0 mm. k_f and liquid column for v_s = 0.5 m/s. (From "Tray efficiencies in distillation columns." Final report of Michigan AIChE, New York, 1960.)

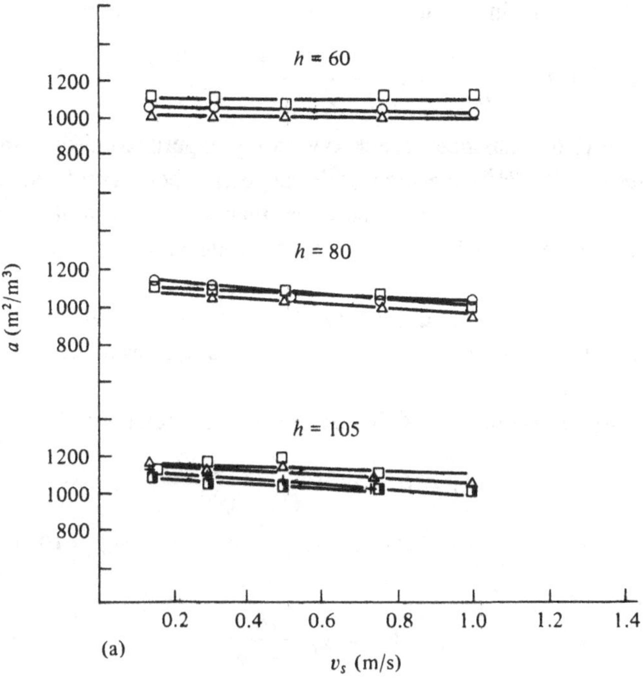

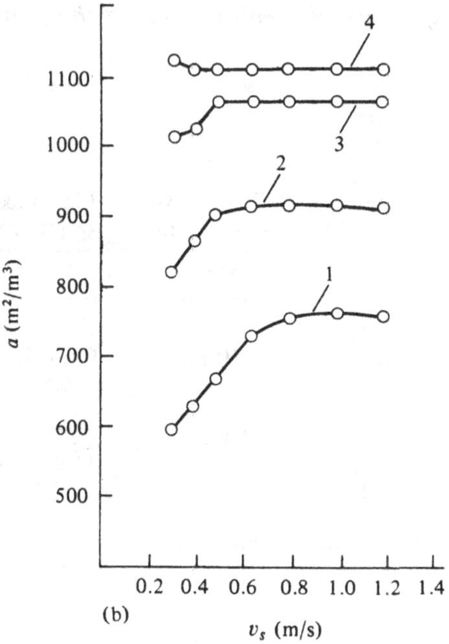

Using experimental values of mass-transfer coefficients k_f, and equations (7.57), (7.58), and (7.62), when appropriate, we can calculate the constant in equation (7.55), and we obtain an equation for the calculation of mass-transfer coefficients in a liquid phase, per unit interface area:

$$k_f = 0.12\left[\frac{D^{1/2}}{(l\nu_f)^{1/4}}\right]\left(\frac{\Delta P}{\rho_f}\right)^{3/8}\frac{(1-\phi)^{1/2}}{(1-\phi^{5/3})^{1/4}} \quad \text{cm/s} \tag{7.65}$$

Equation (7.65) has been checked with the experimental data in several other studies,[32,34,39,93,104] covering a wide range of experimental conditions, as shown in Table 7.3. For all these experiments, the constant in equation (7.65) was equal to 0.12, with a standard deviation equivalent to ± 5 percent.

The good agreement of equation (7.65) with this experimental data obviously supports the validity of the theoretical assumptions made in derivation of the equation.

For air–sodium sulfite and CO_2–water solutions, equation (7.65) becomes

$$k_f = 0.565 h^{0.13}\frac{(1-\phi)^{1/2}}{(1-\phi^{5/3})^{1/4}}$$

because $D_f = 1.77$ cm²/s (0.275 in./s) and $\nu_f = 0.01$ cm²/s (0.002 in./s), while

$$k_{f_s} = k_f\frac{ah}{1-\phi} \tag{7.66}$$

Table 7.3. *Various experimental conditions for which equation (7.65) is found to be valid*

	Plate geometry				
Reference	Orifice diameter, d_0 (mm)	Free area, Fa (%)	Column size, ϕ:dia. (mm)	Liquid load, Q_f (m³/h)	Gas load, Q (m³/s)
39	0.9	2.0	ϕ300.0	—	0.3–0.8
104	0.9–5.0	2.55–10.0	240.130	0.25–5.0	0.3–1.3
32	2.0	10.0	310.110	0.1; 0.2; 0.6	0.5–1.75 with account of mixing
93	2.0	6.95	ϕ200.0	0.35	0.3–1.3
34	2.0	6.85–12.2	ϕ1000.0	11.0; 17.0; 21.0	0.51–1.77 with account of mixing

As we have seen in Chapter 3, the gas energy consumed during bubbling is numerically equal to the height of liquid column, and consequently, the mass-transfer coefficient (being dependent on this height) is to a large degree defined by this energy. In particular, if the specific surface contact area is constant, we have from equation (7.66),

$$k_{f_s} \sim h^{1/13}$$

Note that, in deriving equation (7.65), it was assumed that the hydrodynamic conditions were correctly described by a model of homogeneous isotropical turbulence and, as can be seen in Figure 7.18, the values of the mass-transfer coefficient calculated from equation (7.65) and measured values agree very well.

Figure 7.18 Comparison of calculated and experimental mass-transfer coefficients, k_f.

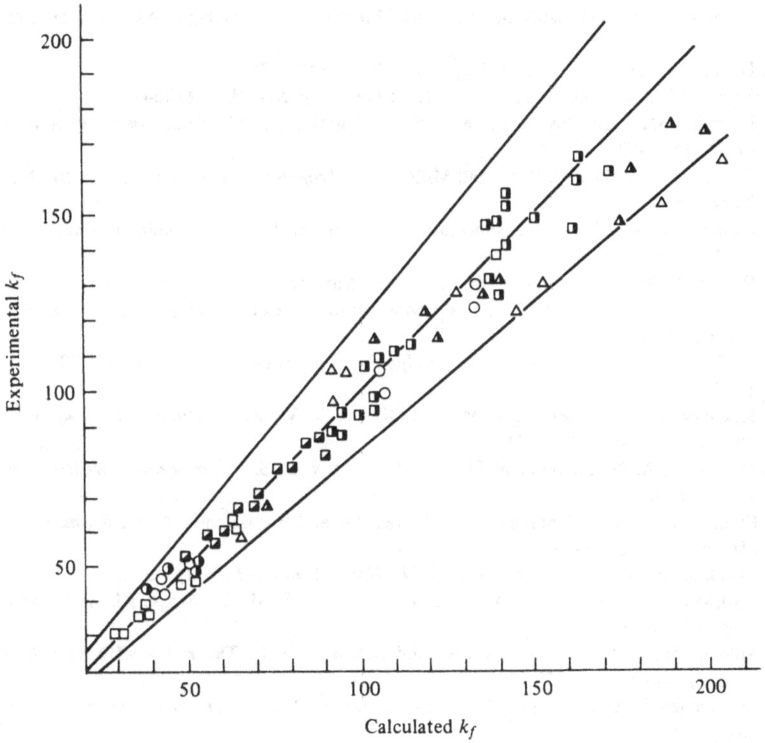

References

1. Nernst, W. Z., *Phys. Chem.*, *47*, 52 (1904).
2. Rosen, A. M., and Krylov, V. S., *Khim. Prom. (Moscow)*, no. 1, 51 (1966).
3. Sherwood, T. K., Digford, R. L., and Wilke, C. R., *Mass Transfer*, McGraw-Hill, New York, 1975.
4. Highbie, R., *Trans. AIChE*, *31*, 365 (1935).
5. Danckwerts, P. V., *Ind. Eng. Chem.*, *23*(6), 1460 (1951).
6. Lamb, W. B., Springer, T. G., and Digford, R. L., *Ind. Eng. Chem. Fund.*, *8*, 823 (1969).
7. Hanratty, T. J., *AIChE J.*, *2*(3), 259 (1956).
8. Toor, H. L., and Marchello, J. M., *AIChE J.*, *4*(1), 97 (1958).
9. Marchello, J. M., and Toor, H. L., *Ind. Eng. Chem. Fund.*, *2*, 8 (1963).
10. Perlmutter, D. D., *Chem. Eng. Sci.*, *16*, 287 (1961).
11. Pinczewski, W. V., and Sideman, S., *Chem. Eng. Sci.*, *29*, 1969 (1974).
12. Rozenshtok, Yu., *Heat and Mass Transfer*, vol. 1, Minsk, 1965. (In Russian.)
13. Malyusov, V. A., Malofeev, N. A., and Zhavoronkov, N. M., *Khim. Prom. (Moscow)*, no. 4, 10 (1953).
14. Popov, D. M., Candidate's dissertation, Moscow Institute of Chemical Technology, 1961. (In Russian.)
15. "Tray efficiencies in distillation columns," Final report of Michigan AIChE, New York 1958.
16. "Tray efficiencies in distillation columns," Final report of Michigan AIChE, New York 1960.
17. DeGoederen, C. N. I., *Chem. Eng. Sci.*, *20*(12), 115 (1965).
18. Sherwood, T. K., and Goodgame, T. K., *Chem. Eng. Sci.*, *3*, 37 (1954).
19. Narman, S. S., Gacolos, T., Frese, A. S., and Sutaliffe, D. M., *Trans. Inst. Chem. Eng.*, *41*(2), 61 (1963).
20. Edwards, D. K., Denny, V. E., and Mills, A. F., *Transfer Processes*, McGraw-Hill, New York, 1973.
21. Kafarov, V. V., *Osnovy massoperedachi*, Izdateltsvo Vysshaya Shkola, Moscow, 1962. (In Russian.)
22. Ramm, V. M., *Absorption of gases*, Chimia, Moscow, 1966. (In Russian.)
23. Akselrod, L. S., Doctoral dissertation, Moscow Institute of Chemical Engineering, 1958. (In Russian.)
24. Levich, V. G., *Physicochemical Hydrodynamics*, Prentice-Hall, Englewood Cliffs, N.J., 1962.
25. Kochergin, N. A., Olevsky, V. M., and Dilman, V. V., *Khim. Prom. (Moscow)*, no. 7, 59 (1960); no. 8, 567 (1961).
26. Planovsky, A. N., Artomonov, D. S., and Chekhov, O. S., *Khim. Prom. (Moscow)*, no. 1, 13 (1960).
27. Planovsky, A. N., Solomakha, G. P., Florea, O., and Surukhanov, A. V., *Khim. Prom. (Moscow)*, no. 2, 129 (1963).
28. Surukhanov, A. V., and Planovsky, A. N., *Khim. Prom. (Moscow)*, no. 4, 49 (1964).
29. Radionov, A. E., Kashnikov, A. M., and Radikovsky, B. M., *Tr. Mosk. Khim.-Tekhnol. Inst.*, *47*, 5 (1964).
30. Solomakha, G. P., Rudevich, G. A., and Nikolaev, P. I., *Theor. Found. Chem. Eng. (USSR)*, *5* (1968).
31. Danilychev, I. A., Planovsky, A. N., and Chekhov, O. S., *Khim. Prom. Moscow*, no. 6, 461 (1964).
32. Danilychev, I. A., Candidate's dissertation, Moscow Institute of Chemical Engineering, 1965. (In Russian.)

33. Rudov, G. Ya., and Chekhov, O. S., *Tr. Mosk. Inst. Khim. Mashinostr.*, *3*, 16 (1968).
34. Zverev, K. G., Candidate's dissertation, Moscow Institute of Chemical Engineers, 1968. (In Russian.)
35. Kasatkin, A. G., Popov, D. P., and Dytnersky, Yu. M., *Khim. Prom. (Moscow)*, no. 4, 49 (1964).
36. Dytnersky, Yu. M., Kasatkin, A. G., and Khalpanov, L. P., *Zh. Fiz. Khim.*, *39*, 1 (1966).
37. Ivanov, V. A., Candidate's dissertation, Moscow Institute of Chemical Engineering, 1968. (In Russian.)
38. Hobler, T., *Mass Transfer and Absorbers*, Pergamon Press, London, 1966.
39. Kusminykh, I. N., and Koval, L. A., *Zh. Fiz. Khim.*, *28*, 1 (1955).
40. Kusminykh, I. N., Akselrod, L. S., Koval, J. A., and Radionov, A. I., *Zh. Fiz. Khim.*, *27*, 2 (1954).
41. Kusminykh, I. N., *Khim. Prom. (Moscow)*, no. 8, 228 (1950).
42. Walter, I. N., and Sherwood, T. K., *Ind. Eng. Chem.*, *33*, 493 (1955).
43. Artomonov, D. S., Candidate's dissertation, Moscow Institute of Chemical Engineering, 1964. (In Russian.)
44. Koval, J. A., Candidate's dissertation, Moscow Institute of Chemical Engineering, 1951. (In Russian.)
45. Chekhov, O. S., Doctoral dissertation, Moscow Institute of Chemical Engineering, 1969. (In Russian.)
46. Gaster, I. A., Bonnet, W. F., and Hess, I., *Chem. Eng. Prog.*, *44*(10), 247 (1952).
47. Foss, A. S., and Gester, I. A., *Chem. Eng. Prog.*, *52*(1), 28 (1956).
48. West, F. S., and Gilbert, I. A., Shiizu, *Ind. Eng. Chem.*, *45*, 358 (1953).
49. Rudevich, G. A., Candidate dissertation, Moscow Institute of Chemical Engineering, 1968. (In Russian.)
50. Griffith, P. M., *Chem. Eng. Sci.*, *12*, 128 (1960).
51. Ruckenstein, E., *Rev. Chim. (Bucharest)*, *10*, 500 (1959).
52. Kruzhilin, G. I., *Zh. Tekh. Fiz.*, *6*, 561 (1936).
53. Boltze, E., Thesis, Göttingen University, 1908.
54. Froessling, N., *Beitr. Geophys.*, *32*, 170 (1938).
55. Lochiel, A. C., Thesis, University of Edinburgh, 1963.
56. Friedlander, S. K., *AIChE J.*, *7*, 317 (1961).
57. Calderbank, P. H., and Moo-Young, M. B., *Chem. Eng. Sci.*, *16*, 39 (1962).
58. Ward, D. M., Trass, O., and Tonson, A. I., *Can. J. Chem. Eng.*, *40*, 164 (1962).
59. Froessling, N., *Acta Univ. Zur. NFAVD*, *36*, 4 (1940).
60. Hadamard, G., *C.R. Acad. Sci.*, *152*, 1735 (1911).
61. Hammerton, D., and Garner, F. H., *Trans. Inst. Chem. Eng.*, *32*, 518 (1954).
62. Chao, B. T., *Phys. Fluids*, *5*, 69 (1962).
63. West, F. B., *Ind. Eng. Chem.*, *43*, 234 (1951).
64. Che, H. Y., and Tobias, Ch. W., *Ind. Eng. Chem. Fund.*, *7*, 1 (1968).
65. Bowman, C. W., and Johnson, A. I., *Can. J. Chem. Eng.*, *40*, 139 (1962).
66. Zahm, A. F., *NACA Rep.*, *153*, 517 (1926).
67. Rosenberg, B., Rep. no. 727, David Taylor Naval Ship R & D Center, Bethesda, Md., 1950.
68. Takadi, T., and Maeda, S., *Chem. Eng. Tokyo*, *25*, 254 (1961).
69. Calderbank, P. H., and Lochiel, A. C., *Chem. Eng. Sci.*, *19*, 485 (1964).
70. Calderbank, P. H., Johnson, D. S. L., and London, J., *Chem. Eng. Sci.*, *25*, 235 (1970).
71. Levich, V. G., Krylov, V. S., and Vorotilin, V. P., *Dokl. Acad. Nauk SSSR*, *161*, 3 (1965).
72. Shebalin, K. N., *Zh. Fiz. Khim.*, *1*, 39 (1939).

73. Baird, M. H. I., and Davidson, J. F., *Chem. Eng. Sci., 17,* 87 (1962).
74. Deindorfer, F. H., and Humphey, A. J., *Ind. Eng. Chem., 53*(9), 755 (1961).
75. Garner, F. H., *Chem.-Ing.-Tech., 29*(1), 28 (1958).
76. Garner, F. H., and Porter, K. E., *Proc. Int. Symp. Distill. 1960, Inst. Chem. Eng.,* p. 43 (1960).
77. Coppock, P. D., and Meilleyohon, *Trans. Inst. Chem. Eng., 29,* 75 (1951).
78. Datta, R. L., Napier, D. N., and Newitt, D. W., *Trans. Inst. Chem. Eng., 28,* 14 (1950).
79. Guyer, A., and Pfister, X., *Helv. Chim. Acta, 29,* 1173 (1964).
80. Leonard, J. H., and Houghton, G., *Chem. Eng. Sci., 18,* 133 (1963).
81. Sherwood, T. K., Evans, F., and Longcor, J. V. A., *Ind. Eng. Chem., 31,* 1146 (1939).
82. Calderbank, P. H., and Patza, R. P., *Chem. Eng., 21,* 719 (1966).
83. Garbarini, G. R., and Tien, C., *Can. J. Chem. Eng., 47,* 35 (1963).
84. Shirotsuka, T., and Hirata, A., *Kagaku Kogaku (Chem. Eng. Japan), 35,* 78 (1971).
85. Kikuchia, A., Tadaki, T., and Maeda, S., *Symp. Soc. Chem. Eng. Jap.,* 1967.
86. Raymond, D. R., and Zieminski, S. A., *AIChE J., 17,* 57 (1971).
87. Hamielec, A. E., Johnson, A. I., and Houghton, W. T., *AIChE J., 13,* 220 (1967).
88. Koide, K., Ori, Y., and Hara, Y., *Chem. Eng. Sci., 29,* 417 (1974).
89. Calderbank, P. H., *Trans. Inst. Chem. Eng., 36,* 443 (1958).
90. Calderbank, P. H., Evans, F., and Rennie, J., (Holland ed.), *Proc. Int. Symp. Distill. 1960, Inst. Chem. Eng.,* p. 51 (1960).
91. Calderbank, P. H., and Moo-Young, M. B., *Proc. Int. Symp. Distill. 1960, Inst. Chem. Eng.,* p. 59 (1960).
92. Calderbank, P. H., and Rennie, F., *Trans. Inst. Chem. Eng., 40,* 3 (1962).
93. Florea, O., Candidate's dissertation, Moscow Institute of Chemical Engineering, 1963. (In Russian.)
94. Lamont, J. C., and Scott, D. S., *AIChE J., 4,* 513 (1970).
95. Harriott, P., *AIChE J., 17,* 149 (1962).
96. Perlmutter, D. O., *Chem. Eng. Sci., 16,* 287 (1961).
97. King, C. J., *Ind. Eng. Chem. Fund., 5,* 1 (1966).
98. King, C. J., and Kozinski, A. A., *AIChE J., 12,* 109 (1961).
99. Harriott, P., *Chem. Eng. Sci., 8,* 93 (1962).
100. Ruckenstein, E., *Chem. Eng. Sci., 21,* 133 (1965).
101. Middleman, S., *AIChE J., 11,* 750 (1965).
102. Chan, Wing-Cheng, Ph.D. thesis, University of Minnesota, 1965.
103. Davies, I. T., *Proc. R. Soc., A290,* 515 (1966).
104. Filatov, L. F., Candidate's dissertation, Moscow Institute of Chemical Engineering, 1969. (In Russian.)

8

Mass transfer in liquid–solid particle systems

8.1 Introduction

Heterogeneous processes, taking place in turbulent flows, occur widely in chemical engineering, and the mass transfer in such processes often takes place, typically, on the surface of active particles when they are dispersed and suspended in a continuous phase. Some examples are the combustion of coal dust blown into a furnace by a stream of air, the transfer of mass from a solution to a growing crystal, the dissolution of fine particles in equipment with agitators, mass transfer in microheterogeneous systems in microbiological processing industries, and so on.

For the determination and quantitative evaluation of the influence of the various parameters that are present in such systems, we need to know the rates of both the chemical transformation and the physical transfer of matter, as well as the effects of interactions between the two processes. Progress in the study of mass transfer, and of heat and momentum transfer (some of which is presented in earlier chapters), allows us to develop a quantitative description of such processes.

In accordance with the objectives of this chapter, we shall limit our discussion to the problem of mass transfer to and from the surfaces of solid particles in turbulent liquid. The nature of such a process involves, of course, both the hydrodynamics of the flow and the character of the motion of discrete particles. The empirical equation

$$m = k(c_s - c_0)At \tag{8.1}$$

was obtained[1] in an investigation of the dissolution of solid particles. This equation gives a relationship between transferred mass m and an overall mass-transfer coefficient k, interface surface area A of the dissolving particle, driving force $\Delta c = c_s - c_0$ (c_s is the concentration of the saturated solution and c_0 is the actual concentration of the solution at a given instant), and time t.

Further studies[2] have shown that k is proportional to the diffusivity D of the substance in the liquid, and consequently equation (8.1) may be rewritten in the form

$$m = D\frac{c_s - c_0}{\delta}At \tag{8.2}$$

According to this theory (which represents a theoretically sound expression for diffusional flow in a stagnant liquid, and which should be regarded as purely empirical when applied to a liquid in motion), there is a thin layer, thickness δ, of stationary liquid immediately adjacent to the surface of the solid particle, in which molecular diffusion takes place. Equation (8.2), then, should be interpreted as relating the rate of mass transfer to the concentration gradient $(c_s - c_0)$, for either a flowing or a stagnant liquid. From the theory alone one cannot calculate the thickness of diffusion layer δ (nor the mass-transfer coefficient) because it depends on the liquid velocity, but the relationship between this thickness δ and the free stream (or characteristic) liquid velocity U is experimentally determined to be

$$\delta \sim 1/U^n$$

where the exponent n has been found to depend on the experimental conditions, ranging from $n = 0.5$ to $n = 1$.

Another theory of mass transfer, generally accepted in chemical engineering technology, is the film theory.[3,4] This theory postulates the existence of a thin film near the interface between the moving phases, the properties of this film differing from those of the bulk of the moving phases, and in which there is no relative motion between the phases. Thus the film at a solid particle–liquid interface is static just as in the first theory, and as before, transport of matter within the film depends solely on molecular diffusion, whereas turbulence of the flow only influences the size of the transfer region, or in other words, the *effective film thickness*. The overall mass-transfer coefficient k in equation (8.1) may then be interpreted as the velocity of molecular diffusion in the stationary layer, and will be given by

$$k = D/\delta \tag{8.3}$$

Now, the concentration distribution in the boundary layer and mass flux at the active surface of the particle are determined by the equation of (steady) convective diffusion:[5]

$$u\frac{\partial c}{\partial x} + v\frac{\partial c}{\partial y} = D\frac{\partial^2 c}{\partial y^2} \tag{8.4}$$

where u and v are components of the flow velocity vector and are functions of Cartesian coordinates x and y (for the two-dimensional case), and c is the local concentration. On the right of equation (8.4) only the y derivative, which is normal to the surface, is given because mass transfer by molecular diffusion in the other directions is insignificant. This equation (or its three-dimensional form) gives us a starting point, for film models, in the study of mass transfer in solid particle–liquid systems.

Another approach to the study of mass transfer was that made by Higbie,[6] in which he proposed the *penetration* (or *surface-renewal*) *theory*. According to this theory, mass transfer was mostly accomplished by the periodic motion of eddies from the bulk of the flow to the interface, this transfer being augmented by periods of unstable molecular diffusion, of short duration, at the boundary. Supposing that all eddies reaching the interface have equal residence times, Higbie obtained the following relationship for calculation of the mass-transfer coefficient:

$$k = 2(D/\pi\tau)^{1/2} \tag{8.5}$$

where τ is the time of contact of each eddy with the interface. It follows from equation (8.5) that k is proportional to the square root of the diffusivity D for Higbie's model. This penetration theory received further development in other studies,[7,8] in which it was postulated that eddies stay at the interface for different times so that, for example, for a short time of contact, mass transfer from one phase to another takes place through molecular diffusion or, under other circumstances, not only by molecular diffusion but also by turbulent diffusion. There is a fundamental similarity between the penetration and film theories[8] even though they indicate different relationships between k and D, as indicated in equations (8.3) and (8.5), because the difference consists only in the choice of coordinate systems. Higbie's model (and modified theories) uses a coordinate system fixed to a liquid element, the eddy, and the film theory uses a coordinate system fixed in space.

Much work has been done on problems associated with mass transfer to the surface of a single moving sphere, but the problem of mass transfer from solid spherical particles has only been solved in some limited cases, because of difficulties in analytically modeling the hydrodynamic conditions. Various studies[9,10] have obtained equations for calculation of a mass-transfer coefficient in the Stokes flow region, for small values of the Peclet number Pe, and also[5,11] for $\mathrm{Pe} \gg 1$. In this latter range, for $\mathrm{Pe} \gg 1$, the solution involved a quite elaborate integral method[11,13] yielding equations that practically coincide with those obtained using a differential method.[14] Other investigations of[13,15] the transport phenomenon in the Reynolds number range $1 \ll$

$\mathrm{Re} \ll 10^4$ using an integral method have obtained, for the diffusional mass $\dot{m}$ on the surface of a sphere, per unit time, the following equations:

1. $\mathrm{Pe} \gg 1$; $\mathrm{Re} \ll 1$,

$$\dot{m} = 8c_0 D^{2/3} U^{1/3} r_p^{4/3}$$

2. $\mathrm{Pe} \ll 1$; $\mathrm{Re} \ll 1$,

$$\dot{m} = 4\pi D c_0 r_p (1 + 0.64\mathrm{Pe}^{1/3})$$

3. $\mathrm{Re} \gg 1$,

$$\dot{m} = 4\pi D^{2/3} U^{1/2} r_p^{3/2} \frac{n c_0}{\nu_f^{1/6}}; \qquad n = 0.5 \text{ to } 1.0$$

Table 8.1. *Effect of particle diameter d_p, viscosity ν_f, and diffusivity D on the mass-transfer coefficient k*

Reference	Particle diameter, d_p: n in $k \sim d_p^n$	Kinematic viscosity, ν_f: n in $k \sim \nu_f^n$	Diffusivity, D: n in $k \sim D^n$
16	—	−0.12 to −0.90	0.45
17	0.26	—	
18	—	−0.50	0.56
19	0.0	−0.17	0.50
20	—	−0.42	0.67
21	0.05 to −0.13	—	—
22	0.0	−0.33	0.50
23	—	0.09	0.50
24	3.08	—	0.45 to 0.95
25	—	−0.37	0.77
26	0.0	−0.42	0.67
27	−0.70 small d_p −0.10 large d_p	−0.22 large d_p −0.06 small d_p	0.60 to 0.80 (depends on d_p)
28	0.03	—	—
29	0.33	—	—
30	—	0.12	0.50
31	0.15	−0.07	0.92
32	−0.19	—	—
33	—	—	0.94
34	−0.17[a] −0.5[b]	−0.26[a] −0.12[b]	0.64[a] 0.62[b]
41	—	−0.21	0.50
42	—	−0.21	0.17

[a]Difference of density close to that for neutral buoyancy.
[b]Significant difference of density exists.

where c_0 is the concentration in the solution (the driving force); U, the flow characteristic velocity, can be set equal to the relative velocity v_r of the particle; r_p is the particle radius; D is the diffusivity; and ν_f is the liquid kinematic viscosity.

8.2 Parameters used for calculation of the mass-transfer coefficient

A survey of the literature on mass transfer in solid particle–liquid systems shows an extremely wide divergence of results and correlations, as is clearly illustrated in Table 8.1, showing effects of particle diameter d_p, kinematic viscosity ν_f, and diffusivity D on the mass-transfer coefficient k and Table 8.2, showing the effect of impeller speed n, impeller diameter d_m, and power

Table 8.2. *Effect of impeller speed n, impeller diameter d_m, and power dissipation ϵ on the mass-transfer coefficient k*

Reference	u: unbaffled; b: baffled	Impeller speed, n: m in $k \sim n^m$	Impeller[a] diameter d_m: n in $k \sim d_m^n$	Power/unit mass, ϵ: n in $k \sim \epsilon^n$
16	u	0.62 to 1.40	0.24 to 1.80	—
18	u	1.00	1.00	—
20	b	0.75	—	—
21	b	0.39 to 0.45	0.39 to 0.45	—
22	b	0.83	0.67	—
23	b	0.59	1.18	—
25	b	0.70	1.40	—
26	b and u	0.75	0.50	0.25
27	b and u	0.30 to 0.50	0.20 to 0.33	0.10 to 0.15
28	u	0.80	—	0.26
30	b	0.38	—	0.13
31	b	0.15	—	—
34	b	0.32	0.17	0.21
35	b	0.60	0.20	—
36	b	0.58 propeller 0.87 turbine	0.16 propeller 0.75 turbine	—
37	u	0.20 to 0.67	0.20 to 0.38	—
38	b	0.20 to 0.38	—	—
39	b	0.27 to 1.22	—	—
40	b	0.63	—	0.23
41	b	0.71	1.42	—
42	b	0.54	1.08	—

[a]Valid only for geometrically similar systems.

dissipation ϵ in agitators. The reported effects[16–42] of these variables, measured or estimated, on the mass-transfer coefficient testifies to the necessity of carrying out a more complete theoretical investigation, in order to obtain a general equation that can hopefully resolve some of the uncertainties in this area of two-phase flow technology.

A very large number of published studies have reported measurements of mass transfer that were correlated by an empirical equation of the form[16,22,23,25,36,41–44]

$$\mathrm{Sh} = a\mathrm{Re}^n\mathrm{Sc}^m \tag{8.6}$$

or by a semiempirical equation[26–28,30,45–51]

$$\mathrm{Sh} = 2 + b\mathrm{Re}^{0.5}\mathrm{Sc}^{0.33} \tag{8.7}$$

where Sh $= kl/D$ is the Sherwood number, Re $= Ul/\nu_f$ the Reynolds number, and Sc $= \nu_f/D$ the Schmidt number (and a, n, and m are constants, derived from experimental data; ν_f the kinematic viscosity; D the diffusivity; k the mass-transfer coefficient related to d_p, the particle diameter; l a length representative of a known contact surface area; and U is, for example, the relative velocity v_r or the slip velocity). For liquids the recommended value of constant b is 0.76 (but it ranges from 0.3 to 1.0).

It is now generally understood that mass transfer is governed by the environment near the particle surface and that a characteristic Reynolds number should be based on the properties of the particle rather than of the flow as a whole. Of primary consideration in applying equations (8.6) and (8.7) is obtaining a representative value for the slip-velocity term appearing in the Reynolds number, and this demands a fundamental knowledge of the structure of turbulence existing in the solid particle–liquid system. However, as an easier alternative,[22] the terminal velocity of the particle falling under the influence of gravity can be adopted[22] as the velocity to be used in the Reynolds number, in order to obtain the mass-transfer coefficient k in these expressions.

In processing and comparing data with equation (8.7), it is found[47] that the value b cannot be considered as a constant, and the following equation was proposed for its determination:

$$b = 0.439 + 0.1807 d_p^{1/2} + 0.234\alpha(\alpha + 0.05)\mathrm{Re}^{1/2} \tag{8.8}$$

Here d_p (in.) is the particle diameter, $\alpha = \overline{u}/v_r$ is the longitudinal turbulence intensity (where $\overline{u}$ is the root-mean-square fluctuating velocity, v_r is the particle relative velocity), and[52] we can take $\alpha \simeq 0.33$. Strictly speaking, equation (8.7) is only correct for calculating mass-transfer coefficients for fixed solid particles in a flowing liquid or settling particles in a still liquid,

cosity for water at 25°C, and d_p the particle diameter. Equations (8.9) and (8.10) were derived for Kolmogoroff's *inertial subrange,* in which[65] the Reynolds number $Re_\lambda = v_e\lambda/\nu_f$ [where v_e (m/s) is the root-mean-square liquid velocity, and λ (m) is the lateral microscale of turbulence] must be above 1500. Let us consider how closely the inertial subrange model approaches a real flow in a stirred vessel, outside the immediate impeller stream region. Following Kolmogoroff's theory, in this region,

$$v_e \sim (\epsilon\lambda)^{1/3}, \qquad \eta \ll \lambda \ll l \tag{8.11}$$

which is equivalent to using

$$\beta\,(\lambda/\eta) \sim (\lambda/\eta)^{1/3}$$

in the equation

$$\frac{v_e}{v} = \beta(\lambda/\eta)$$

where

$$\eta = (\nu_f^3/\epsilon)^{1/4}$$

and

$$v = (\nu_f\epsilon)^{1/4} \tag{8.12}$$

are defined from dimensional reasoning. Here v_e (m/s) is the root-mean-square relative velocity, λ (meters) is now the local turbulence scale, η (meters) the Kolmogoroff length scale, l (meters) the macroscale of turbulence, v (m/s) the velocity scale of turbulence, ν_f the kinematic viscosity, and β a universal function.

The Reynolds number

$$Re_\lambda = v_e\lambda/\nu_f$$

for the inertial subrange in a real flow may be calculated from the microscale of turbulence,[66]

$$\lambda^2 = 15\nu_f v_e^2/\epsilon$$

and typical experimental values, obtained for low-viscosity liquids at moderate agitation conditions, in the flow outside an impeller stream[34] are $v_e \sim 0.12$ m/s (0.39 ft/s), $\epsilon \sim 0.15$ m^2/s^3 (1.6 ft^2/s^3), $\nu_f \sim 10^{-6}$ m^2/s (10^{-5} ft^2/s), resulting in

$$Re_\lambda = v_e^2(15/\epsilon\nu_f)^{1/2} = 144$$

This value of the microscale Reynold's number is far below the requirement of 1500, and consequently the validity of equation (8.11) appears in doubt. This discrepency is increased when, in deriving equations (8.9) and (8.10) for the relative velocity, equation (8.12), which refers to a velocity scale, is used instead of

$$v_e \sim \lambda(\epsilon/\nu_f)^{1/2}, \qquad \lambda \ll \eta$$

The velocity scale thus cannot be correct, because it predicts a nonzero relative velocity even when $\lambda \to 0$.

Now the theoretical approach described above was based on an assumption of isotropic turbulence, but experimental evidence indicates that the turbulence occurring in a vessel with an impeller is far from isotropic. Thus it is found[67] that the local power input varies from 0.26 to 70 times the mean power input, and also a particle traveling about in an agitated vessel is subjected to about an order of magnitude in velocity range. Nonisotropic turbulence has been illustrated still further by measurements of the dissipative energy distribution throughout the volume of the vessel, where it was found that approximately 20 percent of all input energy was dissipated in the immediate vicinity of the impeller, about 50 percent in the flow generated by the impeller, and only 30 percent in the remaining volume of the vessel. The local energy dissipation may differ from the mean dissipation $\bar{\epsilon}$ by two orders of magnitude, and hence the average power input per unit mass approach is not a completely satisfactory method of calculating mass-transfer coefficients in solid particles–liquid systems.

Statistical analysis of a large amount of experimental data obtained with a wide variety of impeller types[34] has established that the variation of the mass-transfer coefficient with agitation conditions is best described by a correlation of the form

$$k = f(\bar{\epsilon}, D/d_m)$$

where d_m is here a typical flow system dimension. However, at the present time, there is no satisfactory theoretical or semiempirical correlation for the calculation of the mass-transfer coefficient of solid particles suspended in a turbulent liquid.

8.3 Mass transfer to a microparticle suspended in a turbulent flow

Mass transfer from (or to) solid particles to (or from) a liquid is of concern in many areas of engineering but is probably more important in conjunction with chemical processes than in any other single area. Many problems of mass transfer from the liquid phase to solid particles are too complicated to permit analytical solution, this being especially due to the mathematical difficulties associated with properly allowing for the effect of hydrodynamic factors on the mass transfer. However, some approximate models may be used in an attempt to predict mass transfer.

In our analysis we will use the model of the diffusion boundary layer, which will permit us, by use of the physically well-founded simplifications involved in the solution of the equations of convective diffusion, to overcome certain of the mathematical difficulties. We will examine mass transfer from solid particles to a moving liquid with a low particle concentration, and with constant diffusivity and constant temperature, and we will call such a mixture a solution. For completeness we start from first principles and consider the flow through a control volume.

Let a Cartesian system of coordinates x, y, z be defined within a moving liquid and consider the mass flowing through an infinitesimal control volume $dV = dx\, dy\, dx$. We assume the fluid to be incompressible, and resolve the arbitrary velocity vector u_i into components u, v, w parallel to the x, y, z directions, respectively. If we denote the concentration of a solute by c, then, according to Fick's first law, the influx of mass into the control volume, parallel to the x direction, is

$$-D \frac{\partial c}{\partial x}\, dy\, dz\, dt$$

where D is the diffusivity; the outflow in the same direction is

$$-D \frac{\partial}{\partial x}\left(c + \frac{\partial c}{\partial x}\, dx \right) dy\, dz\, dt$$

Thus the net outflow of mass parallel to the x direction is the difference of these quantities:

$$-D \frac{\partial^2 c}{\partial x^2}\, dx\, dy\, dz\, dt$$

Similar expressions are easily obtained for the net outflow in the y and z directions,

$$-D \frac{\partial^2 c}{\partial y^2}\, dx\, dy\, dz\, dt$$

and

$$-D \frac{\partial^2 c}{\partial z^2}\, dx\, dy\, dz\, dt$$

and, by addition, we obtain the net outflow of mass as a result of diffusion from and to the control volume in the time interval dt:

$$-D \left(\frac{\partial^2 c}{\partial x^2} + \frac{\partial^2 c}{\partial y^2} + \frac{\partial^2 c}{\partial z^2} \right) dx\, dy\, dz\, dt \tag{8.13}$$

The change in concentration due to convection may be written as

$$\frac{dc}{dt} = \frac{\partial c}{\partial t} + u\frac{\partial c}{\partial x} + v\frac{\partial c}{\partial y} + w\frac{\partial c}{\partial z}$$

and hence mass transfer due to convection during time dt is

$$\frac{dc}{dt}\,dx\,dy\,dz\,dt \tag{8.14}$$

From expressions (8.13) and (8.14), we obtain

$$\frac{dc}{dt}\,dx\,dy\,dz\,dt = D\left(\frac{\partial^2 c}{\partial x^2} + \frac{\partial^2 c}{\partial y^2} + \frac{\partial^2 c}{\partial z^2}\right)dx\,dy\,dz\,dt$$

Thus the amount of solute brought into the differential volume by convection is equal to the amount transferred by diffusion (this corresponding to the law of conservation of mass), and consequently the general equation of convective diffusion, in terms of Cartesian coordinates, may be written as follows:

$$\frac{\partial c}{\partial t} + u\frac{\partial c}{\partial x} + v\frac{\partial c}{\partial y} + w\frac{\partial c}{\partial z} = D\left(\frac{\partial^2 c}{\partial x^2} + \frac{\partial^2 c}{\partial y^2} + \frac{\partial^2 c}{\partial z^2}\right) \tag{8.15}$$

or, in Cartesian tensor form,

$$\frac{\partial c}{\partial t} + u_i\frac{\partial c}{\partial x_i} = D\frac{\partial^2 c}{\partial x_k\,\partial x_k}$$

Equation (8.15) alone is insufficient for determination of the concentration distribution $c(x, y, z, t)$; we must also know the boundary and initial conditions to be satisfied. Assuming steady state, so that the concentration distribution c is independent of time ($\partial c/\partial t = 0$), we can reduce equation (8.15) to the dimensionless form

$$U\frac{\partial C}{\partial X} + V\frac{\partial C}{\partial Y} + W\frac{\partial C}{\partial z} = \frac{1}{\mathrm{Pe}}\left(\frac{\partial^2 C}{\partial X^2} + \frac{\partial^2 C}{\partial Y^2} + \frac{\partial^2 C}{\partial Z^2}\right) \tag{8.16}$$

Here $U = u/U_0$, $V = v/U_0$, and $W = w/U_0$ are dimensionless velocity components; $X = x/l$, $Y = y/l$, and $Z = z/l$ are dimensionless coordinates; and $C = c/c_0$ is the dimensionless concentration, where U_0 is a typical flow velocity, l a typical linear dimension and c_0 the (constant) concentration in the bulk of the solution, and $\mathrm{Pe} = U_0 l/D$ is the Peclet number. Note that as all the dimensionless terms in equation (8.16) are, in general, of order unity, the relationship between convective [left side of equation (8.16)] and diffusional (right side) transfer of matter is now described by a single dimensionless quantity, the Peclet number. Thus for a sufficiently low Pe number, for example small liquid velocities and small particle dimensions for a given diffusivity, mass transfer by convection is negligibly small.

Conversely, when the Peclet number is large, the concentration distribution is determined essentially by convective transfer, and molecular diffusion can be neglected (in regions far from the particle surface).

We will first restrict ourselves to analysis of mass transfer for a solid spherical particle moving in a laminar *creeping flow,* a flow phenomenon that has been treated successfully analytically, and the results of which may be combined with diffusion theory to permit us to obtain values for mass-transfer coefficients.[5,11,13,32] For very high Pe numbers (Pe = ReSc), usually occurring in solid particles–liquid systems, the liquid may be divided into two regions: the region of constant concentration, far from the particle surface (in other words, in the bulk liquid); and a region of rapidly changing concentration, in the immediate vicinity of the particle surface (or in the diffusion boundary layer).

Because the thickness δ of the diffusion boundary layer in the y direction is very small compared to the length l along the particle surface in the x direction, for $y \simeq \delta$ we have

$$\frac{\partial^2 c}{\partial y^2} \gg \frac{\partial^2 c}{\partial x^2} \tag{8.17}$$

and the two-dimensional equation for convective diffusion in a diffusion boundary layer takes the form

$$u\frac{\partial c}{\partial x} + v\frac{\partial c}{\partial y} = D\frac{\partial^2 c}{\partial y^2}$$

or, rewriting in spherical coordinates r and θ, as

$$v_r\frac{\partial c}{\partial r} + \frac{v_\theta}{r}\frac{\partial c}{\partial \theta} = D\left(\frac{\partial^2 c}{\partial r^2} + \frac{2}{r}\frac{\partial c}{\partial r}\right) \tag{8.18}$$

where v_r and v_θ are the radial and tangential velocity components. The angular dependence part of the Laplacian,

$$\frac{1}{r^2 \sin\theta}\frac{\partial}{\partial \theta}\left(\sin\theta\frac{\partial c}{\partial \theta}\right)$$

is omitted from the right side of equation (8.18), because the derivatives along the surface of the spherical particle are small compared to the derivatives along the radius vector [and this approximation corresponds to that in equation (8.17)].

At small distances from the surface of the spherical particle, the stream function ψ is given by[68]

$$\psi = -\frac{u_r}{2}\sin^2\theta\left(r^2 - \frac{3}{2}r_p r + \frac{1}{2}\frac{r_p^3}{r}\right) \simeq -\frac{3}{4}u_r y^2 \sin^2\theta$$

where u_r is the terminal velocity of the particle of radius r_p. Correspondingly, the tangential velocity is

$$u_\theta = -\frac{1}{r \sin\theta}\frac{\partial\psi}{\partial y} \simeq -\frac{1}{r_p \sin\theta}\frac{\partial\psi}{\partial y} = \frac{3}{2}u_r \frac{y}{r_p}\sin\theta \qquad (8.19)$$

Now, noting that for small values of y ($y \ll r_p$),

$$\frac{\partial^2 c}{\partial y^2} \gg \frac{2}{r_p}\frac{\partial c}{\partial y}$$

and taking into account equation (8.19), we can rewrite the equation of convective diffusion in terms of the variables (ψ, θ) as

$$\frac{1}{r_p}\frac{\partial c}{\partial\theta} = Dr_p \sin\theta \frac{\partial}{\partial\psi}\left[(r_p \sin\theta\, v_\theta)\frac{\partial c}{\partial\psi}\right]$$

Expressing v_θ in terms of ψ, we find

$$\frac{\partial c(\psi, \theta)}{\partial\theta} = Dr_p^2 \sin^2\theta\,(3u_r)^{1/2}\frac{\partial}{\partial\psi}\left[(-\psi)^{1/2}\frac{\partial c}{\partial\psi}\right] \qquad (8.20)$$

To solve equation (8.20), a set of boundary conditions is required, and we have

$$c = 0 \qquad \text{at } \psi = 0 \text{ (on the surface of the particle)} \qquad (8.21)$$

$$c = c_0 \qquad \text{for } \psi \to -\infty \text{ (far from particle)} \qquad (8.22)$$

$$c = c_0 \qquad \text{for } \theta = 0,\quad \psi = 0 \qquad (8.23)$$

The point $\theta = 0$, $\psi = 0$ is the point of incidence or the stagnation point of the liquid flow on the spherical particle, and at this point the flow is not depleted by diffusion, its concentration being the same as the concentration in the bulk of the liquid.

Introducing a new variable,

$$\xi \equiv Dr_p^2(3u_r)^{1/2}\int \sin^2\theta\, d\theta = \frac{Dr_p^2(3u_r)^{1/2}}{2}\left(\theta - \frac{\sin 2\theta}{2}\right) + c_1$$

(where c_1 is an integration constant), we can rewrite equation (8.20) as

$$\frac{\partial c}{\partial\xi} = \frac{\partial}{\partial\psi}\left[(-\psi)^{1/2}\frac{\partial c}{\partial\psi}\right] \qquad (8.24)$$

For solving equation (8.24) a similarity method can be used[69] such that

$$\psi \to \alpha\psi' \qquad (8.25)$$

$$\xi \to \beta\xi'$$

provided that the arbitrary changes in the scale of the variables are linked by the relationship

$$\beta = \alpha^{2/3}$$

and that transforming the variables does not require a change in boundary conditions (8.21) through (8.23). In other words,

$$c(\psi, \xi) = c(\alpha\psi, \alpha^{2/3}\xi)$$

A solution $c(\psi, \xi)$ will be sought such that a combination of variables ψ and ξ remains unchanged by transformations (8.25).

Now, consider such a combination of independent variables ψ and ξ (or ψ and θ) in the following form:

$$\eta = \frac{-\psi}{\xi^{2/3}} \tag{8.26}$$

We then have

$$\frac{\partial c}{\partial \xi} = \frac{2\psi}{3\xi^{5/3}} \frac{dc(\eta)}{d\eta} \tag{8.27}$$

$$\frac{\partial c}{\partial \psi} = \frac{-1}{\xi^{2/3}} \frac{dc(\eta)}{d\eta} \tag{8.28}$$

$$\frac{\partial}{\partial \psi}\left[(-\psi)^{1/2} \frac{\partial c}{\partial \psi}\right] = \frac{1}{\xi} \frac{d}{d\eta}\left[\eta^{1/2} \frac{dc(\eta)}{d\eta}\right] \tag{8.29}$$

Taking into account equations (8.26) through (8.29), equation (8.24) is transformed into an ordinary differential equation for c:

$$-\frac{2}{3}\eta \frac{dc}{d\eta} = \frac{d}{d\eta}\left(\eta^{1/2} \frac{dc}{d\eta}\right)$$

Introducing the variable $z \equiv \eta^{1/2}$ for convenience in integration we obtain, finally,

$$\frac{d^2c}{dz^2} + \frac{4}{3} z^2 \frac{dc}{dz} = 0$$

which, after integration, yields

$$c(z) = c_2 \int_0^z \exp\left(-\frac{4}{9} z^3\right) dz + c_3$$

where c_2 and c_3 are constants. Note that z is given by

$$z = \left(\frac{3u_r}{4}\right)^{1/2} \frac{y \sin\theta}{\left[Dr_p^2 \left(\frac{3u_r}{4}\right)^{1/2} \left(\theta - \frac{\sin 2\theta}{2} + c_1\right)\right]^{1/3}} \tag{8.30}$$

The constants c_1, c_2, and c_3 are determined from boundary conditions (8.21) to (8.23). Condition (8.21) gives $c_3 = 0$. From condition (8.22), we obtain

$$c_2 = \frac{c_0}{\int_0^\infty \exp\left(-\frac{4}{9}z^3\right) dz}$$

and the integral in this equation is given by a gamma function as

$$\int_0^\infty \exp\left(-\tfrac{4}{9}z^3\right) dz = \left(\tfrac{9}{4}\right)^{1/3}\Gamma\left(\tfrac{1}{3}\right) \simeq 1.15$$

Finally, in the region of the point of incidence of the flow on the particle, where $\theta = 0$, the concentration c must be a single-valued positive function of the angle θ, but at small values of θ we have

$$z \simeq \left(\frac{3u_r}{4}\right)^{1/2} \frac{y\theta}{[c_1 + \frac{20}{3}\theta^3 Dr_p^2(3u_r)^{1/2}]^{1/3}}$$

Hence for z to be a real, positive, and single-valued quantity over a range of small values of θ, c_1 must equal zero. Finally, we get

$$z = \left(\frac{3u_r}{4Dr_p^2}\right)^{1/3} \frac{y \sin\theta}{(\theta - \sin 2\theta/2)^{1/3}} \tag{8.31}$$

and

$$c = \frac{c_0}{1.15} \int_0^z \exp\left(-\tfrac{4}{9}z^3\right) dz \tag{8.32}$$

The mass diffusion to unit surface of the particle per unit time is then

$$\dot{m}' = D\left(\frac{\partial c}{\partial y}\right)_{y=0} = \frac{Dc_0}{1.15}\left(\frac{3u_r}{4Dr_p^2}\right)^{1/3} \frac{\sin\theta}{(\theta - \sin 2\theta/2)^{1/3}} \tag{8.33}$$

and is seen to be directly proportional to the concentration c_0 of the diffusing matter, to be an increasing function with u_r the velocity of the particle and with the diffusivity D, and a decreasing function of r_p, and to be a function of the angle θ. The latter function has the value $(2/\pi)^{1/3}$ for $\theta = \pi/2$ and is zero for $\theta = \pi$. For $\theta = 0$, equation (8.33) is indeterminate, but by resolving the indeterminacy we find that the function of angle θ has the value unity at $\theta = 0$. Thus the mass flux is highest at the point of incidence, $\theta = 0$, and decreases slowly with increasing θ, until it rapidly goes to zero at the downstream side of the particle. The corresponding effective thickness of the diffusion layer can also be obtained from equations (8.2) and (8.33) and is given by

$$\delta = \frac{1.15(\theta - \sin 2\theta/2)^{1/3}}{\sin\theta}\left(\frac{4Dr_p^2}{3u_r}\right)^{1/3} \tag{8.34}$$

This shows that δ increases with angle θ and that it becomes infinite for $\theta = \pi$. Actually, of course, δ does not become infinite at $\theta = \pi$, and also $\dot{m}'$ will not be zero. [Note that had we not set c_1 in equation (8.30) equal to zero, the value of δ would also be infinite, and the flux $\dot{m}'$ would be zero at the point of incidence ($\theta = 0$), which would obviously be incorrect.] The discrepancy between prediction and experimental results at $\theta = \pi$ is due to the inapplicability of the theory in this region.

Equation (8.34) shows that for a certain value of θ relatively close to π, the thickness of the diffusion layer attains a value comparable to that of the radius of the particle, and so in this range of θ the theory presented above is not applicable, because of the assumption in equation (8.17). It is clear, however, that in the range $\theta \simeq \pi$ there is no significant effect on the total mass transfer m to the particle. The total diffusion flow per unit time is then

$$\dot{m} = \int \dot{m}'\,dA = 2\pi r_p^2 \int_0^\pi \dot{m}' \sin\theta\,d\theta$$

$$= \frac{Dc_0 r_p^{4/3}}{1.15}\left(\frac{3u_r}{4D}\right)^{1/3} 2\pi \int_0^\pi \frac{\sin^2 d\theta}{(\theta - \sin 2\theta/2)^{1/3}}$$

and evaluation of this integral gives

$$\dot{m} = 7.98 c_0 D^{2/3} v_r^{1/3} r_p^{4/3} \tag{8.35}$$

where we have replaced the terminal velocity u_r with the relative velocity v_r.

From equation (8.35) we can determine the average diffusion to unit surface area of the particle,

$$\dot{m}' \simeq \frac{2}{\pi} D^{2/3}\left(\frac{v_r}{r_p^2}\right)^{1/3} c_0$$

This analysis shows that the total rate of mass transfer to a solid particle moving at a small velocity through a solution is thus proportional to c_0, $v_r^{1/3}$, and $r_p^{4/3}$.

Up to this point in our discussion of the diffusion flux to a solid particle, we have confined our treatment to the case of a laminar creeping flow. In actual practice, however, we almost always deal with a liquid in turbulent flow, and therefore the study of convective diffusion phenomena in a turbulent flow regime has practical importance. In this flow regime, which occurs usually as a result of an extremely energetic and random agitation, the concentration of substance in the bulk liquid is essentially constant. In the neighborhood of the interface, in the turbulent buffer layer, matter is transferred by turbulent eddies, and molecular diffusion does not play a significant role. The viscous sublayer, between the turbulent buffer layer and the interface, is a region where mass transfer takes place through both turbulent and molecular diffusion mechanisms. Because the scale of turbulent eddies becomes smaller as we approach the interface, transfer by molecular viscosity is greater than that due to the turbulent eddies near the interface, and in the immediate vicinity of the interface, in the diffusional boundary layer, the molecular diffusion mechanism completely dominates over the turbulent.

In Chapter 6, on the assumption of fully developed turbulence, we derived the relative velocity of particles suspended in a turbulent flow, and this permits us to compute [from equation (8.35)] the diffusion flux to the particle surface. Assuming that, to an order of magnitude, the diffusional flux to the particle is equal to diffusional flux to the surface of a motionless particle around which liquid moves at relative velocity v_r, we obtain the following equations for diffusion to the surface of a particle:

1. For a tubular apparatus (Section 6.7),

$$\dot{m} = \text{const.}\ D^{2/3} r_p^{4/3} \left(\frac{\nu_f}{l}\right)^{1/12} U^{1/4} \left[1 - \frac{V + V_a}{V(\rho_p/\rho_f) + V_a}\right]^{1/6} c_0 \qquad (8.36)$$

2. For a bubbling apparatus (Section 6.8),

$$\dot{m} = \text{const.}\ D^{2/3} r_p^{4/3} \left(\frac{\nu_f}{l}\right)^{1/12} \times \left[1 - \frac{V + V_a}{V(\rho_p/\rho_f) + V_a}\right]^{1/6} c_0 \left(\frac{\Delta P}{\rho_f}\right)^{1/8} \qquad (8.37)$$

3. For a mixer with agitator and baffles (Section 6.9),

$$\dot{m} = \text{const.}\ D^{2/3} r_p^{4/3} \left(\frac{\nu_f}{D_m}\right)^{1/12} \left[1 - \frac{V + V_a}{V(\rho_p/\rho_f) + V_a}\right]^{1/6} \times c_0 \left[\frac{nD_m^2}{(T^2 x_1 q^{0.2})^{1/3}}\right]^{1/4} \qquad (8.38)$$

and the symbols are defined in Chapter 6. In dimensionless form, equation (8.36) may then be written as

$$\frac{kd_p}{D} = \text{const.} \left(\frac{d_p}{l}\right)^{1/3} \left[1 - \frac{V + V_a}{V(\rho_p/\rho_f) + V_a}\right]^{1/6} \left(\frac{Ul}{\nu_f}\right)^{1/4} \left(\frac{\nu_f}{D}\right)^{1/3}$$

or

$$\mathrm{Sh} = \text{const.} \left(\frac{d_p}{l}\right)^{1/3} \left[1 - \frac{V + V_a}{V(\rho_p/\rho_f) + V_a}\right]^{1/6} \mathrm{Re}^{1/4}\mathrm{Sc}^{1/3} \quad (8.39)$$

For the case where the additional mass effect given by term V_a may be neglected, equation (8.39) may in turn be written as

$$\mathrm{Sh} = \text{const.} \left(\frac{d_p}{l}\right)^{1/3} \left(\frac{\Delta\rho}{\rho_p}\right)^{1/6} \mathrm{Re}^{1/4}\mathrm{Sc}^{1/3} \quad (8.40)$$

where $\mathrm{Sh} = kd_p/D$, $\mathrm{Sc} = \nu_f/D$, and $\mathrm{Re} = Ul/\nu_f$ are, respectively, the Sherwood, Schmidt, and Reynolds numbers for the total liquid flow.

The velocity U of large-scale eddies, for a bubbling apparatus, is obtained from equation 6.64, and the eddy scale l is comparable to the height of gas–liquid mixture. For an apparatus with a mixer, the value of U is calculated by equation 6.70, where l is taken to be equal to the diameter D_m of the mixer.

We can use experimental data[34] for verification of the theoretical equation (8.40). Small particles of anion-exchange resin beads, with sizes from 30.8 to 96.1 μm, suspended in a turbulent flow, were mixed with a stirrer and a mass-transfer coefficient was determined for reaction of the ion-exchange resins with aqueous acids. In Table 8.3, a comparison of experimental data[34] with a theoretical equation similar to equation (8.40) is given. The equation is rewritten for a mass-transfer coefficient k for solid particles suspended in agitated tanks:

$$k = \text{const.} \frac{D^{2/3}\,\nu_f^{1/12}}{d_p^{2/3}} \left(\frac{\Delta\rho}{\rho_p}\right)^{1/6} \left[\frac{nD_m^2}{(T^2h)^{1/3}}\right]^{1/4} \frac{1}{D_m^{1/12}}$$

where D is the diffusivity, ν_f the kinematic viscosity, d_p the particle diameter, ρ_p the particle density, $\Delta\rho$ the density difference, n the impeller speed (min^{-1}), T the tank diameter, and h the height of liquid in the tank. Note that the disagreement between experimental and calculated values of k is less than 10 percent in all cases.

8.4 Mass transfer to a macroparticle suspended in a turbulent flow

In many chemical industrial processes, the need arises for the quantitative evaluation of mass transfer in a turbulent flow for particles whose size is considerably greater than the internal scale of turbulence. Such transfer often occurs, for example, in the dissolution of particles, extraction from suspensions, and the heterogeneous reactions occurring on the surface of catalyst particles, and most experimental data are described by criterion equations applicable over a broad range of parameter variation. Many of the theoretical results that have been obtained are in disagreement with experimental data[23,27,35] on mass transfer in liquid–solid systems, and therefore there exists the need for a more comprehensive theoretical description of the phenomenon of mass transfer to a (large) solid particle suspended in a turbulent flow.

Because the size of the particle is much greater than the thickness of diffusion boundary layer, it is possible to reduce the problem of mass transfer to a macroparticle to the problem of diffusion to a flat surface across which there is a turbulent flow. In solving this problem, the hypothesis[5,70] of gradual damping of turbulence in the vicinity of the wall will be used.

Table 8.3. *Comparison of experimental and theoretical values of the mass-transfer coefficient k for solid particles*

Stirrer speed (rpm)	Particle diameter (μm)	Mass-transfer coefficient, k: Exp. ($\times 10^4$ m/s)	Mass-transfer coefficient, k: Calc. ($\times 10^4$ m/s)	$\frac{\text{Exp.} - \text{calc.}}{\text{Exp.}} \times 100$ (%)
692	30.8	5.70	5.88	3.1
692	30.8	5.50	5.88	6.0
818	30.8	5.80	6.10	5.2
832	30.8	5.80	6.16	6.2
926	30.8	6.00	6.32	5.3
271	50.1	3.40	3.36	1.8
274	50.1	3.10	3.37	8.7
359	50.1	3.40	3.61	6.2
409	50.1	3.40	3.73	9.7
692	50.1	4.10	4.25	3.7
692	50.1	3.90	4.25	8.9
925	50.1	4.30	4.57	6.3
279	96.1	2.28	2.20	3.5
416	96.1	2.45	2.43	0.8
471	96.1	2.65	2.50	5.6
482	96.1	2.55	2.52	1.2
492	96.1	2.70	2.53	6.3
537	96.1	2.85	2.59	9.1
815	96.1	3.00	2.87	4.3
832	96.1	3.15	2.88	8.6
1080	96.1	3.15	3.08	2.2
1130	96.1	3.20	3.11	2.8

According to this hypothesis, the turbulent flow has a four-layered structure (Figure 7.2). Far from the surface of the particle (in the bulk flow) there is a zone of developed turbulence and a constant concentration. Closer to the particle, in a turbulent buffer layer, both the average velocity and the average concentration decrease slowly according to a logarithmic law, and matter is here transferred by means of turbulent eddies, with molecular viscosity and diffusion not playing a noticeable part in the process. Still closer to the particle surface, in the viscous sublayer, momentum transferred by molecular viscosity exceeds that transferred by turbulent eddies. However, since the diffusivity D is about three orders of magnitude smaller than kinematic viscosity ν_f, the remaining turbulent eddies still transfer substantially more solute than molecular diffusion. Only in the innermost portion of the viscous sublayer, at a distance y from the particle surface that is less than the thickness of the diffusional sublayer δ, does the molecular diffusion mechanism predominate over the turbulence mechanism.

By combining solutions for the regions within and outside the viscous sublayer, it is possible to find the limiting diffusional flux passing through the boundary layer and hence to calculate a mass-transfer coefficient for the solid particle. In a turbulent flow regime, in spite of random agitation of the liquid, it may be assumed that turbulent eddies transport the matter dissolved in the liquid, and that this process is characterized by a turbulent diffusion gradient leading to transfer of mass in the direction of decreasing concentration; the net flow of the solute may be assumed proportional to the gradient of average concentration $\partial c/\partial y$, with the concentration of the subsurface at the particle surface being maintained at $c = 0$, whereas in the bulk flow of the solution it has the value c_0.

Denoting $\dot{m}'_{\text{turb}}$ as the average flow of the substance transported by turbulent eddies to unit area of the particle surface per unit time, we may write

$$\dot{m}'_{\text{turb}} = D_{\text{turb}} \frac{\partial c}{\partial y} \tag{8.41}$$

where $\partial c/\partial y$ is the average concentration gradient and D_{turb} (with dimensions $[L^2T^{-1}]$) is the turbulent diffusivity. D_{turb}, characterizing the transfer of substance by the turbulent motion, varies with the distance from the solid surface and may be related to the quantities that characterize the turbulence, such as density ρ, average velocity difference ΔU, and length scale l. From dimensional considerations, taking into account that the only composite quantity with dimensions $[L^2T^{-1}]$ is the product ΔUl, we obtain

$$D_{\text{turb}} \simeq \Delta Ul$$

Since the velocity v_e of the turbulent eddies is of the same order of magnitude as the change in the average flow velocity, over distances l, of the same order of magnitude as the scale of the turbulent eddies, we have

$$v_e \simeq \Delta U \simeq \frac{\partial U}{\partial l}\, l$$

and we may write

$$D_{\text{turb}} \simeq v_e l \simeq l^2 \frac{\partial U}{\partial l} \tag{8.42}$$

From equations (8.41) and (8.42), we find the diffusional flux of solute transported by turbulent eddies in the main turbulent stream (zone I; see Figure 7.2) to be

$$\dot{m}'_{\text{turb}} = \alpha y^2 \frac{\partial U}{\partial y}\frac{\partial c}{\partial y} \tag{8.43}$$

where α is a constant. Within the bulk turbulent flow (zone I), both the velocity and concentration have constant values and they may be regarded as independent of the distance from the particle.

We now need to evaluate the flow of solute transported by turbulent eddies in the buffer layer (zone II), in other words in the interval $\delta_0 < y < \delta_b$, where δ_b is the thickness of the turbulent buffer layer determined by

$$\delta_b \simeq \frac{v^* x}{U_0} \tag{8.44}$$

and δ_0 is the thickness of the viscous sublayer given by

$$\delta_0 = a\frac{\nu_f}{v^*} \tag{8.45}$$

(where a is a proportionality factor $\simeq 10$). Here v^* is the shear velocity, U_0 is the mean flow velocity, and x is the distance along the surface. Substituting the expression for the average velocity[71] distribution,

$$U = \frac{v^*}{\epsilon^{1/2}} \ln\left(\frac{v^* y}{a\nu_f}\right)$$

where ϵ is a constant,

$$v^* = \left(\frac{\tau}{\rho_f}\right)^{1/2} = \left(\frac{K_f}{2}\right)^{1/2} U_0 \simeq \frac{K_f^{1/2}}{1.41} U_0 \tag{8.46}$$

τ is the shear stress, and K_f is a friction factor, into equation (8.43), we find

$$\dot{m}'_{\text{turb}} = \beta v^* y \frac{dc}{dy} \qquad \left(\beta = \frac{1}{\epsilon^{1/2}}\alpha\right) \tag{8.47}$$

Integrating (8.47), we obtain the average concentration distribution in the turbulent buffer layer:

$$c_{II} = \frac{\dot{m}'_{turb}}{\beta v^*} \ln(y) + \alpha_1$$

where α_1 is a constant of integration.

At $y = \delta_b$ the equality

$$c_{II} = \frac{\dot{m}'_{turb}}{\beta v^*} \ln(\delta_b) + \alpha_1 = c_0$$

permits us to determine constant α_1, and consequently the concentration in the turbulent buffer layer may be written in the form

$$c_{II} = \frac{\dot{m}'_{turb}}{\beta v^*} \ln\left(\frac{y}{\delta_b}\right) + c_0 \tag{8.48}$$

The concentration c_{II} is a function not only of the distance from the surface of the particle (coordinate y) but also of coordinate x, this latter dependence due to the dependence of δ_b and v^* on x, given by equations (8.44) and (8.46).

For the viscous sublayer (zone III), it is assumed that mass transfer by turbulence eddies is significantly greater than the transfer by molecular diffusion, and for $y < \delta_0$ the diffusional flux is given by equation (8.41), as is the case in the outer portion of boundary layer. The turbulent diffusivity in the viscous sublayer may be found from equation (8.42) by expressing the quantities on the right-hand side as functions of v^*, δ_0, and y, and we find

$$D_{turb} \simeq \frac{\gamma v^* y^4}{\delta_0^3} \tag{8.49}$$

where γ is an unknown numerical coefficient. Substituting D_{turb} from equation (8.49) into equation (8.41), we arrive at an expression for the diffusional flux in the viscous sublayer

$$\dot{m}'_{turb} = \gamma v^* \frac{y^4}{\delta_0^3} \frac{\partial c}{\partial y} \tag{8.50}$$

Integrating (8.50), we find the concentration distribution in zone III ($y < \delta_0$) to be given by

$$c_{III} = \frac{-\dot{m}'_{turb}\delta_0^3}{3\gamma v^* y^3} + \alpha_2 \tag{8.51}$$

where α_2 is an integration constant.

In the diffusion sublayer (zone IV), for $y < \delta$, the turbulent diffusivity becomes smaller than the molecular diffusivity, so molecular diffusion becomes the mechanism of mass transfer and

$$\dot{m}' = D\frac{\partial c}{\partial y}$$

leading to

$$c_{IV} = \frac{\dot{m}'}{D}y \tag{8.52}$$

The constant α_2 in equation (8.51) allows for the condition that at the point $y = \delta$, equations (8.51) and (8.52) will give the same value for the concentration, so, with $\dot{m}' = \dot{m}'_{turb}$ by mass continuity,

$$\alpha_2 = \frac{\dot{m}'\delta_0^3}{3\gamma v^*\delta^3} + \frac{\dot{m}'\delta}{D}$$

so that

$$c_{III} = \frac{\dot{m}'\delta}{D} + \frac{\dot{m}'\delta_0^3}{3\gamma v^*}\left(\frac{1}{\delta^3} - \frac{1}{y^3}\right) \tag{8.53}$$

Now, since c_{II} and c_{III} are equal at $y = \delta_0$, we find the flux per unit area and time, $\dot{m}'$, of the solute to be given by a combination of equations (8.48) and (8.53):

$$\frac{\dot{m}'}{\beta v^*}\ln\left(\frac{\delta_0}{\delta_b}\right) + c_0 = \frac{\dot{m}'\delta}{D} + \frac{\dot{m}'\delta_0^3}{3\gamma v^*}\left(\frac{1}{\delta^3} - \frac{1}{\delta_0^3}\right)$$

Therefore,

$$\dot{m}' = \frac{Dc_0}{\delta - \dfrac{D}{\beta v^*}\ln\left(\dfrac{\delta_0}{\delta_b}\right) + \dfrac{\delta_0^3 D}{3\gamma v^*}\left(\dfrac{1}{\delta^3} - \dfrac{1}{\delta_0^3}\right)} \tag{8.54}$$

We have now to obtain the thickness δ of the diffusion sublayer in order to make equation (8.54) of use. As we have shown in equation (8.49), the turbulent diffusivity, D_{turb}, in the viscous sublayer is proportional to y^4, and therefore decreases rapidly as the surface of the particle is approached. At a certain distance (at which we define $y = \delta$) from the particle, D_{turb} must therefore become so small that the following equality will be valid:

$$D_{turb}(\delta) = \gamma v^*\frac{\delta^4}{\delta_0^3} = D \tag{8.55}$$

Equation (8.55) permits us to derive the value of δ, taking into account (8.45):

$$\delta = \left(\frac{D\delta_0^3}{\gamma v^*}\right)^{1/4} \simeq \frac{\delta_0}{\mathrm{Sc}^{1/4}(10\gamma)^{1/4}} = \frac{10^{3/4}\nu_f}{\mathrm{Sc}^{1/4}\gamma^{1/4}v^*} \tag{8.56}$$

We have now achieved our objective of obtaining a formula for δ, and substituting equation (8.56) into equation (8.54), we obtain

$$\dot{m}' = \frac{Dc_0}{\delta\left[1 + \dfrac{1}{30\gamma \mathrm{Sc}}\left(\dfrac{\delta_0^4}{\delta^4} - \dfrac{\delta_0}{\delta}\right) - \dfrac{D}{\beta v^* \delta}\ln\left(\dfrac{\delta_0}{\delta_b}\right)\right]}$$

$$= \frac{Dc_0}{\delta\left[1 + \dfrac{1}{30\gamma \mathrm{Sc}}\left[10\gamma \mathrm{Sc} - (10\gamma \mathrm{Sc})^{1/4}\right] - \dfrac{\delta_0}{10\beta \mathrm{Sc}\,\delta}\ln\left(\dfrac{\delta_0}{\delta_b}\right)\right]}$$

Note that for $\mathrm{Sc} \gg 1$,

$$\dot{m}' \simeq \frac{Dc_0}{\delta[\frac{4}{3} - (\gamma^{1/4}/10^{3/4}\beta \mathrm{Sc}^{3/4})\ln(\delta_0/\delta_b)]} \tag{8.57}$$

Equation (8.57) may be rewritten in the form

$$\dot{m}' = \frac{Dc_0}{\frac{4}{3}\delta} = \frac{c_0 v^*}{\lambda \mathrm{Sc}^{3/4}} \tag{8.58}$$

by using equations (8.55) and (8.56), where $\lambda = \frac{4}{3}10^{3/4}\gamma^{-1/4}$, because the constants β and γ are close to unity (as indicated by experimental data) and the second term in the denominator in (8.57) is thus substantially smaller than the first for $\mathrm{Sc} \gg 1$.

The diffusional flux given by equation (8.58) contains the characteristic velocity of the turbulent flow and the thickness of the turbulent buffer layer. Taking into account equation (8.46), we can rewrite equation (8.58) as

$$\dot{m}' = \frac{c_0 U_0 K_f^{1/2}}{1.41\lambda \mathrm{Sc}^{3/4}}$$

and then obtain the total diffusional flow reaching the entire surface of the particle as

$$\dot{m} = \frac{c_0 U_0 S}{1.41\lambda \mathrm{Sc}^{3/4} l}\int_0^l K_f^{1/2}\,dx$$

where S is the particle-surface area.

The integral friction factor per unit length along a surface is related to the drag coefficient C_D by the equation defining the coefficient as the ratio of the total force exerted on one side of the plate to the product of the dynamic pressure and the length of the surface:

$$C_D = \frac{F_D}{\rho_f U_0 l/2} = \frac{\int_0^l K_f^{1/2}\,dx}{l} \tag{8.59}$$

Table 8.4. *Comparison of experimental and theoretical values of the mass-transfer coefficient k for various solid particle–liquid systems*

Particle diameter (μm)	Stirrer speed (rpm)	Mass-transfer coefficient: Experimental ($\times 10^4$ m/s)	Mass-transfer coefficient: Calculated ($\times 10^4$ m/s)	$\frac{\text{Exp.} - \text{calc.}}{\text{Exp.}} \times 100$ (%)	Ref.
Anion-Exchange Resin Particles in Aqueous Acids[a]					
231	360	1.53	1.39	9.1	34
231	384	1.60	1.46	8.8	
231	460	1.85	1.67	9.7	
231	537	2.09	1.88	10.0	
231	600	2.10	2.03	3.3	
231	649	2.40	2.16	10.0	
231	666	2.30	2.20	4.3	
231	756	2.45	2.62	6.9	
231	773	2.45	2.42	1.2	
231	792	2.65	2.47	6.8	
231	814	2.75	2.51	8.7	
231	1130	3.25	3.27	0.6	
593	497	1.40	1.51	7.9	
593	540	1.45	1.55	6.9	
593	597	1.65	1.77	7.3	
593	597	1.80	1.88	4.5	
593	658	1.85	2.03	9.7	
593	661	2.00	2.18	9.0	
593	697	2.03	2.19	7.9	
593	702	2.10	2.28	8.6	
593	947	2.12	2.29	8.0	
Cation-Exchange Resin Particles in Aqueous Bases[b]					
486	350	1.010	0.965	4.5	34
486	433	1.150	1.130	1.3	
486	556	1.350	1.362	0.9	
486	671	1.550	1.570	1.3	
486	730	1.650	1.672	1.3	
486	780	1.730	1.758	1.6	
Ammonium Nitrate Particles in Ethanol[c]					
1950	423	0.390	0.351	10.0	34
1950	533	0.450	0.429	4.7	
1950	692	0.500	0.508	1.6	
1950	697	0.480	0.511	6.4	
1950	846	0.538	0.590	9.7	
1950	991	0.600	0.660	10.0	
Benzoic Acid Particles in Water[d]					
130	300	0.60	0.55	8.3	27
230	300	0.51	0.55	7.8	
1000	300	0.52	0.55	5.8	

Taking into account equation (8.59), we then rewrite the diffusional flux to the solid particle in the form

$$\dot{m} = \frac{c_0 C_D^{1/2} U_0 S}{1.41\lambda \mathrm{Sc}^{3/4}} \tag{8.60}$$

To calculate the diffusion given by equation (8.60), we can now use equations derived in Chapter 6 for the particle relative velocity. It then remains to determine the drag coefficient C_D, and to do this we will use an approach described elsewhere,[5,68] in which we can evaluate the viscous drag coefficients by calculating the energy dissipation in the boundary layer and, in addition, we can take into account the turbulent resistance.

Now for $v^i \gg v_e$ (where v^i is the ideal fluid flow velocity and v_e is the eddy velocity) we can assume that we are in the range of viscous flow, and the rate of energy dissipation here will be

$$-\frac{\partial E}{\partial t} = \mu_f \int \left[\frac{\partial}{\partial r}(v_r^i)^2 + \frac{\partial}{\partial r}(v_\theta^i)^2 \right] 2\pi r_p^2 \sin\theta \, d\theta = 8\pi\mu_f u_r^2 \, \Psi^2(\gamma)$$

where v_r^i and v_θ^i are the components of v^i in the spherical coordinate directions r and θ, respectively; μ_f is the dynamic viscosity; and $\Psi(\gamma)$ is defined in Chapter 6.

This energy of dissipation then yields force F_D acting on a particle as

$$F_D = -\frac{1}{2}\frac{\partial}{\partial v}\left(\frac{\partial E}{\partial t}\right) = 8\pi\mu_f u_r \Psi^2(\gamma)$$

and the drag coefficient becomes

$$C_D = \frac{32}{\mathrm{Re}} \Psi^2(\gamma) \tag{8.61}$$

Substituting into equation (8.60) the values of u_r from equation (6.79), ϵ from equation (6.61), and C_D from equation (8.61), the diffusion flux per unit surface area of the particle will be

---

Notes to Table 8.4

[a] $\rho_p = 1100$ kg/m³; $\Delta\rho = 100$ kg/m³; $\nu_f = 1 \times 10^{-6}$ m²/s; $D = 1.49 \times 10^9$ m²/s; $D_m = 0.075$ m; $T = 0.25$ m; $h = 0.216$.
[b] $\rho_p = 1280$ kg/m³; $\Delta\rho = 280$ kg/m³; $\nu_f = 1 \times 10^{-6}$ m²/s; $D = 2.13 \times 10^9$ m²/s; $D_m = 0.075$ m; $T = 0.25$ m; $h = 0.216$ m.
[c] $\rho_p = 1490$ kg/m³; $\Delta\rho = 700$ kg/m³; $\nu_f = 1.6 \times 10^{-6}$ m²/s; $D = 0.541 \times 10^9$ m²/s; $D_m = 0.075$ m; $T = 0.25$ m; $h = 0.216$ m.
[d] $\rho_p = 1280$ kg/m³; $\Delta\rho = 280$ kg/m³; $\nu_f = 1.12 \times 10^{-6}$ m²/s; $D = 0.77 \times 10^9$ m²/s; $D_m = 0.0508$ m; $T = 0.1016$ m; $h = 0.127$ m.

$$\dot{m}' = \text{const.} \frac{D^{3/4}}{\nu_f^{1/2}} \left(\frac{\Delta\rho}{\rho_p + 0.5\rho_f} \right)^{1/2} \frac{U^{3/4}}{l^{1/4}} \Psi(\gamma) c_0$$

or, in dimensionless form,

$$\text{Sh} = \text{const. Re}^{3/4}\text{Sc}^{1/4} \left(\frac{a}{l} \right) \left(\frac{\Delta\rho}{\rho + 0.5\rho_0} \right)^{1/2} \Psi(\gamma) \quad (8.62)$$

In Table 8.4, a comparison of experimental data[27,60] is made with a theoretical equation similar to equation (8.62). When it is rewritten for a mass-transfer coefficient k for solid particles suspended in agitated tanks, equation (8.62) becomes

$$k = \text{const.} \frac{D^{3/4}}{\nu_f^{1/2}} \left(\frac{\Delta\rho}{\rho_p + 0.5\rho_f} \right)^{1/2} \left[\frac{n^3 D_m^5}{(T^2 h)^{1/3}} \right]^{1/4} \quad (8.63)$$

In Table 8.5, a comparison of experimental data with theoretical equation (8.63) is made.

As is seen from the tables, the theoretical equations are in good agreement with experimental data.

Table 8.5. *Comparison of values of the mass-transfer coefficient k obtained from equation (8.63) with experimental data*

System	d_p (cm)	Re	Sc	Sh Exp.	Sh Calc.	$\frac{\text{Exp.} - \text{calc.}}{\text{Exp.}} \times 100$ (%)	Ref.
Sublimation of naphthalene	1.59	239	2.57	16	16.93	5.8	50
particles in air	3.81	228	2.55	16	16.31	2.0	
Dissolution	1.59	181	1230	105	95.3	9.2	50
of benzoic	1.59	168	1230	100	90.1	9.9	
acid	1.59	200	1230	108	102.7	4.9	
particles in	1.59	160	1250	95	87.2	8.2	
water	1.59	263	1230	121	126.12	4.2	
	1.59	285	1230	131	134	2.3	
	1.59	299	1230	138	139	0.6	
	1.59	287	1220	124	134.4	8.4	
	3.81	380	1230	160	166.2	3.9	
	3.81	376	1230	158	164.9	4.4	
	3.81	229	1260	114	114.4	0.3	
	3.81	294	1250	140	137.7	1.6	
	3.81	458	1260	184	192.4	4.6	
	3.81	304	1220	136	140.3	3.2	
	3.81	211	1220	117	106.7	8.8	
	3.81	467	1250	184	194.8	5.9	

References

1. Chukarev, A. N., *Zh. Russ. Fiz. Ova.*, *28*, 604 (1896). (In Russian.)
2. Nernst, W. T., *Phys. Chem.*, *47*, 52 (1904).
3. Lewis, W. K., and Whitman, W. G., *Ind. Eng. Chem.*, *16*, 1215 (1924).
4. Whitman, W. G., *Chem. Met. Eng.*, *29*, 23 (1923).
5. Levich, V. G., *Physicochemical Hydrodynamics*, Prentice-Hall, Englewood Cliffs, N.J., 1962.
6. Higbie, R., *Trans. Am. Inst. Chem. Eng.*, *31*, 365 (1935).
7. Danckwerts, P. V., *Trans. Faraday Soc.*, *46*(300), 701 (1950).
8. Kishinevsky, M., *Zh. Prikl. Khim. (Moscow) 24*, 542 (1951); *27*(382), 450 (1954).
9. Kraunig, R., and Brunstain, *J. Appl. Sci. Res.*, *A2*, 439 (1951).
10. Friss, H. L., *J. Chem. Phys.*, *22*, 129 (1954).
11. Fridlander, S. K., *Am. Inst. Chem. Eng. J.*, *7*, 347 (1961).
12. Kruzhilin, G. I., *Zh. Tekh. Fiz.*, *6*, 561 (1936).
13. Akselrud, G. A., Candidate's dissertation, Lvov, USSR, 1955. (In Russian.)
14. Froessling, N., *Acta Univ. Zur. NFAVD*, *36*, 4 (1940).
15. Garner, F. H., and Kely, R. B., *Chem. Eng. Sci.*, *36*, 4 (1958).
16. Hixson, A. W., and Baum, S. J., *Ind. Eng. Chem.*, *33*, 478 (1941).
17. Wihelm, R. H., Conklin, L. H., and Sauer, P. C., *Ind. Eng. Chem.*, *34*, 120 (1942).
18. Hixson, A. W., and Baum, S. J., *Ind. Eng. Chem.*, *34*, 120 (1942).
19. Kneule F., *Chem.-Ing.-Tech.*, *28*, 221 (1956).
20. Johnson, D. L., Saito, H., Polejes, J., and Hougen, O., *AIChE J.*, *3*, 411 (1957).
21. Kolar, V., *Collect. Czech. Chem. Commun.*, *24*, 3309 (1959).
22. Barker, J. J., and Treybal, R. E., *AIChE J.*, *6*, 289 (1960).
23. Din-vei, Candidate's dissertation, Moscow Institute of Chemical Technology, Moscow, 1959. (In Russian.)
24. Nagata, S., Yamaguchi, I., Yabuta, S., and Harada, M., *Mem. Fac. Eng. Kyoto Univ.*, *22*, 86 (1960).
25. Marangozis, J., and Johnson, A., *Can. J. Chem. Eng.*, *40*, 231 (1962).
26. Calderbank, P. H., and Moo-Young, M. B., *Chem. Eng. Sci.*, *16*, 39 (1961).
27. Harriott, P., *AIChE J.*, *8*, 93 (1962).
28. Keey, R. B., and Glen, J. B., *AIChE J.*, *12*, 401 (1966).
29. Lastorchkin, V., Sokolova, A. D., Vil'nkts, E. L., and Baram, A. A., *Zh. Prikl. Khim. (Moscow)*, *40*, 849 (1967).
30. Sykes, P., Gomesplata, A., *Can. J. Chem. Eng.*, *45*, 189 (1967).
31. Weinspach, P. M., *Chem.-Ing.-Tech.*, *39*, 231 (1967).
32. Brian, P. L., Hales, H. B., and Sherwood, T. K., *AIChE J. 15*, 419 (1969).
33. Brown, D. E., and Coulson, J. M., in J. N. Sherwood et al., *Diffusion Processes*, vol. 2, Gordon and Breach, New York, 1971, p. 573.
34. Levins, D. M., and Glustonbury, J. R., *Trans. Inst. Chem. Eng.*, *50*, 132 (1972).
35. Mack, D. E., and Marriner, R. A., *Chem. Eng. Prog.*, *45*(4), 545 (1949).
36. Humphrey, D. W., and Van Ness, H. C., *AIChE J.*, *3*(2), 283 (1957).
37. Nagata, S., Adachi, M., and Yamaguchi, I., *Mem. Fac. Eng., Kyoto Univ.*, *20*, 72 (1958).
38. Madden, A., and Nelson, D. G., *AIChE J.*, *3*, 415 (1964).
39. Nienow, A. W., *Can. J. Chem. Eng.*, *47*(3), 248 (1969).
40. Miller, D. N., *Ind. Eng. Chem. Process Des. Dev.*, *10*, 365 (1971).
41. Johnson, A. I., and Chang-Young Huang, *AIChE J.*, *2*, 412 (1956).
42. Liachshenko, P. V., *Gravitatzionie metodi obogachshenia*, Gostoptekhizdat, Moscow, 1940.
43. Chiloly, T. H., and Colburn, A. P., *Ind. Eng. Chem.*, *26*, 1183 (1934).
44. Gilliland, E. K., and Sherwood, T. K., *Ind. Eng. Chem.*, *26*, 516 (1934).

45. Ranz, W. E., and Marshall, W. R., *Chem. Eng. Prog., 48,* 141 (1952).
46. Froessling, N., *Beitr. Geophys., 32,* 170 (1938).
47. Galloway, T. R., and Sage, B. H., *Int. J. Heat Mass Transfer, 7,* 283 (1964).
48. Nienow, A. W., Unahabhokha, R., and Mullin, J. W., *J. Appl. Chem., 18,* 154 (1968).
49. Nienow, A. W., Unahabhokha, R., and Mullin, J. W., *Ing. Eng. Chem. Fund., 5,* 597 (1966).
50. Rowe, P. N., et al., *Trans. Inst. Chem. Eng., 43,* T14 (1965).
51. Rowe, P. N., and Claxton, K. T., Rep. AERE R-4673, R-4675 (1964).
52. Schwartzberg, H. G., and Treybal, R. E., *Ind. Eng. Chem. Fund. 7*(1), 1 (1968).
53. Zundulevich, Y. V., and Vigdorchik, E. M., *Tr. Inst. Gipronikel, 47,* 111 (1970).
54. Middleman, S., *AIChE J., 11,* 750 (1965).
55. Hughmark, G. A., *Chem. Eng. Sci., 24,* 291 (1969).
56. Keey, R. B., Mandeno, P., and Trink Khank Tuoc, Chemeca 1970, Sess. 5.53.
57. Shinnar, R., and Church, J. M., *Ind. Eng. Chem., 52,* 253 (1960).
58. Oyama, Y., and Endoh, K., *Chem. Eng. (Tokyo), 42,* 227 (1967).
59. Calderbank, P., and Jones, S. J. R., *Trans. Inst. Chem. Eng., 39,* 363 (1961).
60. Levins, D. M., and Glastonbury, J. R., *Chem. Eng. Sci., 27,* 537 (1972).
61. Van Den Berg, H. J., in *Chemical Reaction Engineering,* Advances in Chemistry Series 109, American Chemical Society, Washington, D.C., 1972.
62. Kolmogoroff, A. N., *C. R. Acad. Sci., USSR, 30,* 301 (1941).
63. Din-Vei et al., *Khim. Prom. Moscow,* no. 4, 46 (1963).
64. Ishii, T., and Fujita, S., *Kagku Kogaku* (abridg.), *3,* 237 (1965).
65. Stewart, R. W., and Townsend, A. A., *Trans. Phil. R. Soc., 243A,* 359 (1951).
66. Taylor, G. I., *Proc. R. Soc., 151A,* 421 (1935).
67. Cutter, L. A., *AIChE J., 12,* 35 (1966).
68. Kochin, N. E., Kibel, I. A., and Roze, N. V., *Theoretische Hydromechanik,* Akademie-Verlag, Berlin, 1954.
69. Tikhanov, A. N., and Samarskiy, A. A., *Differentialgleichungen der mathematischen Physiks,* Deutscher Verlag der Wissenschaften, Berlin, 1959.
70. Landau, L. D., and Lifshits, L., *Fluid Mechanics,* Pergamon Press, London, 1959.
71. Hinze, J. O., *Turbulence,* McGraw-Hill, New York, 1959.

PART III

Application to chemical and biochemical processes

9

Chemical applications

9.1 Introduction

Bubble processes are widely employed in industry, for example for carrying out oxidation, chlorination, sulfochlorination, and so on. The rates of these processes are primarily limited by the mass-transfer process in the liquid phase, the concentration gradient at the gas–liquid (or solid–liquid) interface determining the driving force for mass transfer from one phase to another.[1,2] The occurrence of reactions in a flow consisting of a liquid phase with a low-soluble gas introduces certain changes in this driving force. However, the concentration at even a short distance from the interface is effectively constant for flows with intensive mixing, and therefore the volume in which the concentration gradient has a significant value is only a small part of the total reaction volume. This means that the deviation of measured gas concentrations from those calculated under the assumption of mass transfer without chemical reaction has an insignificant influence on the macrokinetics of the process, and therefore we will consider the concentration of dissolving gas to be uniform in the whole volume of the apparatus.

As a model for reactions commonly occurring in bubble equipment, oxidation of liquid-phase hydrocarbons can be studied, with possible application to such processes as synthesis of fat alcohols for detergent manufacturing, dimethyl phthalate for manufacturing of nylon, cyclohexane from carbon, phenol, acetone, and many other products.

The total rate of oxidation of liquid-phase hydrocarbons is determined by the hydrocarbon reactivity and by the oxygen feed rate into the reaction zone, and the interaction of these two effects also influences the kinetic, or diffusional, regime of oxidation. We will consider quantitatively each of these effects and using equations derived in Chapters 4 and 7, an analysis will be made of the influence of various parameters on the total rate of the process.

Now, the theory of chain-branched termination reactions with a square termination forms the basis of most modern theories of liquid-phase oxidation of hydrocarbons,[3–6] and this enables us to describe the complex chemical process of hydrocarbon oxidation in terms of a combination of elementary reactions, which can in turn be characterized by the numerical values of various constants. Using this approach to the study of the kinetics of the process, and also by taking into account hydrodynamic and diffusional effects, it becomes possible to determine the optimal conditions for the liquid-phase oxidation process, ultimately leading to appropriate design criteria for process equipment. It is known[7] that the oxidation of hydrocarbons in a liquid phase can be broken down into three steps: originating, continuing, and chain breaking.

As a result of the initiation reaction, the radicals R· are generated (and it is worth noting that in the steady regime the initiation rate is equal to chain-breaking rate). Interacting with oxygen molecules, these radicals are transformed into hydroperoxide radicals as follows:

$$R^{\cdot} + O_2 \xrightarrow{k_1} ROO^{\cdot} \tag{9.1}$$

which in turn react with hydrocarbon RH to produce hydroperoxide ROOH, leading to a restoration of the radicals:

$$ROO^{\cdot} + RH \xrightarrow{k_2} ROOH + R^{\cdot} \tag{9.2}$$

Now, as a result of the radical recombination, chain breaking takes place:

$$R^{\cdot} + R^{\cdot}\ (k_{10}) \tag{9.3}$$

$$R^{\cdot} + ROO^{\cdot}\ (k_9) \tag{9.4}$$

$$ROO^{\cdot} + ROO^{\cdot}\ (k_6) \tag{9.5}$$

In expressions (9.1) through (9.5), in which k_i are reaction-rate constants, we can determine the oxidation rate W and the radical concentration [R] as functions of the concentration of initial reagents [RH], $[O_2]$, the initiation rate W_i, and the rate constants[5,7] k_i:

$$\begin{aligned} W &= \frac{k_1[O_2](W_i)^{1/2}}{\left\{k_{10}\left[\dfrac{k_6k_1^2[O_2]^2}{k_{10}k_2^2[RH]^2} + \dfrac{k_9[O_2]}{k_{10}[RH]} + 1\right]\right\}^{1/2}} \\ [R^{\cdot}] &= \left\{\frac{W_i}{k_{10}\left[\dfrac{k_6k_1^2[O_2]^2}{k_{10}k_2^2[RH]^2} + \dfrac{k_9k_1[O_2]}{k_{10}k_2[RH]} + 1\right]}\right\}^{1/2} \end{aligned} \tag{9.6}$$

The oxygen-containing products are composed of hydroperoxide radical recombinations[8] and high-molecular and unsaturated substances, which give rise to resins, and are formed by alkyl radical recombination. The latter may become inhibitors of the oxidation chain reaction. For low oxygen concentrations, there will be an accumulation of R· radicals and, consequently, higher resin formation. However, in rationally arranged processes of liquid-phase oxidation, we seek conditions that exclude resin formation, and these conditions may be found by investigating the effect of oxygen partial pressure P^2_O on the rate of oxidation reaction and on the concentration of radicals. There are two cases of practical importance: initiation through hydroperoxide decomposition in noncatalytic oxidation, and initiation through reaction of metals of changeable valency with hydrocarbon in catalytic oxidation. We consider each case in turn.

9.2 Noncatalytic oxidation

For a low hydroperoxide concentration, the initiation rate can be described by the equation[3,5]

$$W_i = k_4[\mathrm{RH}][\mathrm{ROOH}] \tag{9.7}$$

Now, hydroperoxide may be decomposed as a result of molecular reactions,

$$\mathrm{ROOH} \xrightarrow{k_5} \mathrm{RO\cdot} + \mathrm{OH}$$

and also may be consumed in a chain reaction with carbonyl compounds[3,5] and alcohols[9]

$$\left.\begin{aligned} \mathrm{ROOH} + \mathrm{ROO\cdot} &\xrightarrow{k_{11}} \mathrm{R{=}O} + \mathrm{\cdot OH} + \mathrm{ROOH} \\ \mathrm{RH} + \mathrm{\cdot OH} &\xrightarrow{k'_{11}} \mathrm{R\cdot} + \mathrm{H_2O} \end{aligned}\right\}$$

$$\left.\begin{aligned} \mathrm{ROOH} + \mathrm{R\cdot} &\xrightarrow{k_{12}} \mathrm{RO\cdot} + \mathrm{ROH} \\ \mathrm{RO\cdot} + \mathrm{RH} &\xrightarrow{k'_{12}} \mathrm{ROH} + \mathrm{R\cdot} \end{aligned}\right\}$$

The change of hydroperoxide concentration with time may then be represented by a differential equation:

$$\frac{d[\mathrm{ROOH}]}{dt} = k_2[\mathrm{RH}][\mathrm{ROO\cdot}] - k_{12}[\mathrm{ROOH}][\mathrm{R\cdot}] - k_4[\mathrm{RH}][\mathrm{ROOH}] - k_5[\mathrm{ROOH}] \tag{9.8}$$

Let us consider the conditions for [ROOH] to reach a maximum value, in other words the case

$$\frac{d[\mathrm{ROOH}]}{dt} = 0$$

and, for simplicity, let us also ignore consumption of hydroperoxide in the chain mechanism. Equation (9.8) then leads to

$$k_2[\mathrm{RH}][\mathrm{ROO}\cdot] = (k_4[\mathrm{RH}] + k_5)[\mathrm{ROOH}] \tag{9.9}$$

In a steady oxidation process, when chains are sufficiently long, the reaction rate for formation of peroxide radicals is approximately equal to the rate of their interchange with hydrocarbon,[1]

$$k_1[\mathrm{R}\cdot][\mathrm{O}_2] \simeq k_2[\mathrm{ROO}\cdot][\mathrm{RH}]$$

and it follows from this that the steady concentration of hydroperoxide radicals is given by

$$[\mathrm{ROO}\cdot] = \frac{k_1[\mathrm{O}_2]}{k_2[\mathrm{RH}]}[\mathrm{R}\cdot]$$

Substituting this into equation (9.9), we obtain

$$[\mathrm{ROOH}] = \frac{k_1[\mathrm{O}_2][\mathrm{R}\cdot]}{k_4[\mathrm{RH}] + k_5} \tag{9.10}$$

Now, using equation (9.10) in equation (9.7), the chain initiation rate becomes

$$W_i = \frac{k_1[\mathrm{O}_2][\mathrm{R}\cdot]}{1 + k_5/k_4[\mathrm{RH}]} \tag{9.11}$$

Combining equations (9.6) and (9.11), we can determine the concentration of alkyl radicals for maximum hydroperoxide concentration:

$$[\mathrm{R}\cdot] = \frac{k_1[\mathrm{O}_2]}{k_{10}\left(1 + \dfrac{k_5}{k_4[\mathrm{RH}]}\right)\left(\dfrac{k_6 k_1^2[\mathrm{O}_2]}{k_{10}k_2^2[\mathrm{RH}]^2} + \dfrac{k_9 k_1[\mathrm{O}_2]}{k_{10}k_2[\mathrm{RH}]} + 1\right)} \tag{9.12}$$

For future use, let us introduce the dimensionless oxygen concentration

$$\omega = \frac{k_1[\mathrm{O}_2]}{k_2[\mathrm{RH}]}\left(\frac{k_6}{k_{10}}\right)^{1/2} \tag{9.13}$$

and a dimensionless constant

$$\chi = \frac{k_9}{(k_6 k_{10})^{1/2}} \tag{9.14}$$

and also an assumption:

$$\frac{[\mathrm{ROO}\cdot]}{[\mathrm{R}\cdot]} = \frac{k_1[\mathrm{O}_2]}{k_2[\mathrm{RH}]} \tag{9.15}$$

Now, for a steady oxidation process, and with sufficiently long chains, the consumption of R· radicals is equal to the consumption of ROOH radicals:

$$k_1[O_2][R\cdot] + k_{10}[R\cdot]^2 + k_9[R\cdot][ROO\cdot]$$
$$= k_2[RH][ROO\cdot] + k_6[ROO\cdot]^2 + k_9[R\cdot][ROO\cdot]$$

or

$$k_1[O_2] + k_{10}[R\cdot] = k_2[RH]\frac{[ROO\cdot]}{[R\cdot]} + k_6[ROO\cdot]\frac{[ROO\cdot]}{[R\cdot]}$$

and it follows that

$$\frac{[ROO\cdot]}{[R\cdot]} = \frac{k_1[O_2] + k_{10}[R\cdot]}{k_2[RH] + k_6[ROO\cdot]}$$

and

$$\frac{[ROO\cdot]}{[R\cdot]} = \frac{k_1[O_2]}{k_2[RH]}\left(\frac{1 + (k_{10}[R\cdot]/k_1[O_2])}{1 + (k_6[ROO\cdot]/k_2[RH])}\right)$$

Denoting

$$\alpha = \frac{1 + (k_{10}[R\cdot]/k_1[O_2])}{1 + (k_6[ROO\cdot]/k_2[RH])} \tag{9.16}$$

we can now evaluate the limits of variation of factor α.

At low concentrations of radicals [R·], chain breaking will take place through the reaction ROO· + ROO· and therefore the oxidation rate must be given by

$$W = k_6[ROO\cdot]^2 + k_2[RH][ROO\cdot] \tag{9.17}$$

As a result of the reaction RH + ROO·, hydroperoxide is formed at decomposition, and two radicals are generated, the rate of chain breaking of radical [ROO·] being

$$W_b = k_6[ROO\cdot]^2 = \tfrac{1}{2}W \tag{9.18}$$

Hence, from equations (9.17) and (9.18),

$$k_6[ROO\cdot]^2 = k_2[RH][ROO\cdot]$$

and equation (9.16) then yields

$$\alpha = \tfrac{1}{2}$$

For limited rate of oxidation, in the case of oxygen shortage,

$$W_b = k_{10}[R\cdot]^2 = \tfrac{1}{2}W$$

and consequently,

$$k_{10}[\mathrm{R}\cdot] = k_1[\mathrm{O}_2]$$

and here

$$\alpha = 2$$

Thus, for a surplus oxygen concentration, in equation (9.13),

$$\omega \simeq \frac{k_1 k_6^{1/2}[\mathrm{O}_2]}{2k_2 k_{10}^{1/2}[\mathrm{RH}]}$$

and for a low oxygen concentration,

$$\omega \simeq \frac{2k_1 k_6^{1/2}[\mathrm{O}_2]}{k_2 k_{10}^{1/2}[\mathrm{RH}]}$$

For the sake of simplicity, we will now consider, as reasonably representative of most systems, an average of these:

$$\omega \simeq \frac{k_1 k_6^{1/2}[\mathrm{O}]}{k_2 k_{10}^{1/2}[\mathrm{RH}]}$$

Let us now evaluate the constant χ. Reaction-rate constants[10] are given by equations of the form

$$k = PZe^{-\Delta E/RT}$$

where P is the steric factor and Z is the collision factor. Note that for species p and q,

$$P_{pq} = P_p P_q$$

Hence, from equation (9.14),

$$\chi = \frac{k_9}{(k_6 k_{10})^{1/2}} \tag{9.19}$$

$$= \frac{P_{\mathrm{R}\cdot} P_{\mathrm{ROO}\cdot} Z_{(\mathrm{R}\cdot+\mathrm{ROO}\cdot)} \exp[-\Delta E_9/RT]}{(P_{\mathrm{R}\cdot} P_{\mathrm{ROO}\cdot} P_{\mathrm{R}\cdot} P_{\mathrm{ROO}\cdot} Z_{(\mathrm{R}\cdot+\mathrm{R}\cdot)} Z_{(\mathrm{ROO}\cdot+\mathrm{ROO}\cdot)} \exp[(-\Delta E_{10} - \Delta E_6)/RT])^{1/2}}$$

Because the reactions in equations (9.3), (9.4), and (9.5) are radical recombination reactions, we have

$$\Delta E_9 \simeq \Delta E_{10} \simeq \Delta E_6 \simeq 0$$

and equation (9.19) may then be rewritten as

$$\chi = \frac{Z_{(\mathrm{R}\cdot+\mathrm{ROO}\cdot)}}{[Z_{(\mathrm{R}\cdot+\mathrm{R}\cdot)} Z_{(\mathrm{ROO}\cdot+\mathrm{ROO}\cdot)}]^{1/2}} \tag{9.20}$$

Now, from collision theory we know that

$$Z_{pq} = \pi(r_p^* + r_q^*)^2 \left[\frac{8KT(m_p + m_q)}{m_p m_q}\right]$$

where r^* is the radical size and m the molecular mass. Substituting this into equation (9.20) yields

$$\chi = \frac{(r^*_{R\cdot} + r^*_{ROO\cdot})^2}{4r^*_{R\cdot}r^*_{ROO\cdot}} \frac{m_{R\cdot} + m_{ROO\cdot}}{(m_{R\cdot}m_{ROO\cdot})^{1/2}}$$

From this equation we see that, because $r^*_{R\cdot} \simeq r^*_{ROO\cdot}$ and $m_{R\cdot}$ is somewhat less than $m_{ROO\cdot}$, the factor $\chi \simeq 1$ (and with increase in radical size χ tends to $\sqrt{2}$), and consequently, according to equation (9.14), we obtain

$$k_9 \simeq (k_6 k_{10})^{1/2} \qquad (9.21)$$

The process of liquid-phase oxidation defined by reactions (9.1) through (9.5) has the rate

$$W = k_1[O_2][R\cdot]$$

where W is the rate of oxygen consumption, k_1 is the reaction-rate constant of equation (9.1), and $[O_2]$ is the oxygen concentration in the liquid phase, proportional to the oxygen partial pressure P_{O_2} in the gas phase. Now, introducing a dimensionless oxidation rate

$$\nu = \frac{k_6(1 + k_5/k_4[RH])}{k_2^2[RH]^2} W \qquad (9.22)$$

we may obtain, as a result of the combined solution of equations (9.12), (9.13), (9.21), and (9.22), a relation between the dimensionless oxidation rate ν and the dimensionless concentration of oxygen, ω:

$$\nu = \frac{\omega^2}{\omega^2 + \omega + 1} \qquad (9.23)$$

Equation (9.23) is the principal equation for obtaining the oxidation rate in a liquid phase, as it is affected by the quadratic character of chain breaking.[10]

It follows from equation (9.23) that at low values of ω,

$$\nu \simeq \omega^2 \qquad (9.24)$$

and according to equation (9.13), (9.22), and (9.24), the oxidation rate in this case is

$$W = \frac{k_1^2[O_2]^2}{k_{10}(1 + k_5/k_4[RH])} \qquad (9.25)$$

For total decomposition of hydroperoxide into radicals, $k_4[RH] \gg k_5$, and equation (9.25) takes the form

$$W = \frac{k_1^2[O_2]^2}{k_{10}}$$

This indicates that, for low oxygen concentration, the order of the reaction for oxygen is two and with respect to hydrocarbon is zero.

For $\omega \to \infty$ we obtain from equation (9.23), $\nu \simeq 1$ and consequently from equation (9.22), when $k_4[\mathrm{RH}] \gg k_5$, the maximum rate of oxidation is obtained:

$$W_{\max} = \frac{k_2^2[\mathrm{RH}]^2}{k_6} \tag{9.26}$$

This means that at the maximum oxidation rate (with abundant oxygen) the reaction is of zero order with respect to oxygen and second order with respect to hydrocarbon.[11]

Figure 9.1 shows the change of dimensionless oxidation rate ν with dimensionless oxygen concentration ω, and we see that ν increases uniformly with ω, and as $\omega \to \infty$, ν asymptotically approaches the maximum value of unity. It follows that

$$\nu = W/W_{\max}$$

Recombination of R· radicals is known to result in the formation of recombination products, or in other words, in tar formation, and both the volume of the tar formed relative to product transformed and the net output are significant factors when the technological effectiveness of a process is being estimated. Because stationary radical and oxygen concentrations set in very soon during the oxidation process, we can consider the product formation rates rather than output of the system. Tar formation can take place

Figure 9.1 Dimensionless oxidation rate of hydrocarbon ν versus dimensionless oxygen concentration ω.

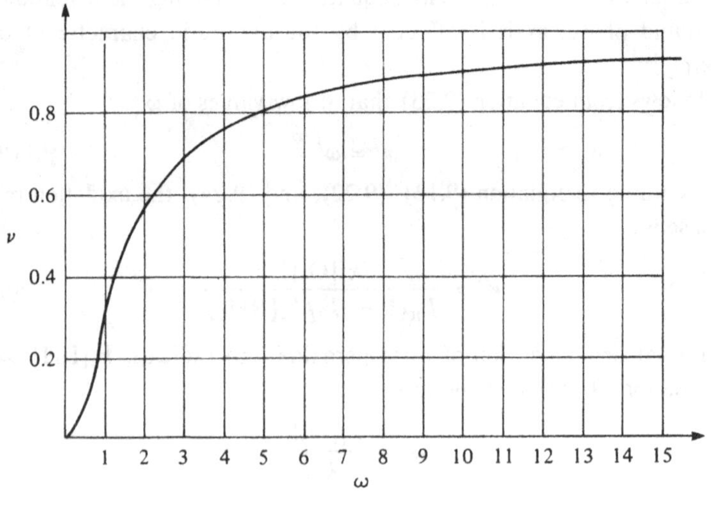

not only as a result of radical recombination, but also as a result of condensation of intermediate products,[12] but in processes of high efficiency (especially catalytical) the concentration of intermediate products is low, and we can thus use the following formula for calculation of the concentration of intermediate product:

$$W_{tf} = \beta k_{10}[\mathrm{R}^{\cdot}]^2 \tag{9.27}$$

Here β is a measure of the radical spent in tar formation ($\beta < 1$). Let us now introduce the dimensionless concentration of radicals $[\mathrm{R}^{\cdot}]$ as

$$\rho = \frac{(k_6 k_{10})^{1/2}(1 + k_5/k_4[\mathrm{RH}])[\mathrm{R}^{\cdot}]}{k_2[\mathrm{RH}]} \tag{9.28}$$

For $k_4[\mathrm{RH}] \gg k_5$, we find that

$$\rho \simeq \frac{(k_6 k_{10})^{1/2}}{k_2[\mathrm{RH}]}[\mathrm{R}^{\cdot}] \tag{9.29}$$

According to equations (9.12), (9.13), (9.14), and (9.21), the relationship between this dimensionless radical concentration ρ and the dimensionless concentration of oxygen ω may be written

$$\rho = \frac{\omega}{\omega^2 + \omega + 1} \tag{9.30}$$

In Figure 9.2 the relationship $\rho = \rho(\omega)$ is shown, and it is seen from this figure that ρ reaches a maximum for $\omega = 1$.

Figure 9.2 Dimensionless concentration of radicals ρ versus dimensionless oxygen concentration ω.

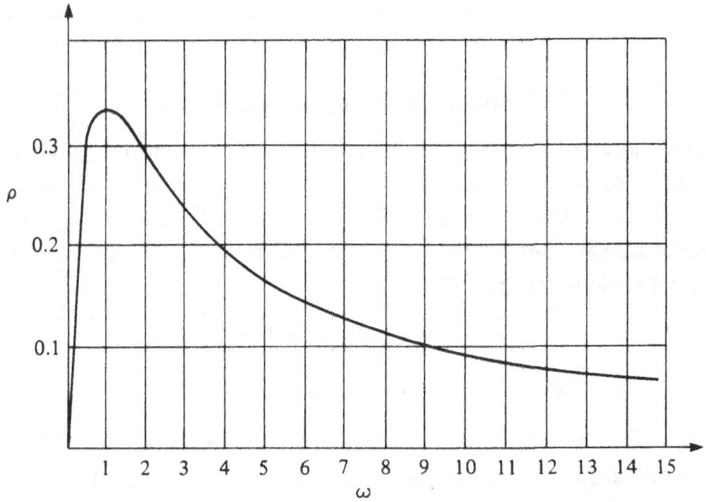

Taking into account equation (9.29), equation (9.27) may be rewritten as

$$W_{tf} = \beta \frac{k_2^2[\mathrm{RH}]^2}{k_6} \rho^2 \tag{9.31}$$

and introducing a dimensionless tar formation rate,

$$\nu_{tf} = \frac{k_6}{k_2^2[\mathrm{RH}]^2} W_{tf}$$

and using this in equation (9.31), we obtain

$$\nu_{tf} = \beta\rho^2$$

Hence ν_{tf} reaches a maximum of $\frac{1}{9}$ times the given value of β at the maximum value of ρ. In a real oxidation process (taking the worst case, $\beta = 1$) we may say that $\nu_{tf} \ll 0.01$. It is also possible to demonstrate that the ν_{tf} maximum corresponds to the case of $\omega \ll 10$.

Let us define the relative dimensionless tar formation rate as

$$\eta = \frac{k_{10}[\mathrm{R\cdot}]^2}{k_1[\mathrm{R\cdot}][\mathrm{O_2}]}$$

Combining equations (9.13) and (9.29), we obtain

$$\eta = \rho/\omega$$

and using this in equation (9.30) then yields

$$\eta = \frac{1}{\omega^2 + \omega + 1}$$

For $\omega \gg 10$ we find $\eta < 0.01$, or the relative tar formation rate for $\omega = 10$ is less than 1.0 percent. For specific mechanisms of oxidation, other values of η and ν_{tf} may be taken as a criterion of technological efficiency.

9.3 Catalytic oxidation of hydrocarbons

The acceleration of the process of liquid-phase oxidation of hydrocarbons due to the presence of catalysts (which are dissolved salts of metals of changeable valency) is determined by the catalysts' capacity for reaction with hydrocarbons and the generation of free radicals. As a result of reactions of catalytic initiation,[13,14]

$$\mathrm{K}_{t_{ox}} + \mathrm{RH} \rightarrow \mathrm{K}_{tb} + \mathrm{H}^+ + \mathrm{R\cdot}$$

$$\mathrm{K}_{tb} + \mathrm{ROOH} \rightarrow \mathrm{K}_{t_{ox}} + \mathrm{OH}^- + \mathrm{RO\cdot}$$

where $\mathrm{K}_{t_{ox}}$ is the oxidized catalyst, K_{tb} the recuperated catalyst, and RH the initial hydrocarbon or intermediate reaction product. R· and RO· radicals are generated, oxidized catalyst interchanging with the weakest bond. C—H leads to R· radicals, and the recuperated catalyst, interchanging with hydroperoxide ROOH, leads to RO· radicals.

It has been shown[9] that the rate of catalytic initiation for the case of steady oxidation is given by

$$W_i = 2k_0(\mathrm{Me}^{+N+1})[\mathrm{RH}]$$

where Me is the catalyst mass and in the exponent N is the valency. Substituting into this equation the value[9] of (ME^{+N+1}), we find that

$$W_i = \frac{2k_0k_7k_1[\mathrm{O}_2]c_k[\mathrm{R}\cdot]}{k_0k_7c_k + k_5k_2 + k_7k_1([\mathrm{O2}]/[\mathrm{RH}])[\mathrm{R}\cdot]}$$

where c_k is the catalyst concentration, and combining this with equations (9.6), (9.13), and (9.21), we obtain a relationship for the R· radical concentration:

$$[\mathrm{R}\cdot] = \left\{\frac{2k_0k_7k_1c_k[\mathrm{O}_2][\mathrm{R}\cdot]}{k_{10}(\omega^2+\omega+1)[k_0k_7c_k + k_5k_2 + k_7k_1([\mathrm{O}_2]/[\mathrm{RH}])[\mathrm{R}\cdot]]}\right\}^{1/2}$$

or

$$[\mathrm{R}\cdot]^2 + \left(\frac{k_0}{k_1}c_k\frac{[\mathrm{RH}]}{[\mathrm{O}_2]} + \frac{k_2k_5[\mathrm{RH}]}{k_1k_7[\mathrm{O}_2]}\right)[\mathrm{R}\cdot] - \frac{2k_0c_k[\mathrm{RH}]}{k_{10}(\omega^2+\omega+1)} = 0 \tag{9.32}$$

Now, solving equation (9.28) with respect to [R·], and substituting its value into equation (9.32), taking into account that radicals formation is a catalytic process, and using equation (9.11) with k_4[RH] replaced by k_7c_k, we find that

$$\left[\frac{k_2^2[\mathrm{RH}]^2}{k_6k_{10}(1+k_5/k_7c_k)^2}\right]\rho^2 + \left(\frac{k_0}{k_1}c_k\frac{[\mathrm{RH}]}{[\mathrm{O}_2]} + \frac{k_2k_5[\mathrm{RH}]}{k_1k_7[\mathrm{O}_2]}\right)$$
$$\times\left[\frac{k_2[\mathrm{RH}]}{(k_6k_{10})^{1/2}(1+k_5/k_7c_k)}\right]\rho - \frac{2k_0c_k[\mathrm{RH}]}{k_{10}(\omega^2+\omega+1)} = 0$$

Introducing the dimensionless rate of molecular hydroperoxide breaking as

$$k_5' = k_5/c_k \tag{9.33}$$

in this equation leads to

$$\rho^2 + \left(\frac{k_0}{k_1}c_k\frac{[\mathrm{RH}]}{[\mathrm{O}_2]} + \frac{k_2k_5[\mathrm{RH}]}{k_1k_7[\mathrm{O}_2]}\right)\frac{k_2[\mathrm{RH}]k_6k_{10}(1+k_5'/k_7)^2}{(k_6k_{10})^{1/2}(1+k_5'/k_7)k_2^2[\mathrm{RH}]^2}\rho$$
$$- \frac{2k_0c_k[\mathrm{RH}]k_6k_{10}(1+k_5'/k_7)^2}{k_{10}(\omega^2+\omega+1)k_2^2[\mathrm{RH}]^2} = 0$$

and after substituting

$$\frac{1}{\omega} = \frac{k_2}{k_1} \frac{[\mathrm{RH}]}{\mathrm{O}_2} \frac{k_{10}^{1/2}}{k_6^{1/2}}$$

from equation (9.13), we find

$$\rho^2 + \frac{(k_6 k_{10})^{1/2}(1 + k_5'/k_7)}{k_2[\mathrm{RH}]\,\omega} \left(\frac{k_0}{k_2} c_k \frac{k_6^{1/2}}{k_{10}^{1/2}} + \frac{k_5 k_6^{1/2}}{k_7 k_{10}^{1/2}} \right) \rho - \frac{2 k_0 c_k k_6 (1 + k_5'/k_7)}{k_2^2[\mathrm{RH}](\omega^2 + \omega + 1)} = 0 \quad (9.34)$$

We now define the dimensionless catalyst concentration:

$$\sigma = \frac{k_0 k_7}{k_2 k_5} c_k$$

and the dimensionless constants

$$\pi^* = \frac{k_7 + k_5'}{k_7}$$

$$k^* = \frac{k_5 k_6 (k_7 + k_5')}{k_2 k_7^2 [\mathrm{RH}]}$$

and we can then rewrite equation (9.34) in the form

$$\rho^2 + \frac{k^*(\sigma + 1)}{\omega} \rho - \frac{2k^* \sigma \pi^*}{\omega^2 + \omega + 1} = 0$$

Solving with respect to ρ, we obtain

$$\rho = \frac{k^*(\sigma + 1)}{2\omega} \left\{ \left[1 + \frac{8\omega^2 \sigma \pi^*}{k^*(\sigma + 1)^2(\omega^2 + \omega + 1)} \right]^{1/2} - 1 \right\} \quad (9.35)$$

For catalytical liquid-phase oxidation, initial hydrocarbon is transformed into the final product at the rate

$$W' = 2k_1[\mathrm{R}\cdot][\mathrm{O}_2] \quad (9.36)$$

where the prime denotes a process. For a constant initiation rate and square chain breaking, the accumulation of hydroperoxide and final products are equal, so substituting into equation (9.36) the value of $[\mathrm{O}_2]$ from (9.13) and $[\mathrm{R}]$ from equation (9.28), taking equation (9.33) into account, we obtain

$$W' = 2k_1 \left[\frac{k_2[\mathrm{RH}]}{(k_6 k_{10})^{1/2}(1 + k_5'/k_7)} \right] \rho \left(\frac{k_2[\mathrm{RH}]\, k_{10}^{1/2}}{k_1 k_6^{1/2}} \right) \omega = \frac{2k_2^2[\mathrm{RH}]^2}{k_6} \frac{k_7}{k_7 + k_5'} \rho\omega \quad (9.37)$$

Now, let us introduce a dimensionless rate of catalytic oxidation:

$$\nu' = \frac{W' k_6 (k_7 + k_5')}{k_7 k_2^2 [\mathrm{RH}]^2} \tag{9.38}$$

Substituting in this equation the value of W from equation (9.37), with the value of ρ being obtained from equation (9.35), the rate is

$$\nu' = k^*(\sigma + 1)\left\{\left[\frac{2\omega^2 \sigma \pi^*}{k^*(\sigma + 1)^2(\omega^2 + \omega + 1)}\right]^{1/2} - 1\right\} \tag{9.39}$$

Now, at sufficiently large values of σ,

$$\left[1 + \frac{2\omega^2 \sigma \pi^*}{k^*(\sigma + 1)^2(\omega^2 + \omega + 1)}\right]^{1/2} \simeq 1 + \frac{\omega^2 \sigma \pi^*}{k^*(\sigma + 1)^2(\omega^2 + \omega + 1)}$$

and consequently in equation (9.39)

$$\nu' \simeq \frac{\sigma \pi^* \omega^2}{(\sigma + 1)(\omega^2 + \omega + 1)}$$

Also, as $k_7 \gg k_5'$ (thus $\pi^* \simeq 1$), from equation (9.38) we find that

$$\nu' \simeq \frac{W' k_6}{k_2^2 [\mathrm{RH}]^2} \simeq \frac{\sigma \omega^2}{(\sigma + 1)(\omega^2 + \omega + 1)}$$

and from this it follows that for $\sigma \to \infty$,

$$\nu' \simeq \frac{\omega^2}{\omega^2 + \omega + 1} \tag{9.40}$$

Then, for $\omega \to \infty$, the maximum rate of catalytic oxidation, as given by equation (9.38), will be

$$W'_{\max} \simeq k_2^2 [\mathrm{RH}]^2 / k_6$$

which is the same as in case of noncatalytical oxidation for $k_4 \gg k_5$, as a comparison with equation (9.26) shows.

9.4 Macrokinetics of liquid-phase oxidation of hydrocarbons

As we have seen, the maximum rate of hydrocarbon oxidation is affected if adequate quantities of oxidizing agent are supplied to the reaction zone. The oxygen supply process, which consists of three consecutive stages (transport from the gas phase to the gas–liquid interface, solution in the liquid, and diffusion into the liquid phase), is limited by the slowest of these, the convective diffusion of oxygen to the hydrocarbon. For instance, for a constant oxygen concentration near the surface of a gas bubble, a pseudo-steady-state condition can be said to occur as a result of convective transport of the molecules. This is a realistic model because, for example, under the conditions of industrial sparging, the renewal time of the bubble surface is negligibly small.

The theory we developed in Chapters 4 and 7 will be used to compute the diffusional flux of oxygen for the hydrocarbon, and in this theory, based on the existence of a diffusion boundary layer (which is dependent on the hydrodynamic conditions and physical properties of the phase components), the movement of the liquid and the convective mass transfer generated by this movement are taken into account.

Under typical conditions of industrial sparging, at moderate and large Reynolds numbers, the diffusion flux of oxygen from one bubble is given by equation (7.19),

$$\dot{m} = 8\left(\frac{\pi D v_r}{d_b}\right)^{1/2} r_b^2(c_0 - c_1) \tag{9.41}$$

where D is the diffusivity, v_r the bubble rise (or relative) velocity, $r_b = \frac{1}{2}d_b$ the radius of the bubble, and c_0 and c_1 are the concentrations of oxygen in the gas and liquid phases, respectively. If we denote the gas void fraction in the gas–liquid mixture by ϕ (at the gas interface its value being unity), we can write [equation (4.4)]

$$v_r = v_s/\phi \tag{9.42}$$

Here v_s is the gas superficial velocity, based on the total cross-sectional area of the reactor. Thus equation (9.41) assumes the form

$$\dot{m} = 8\left(\frac{\pi D v_s}{d_b \phi}\right)^{1/2} r_b^2(c_0 - c_1)$$

For all of the bubbles at a section of the flow, per unit area of interfacial contact, we have

$$\dot{m}' = \frac{\dot{m}}{4\pi r_b^2} = \left(\frac{4 D v_s}{\pi d_b \phi}\right)^{1/2}(c_0 - c_1)$$

and knowing the specific surface area[15] from Section 3.5,

$$a = 3\phi/r_b$$

we can determine the diffusional flux per unit system volume:

$$\dot{m}'a = 2.395\left(\frac{D v_s \phi}{r_b^3}\right)^{1/2}(c_0 - c_1)$$

We can now consider a material balance for oxygen, assuming that the oxidation rate W'per unit system volume is equal to the diffusional flux $\dot{m}'a$,

$$W' = 2.395\left(\frac{D v_s \phi}{r_b^3}\right)^{1/2}(c_0 - c_1)$$

or

$$W' = 2.395 \left(\frac{D v_s \phi}{r_b^3} \right)^{1/2} \left(\frac{P_{O_2}}{H_{O_2}} - [O_2]_{\text{liq}} \right) \tag{9.43}$$

where P_{O_2} is the oxygen partial pressure and H_{O_2} the Henry constant for the oxygen–hydrocarbon system.

Now, from equation (9.40) and Figure 9.1, for a value of the dimensionless oxygen concentration $\omega \simeq 10$, the oxidation conditions are close to the maximum, and in this case the dimensionless oxidation rate $\nu' = 0.91$ and the dimensionless tar formation rate ν_{tf} falls quickly (as we see in Figure 9.2). Thus we conclude that the maximum oxidation rate and minimum tar formation rate conditions require oxidation at $\omega > 10$, and hence $\omega = 10$ is adopted as the borderline for such an adequate process.

Equation (9.43) enables us to calculate the appropriate cross section A of the equipment and corresponding air consumption Q. To determine air consumption, we substitute in the equation the values of W' from equation (9.38), ϕ from equation (9.42), and $[O_2]$ from equation (9.13), to find

$$\frac{k_7 k_2^2 [\text{RH}]^2}{k_6(k_7 + k_5')} \nu' = 2.395 \frac{D^{1/2} v_s}{r_b^{3/2} v_r^{1/2}} \left(\frac{P_{O_2}}{H_{O_2}} - \frac{k_2 k_{10}^{1/2} [\text{RH}]}{k_1 k_6^{1/2}} \omega \right) \tag{9.44}$$

Now, the gas volumetric flow rate is

$$Q = A v_s$$

where A is the cross-sectional area of the apparatus, so equation (9.44) may be rewritten as

$$\nu' = C_1 Q(\omega_{\text{max}} - \omega) \tag{9.45}$$

with

$$C_1 = \frac{2.395 D^{1/2} (k_6 k_{10})^{1/2} (k_7 + k_5')}{v_r^{1/2} r_b^{3/2} A k_1 k_2 k_7 [\text{RH}]}$$

characterizing reaction kinetics, mass-transfer conditions, and apparatus size. Also,

$$\omega_{\text{max}} = \frac{k_1 k_6^{1/2} P_{O_2}}{k_2 k_{10}^{1/2} [\text{RH}] H_{O_2}} = a P_{O_2}$$

is the maximum dimensionless oxygen concentration (proportional to oxygen partial pressure). Substituting into equation (9.45) the value of ν' from equation (9.40), we obtain the basic equation

$$C_1 Q(\omega_{\text{max}} - \omega) = \frac{\omega^2}{\omega^2 + \omega + 1} = C_2 W' \tag{9.46}$$

where

$$C_2 = \frac{1}{W'_{\max}} = \frac{k_6}{k_2^2[\mathrm{RH}]^2}$$

is a term inversely proportional to the oxidation rate (because of the definition of ν).

Rewriting equation (9.46), we obtain

$$\omega = \omega_{\max} - \frac{C_2 W'}{C_1 Q} \tag{9.47}$$

and also

$$\omega = \frac{C_2 W'}{2(C_2 W' - 1)} \left\{ \left[1 - \frac{4(C_2 W' - 1)}{C_2 W'} \right]^{1/2} + 1 \right\} \tag{9.48}$$

obtained from equation (9.23).

Thus, from equation (9.47), a plot of ω against W'/Q will yield a straight line with slope C_2/C_1 and the intersection of the line with the ω axis will give $\omega_{\max}$. Equation (9.46) is solved by finding a value for coefficient C_2 giving optimum agreement of experimental data with the straight line.[16] An example of this procedure is given in the next section.

9.5 Estimation of air supply requirements for processes in the kinetic zone

As the air supply has a considerable influence on liquid-phase oxidation of hydrocarbons, experimental and theoretical investigations to estimate the optimum air supply for such oxidation processes in the kinetic zone, on an industrial scale, are of importance.

We now consider the catalytic oxidation of *o*-xylene, with atmospheric oxygen, as a model reaction, and to estimate the dimensionless oxygen concentration ω from equation (9.48), the results of three experiments of *o*-xylene oxidation at various volumetric air flow rates are presented. The conditions for the tests were $Q_{\mathrm{I}} = 1000$ reduced liters per hour, $Q_{\mathrm{II}} = 1500$, and $Q_{\mathrm{III}} = 2000$, pressure $= 5$ atm, reactor temperature $T = 160\,^{\circ}\mathrm{C}$, and catalyst expenditure 0.02 percent by weight.

The experimental data are given in Table 9.1, according to which (concentration–time) curves are constructed for *o*-toluyl acid, as shown in Figure 9.3. The average value of the oxidation rate is determined from the curves in Figure 9.4 plotted in ($W' - c$) coordinates, and the average rates for the first, second, and third experiments, respectively, were

$$W'_{\mathrm{I}} = 3.4 \frac{\text{kg mol}}{\text{m}^3 \cdot \text{h}}$$

$$W'_{\mathrm{II}} = 5.03 \frac{\text{kg mol}}{\text{m}^3 \cdot \text{h}}$$

$$W'_{\mathrm{II}} = 6.25 \frac{\text{kg mol}}{\text{m}^3 \cdot \text{h}}$$

Further for $C_2 = 0.15$ (a guess), from equation (9.48) we find

$$\omega_{\mathrm{I}} = 1.66$$

$$\omega_{\mathrm{II}} = 4.0$$

$$\omega_{\mathrm{III}} = 16.1$$

Figure 9.3 Kinetic curves for accumulation of *o*-toluyl acid at $P = 5.0$ atm. Air flow rate (reduced liters per hour): I, $Q = 1000$; II, $Q = 1500$; III, $Q = 2000$.

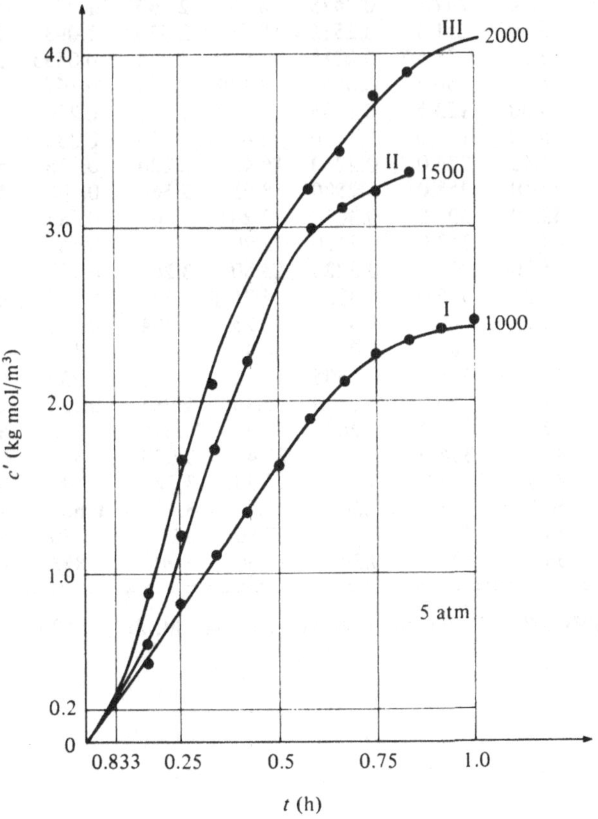

Table 9.1. *Data on catalytic oxidation of o-xylene*

(1)	(2)	(3)	(4)	(5)	(6)	(7)	(8)	(9)	(10)
		Concentration of *o*-toluyl acid							
Run	Point	wt % of acid	grams of acid	kg of acid	kg mol $\times 10^{-4}$	kg mol $\cdot$ m³	Time (h)	W' (kg mol $\cdot$ m³/h)	Q (m³/h)
I	1	3.70	24.00	0.024	1.765	0.239	0.0833	2.87	
	2	7.00	45.80	0.0458	3.360	0.454	0.167	2.825	
	3	12.70	83.60	0.0836	6.140	0.831	0.25	3.430	
	4	16.83	111.1	0.1111	8.170	1.108	0.333	3.424	
	5	20.50	135.5	0.1355	9.950	1.349	0.416	3.340	
	6	24.70	163.5	0.1635	12.00	1.623	0.500	3.346	
	7	28.70	190.0	0.1900	13.95	1.890	0.585	3.340	
	8	32.00	212.1	0.2121	15.61	2.115	0.667	3.270	
	9	34.70	228.0	0.2280	16.75	2.270	0.750	3.030	1.0
	10	37.00	242.3	0.2423	17.86	2.420	0.916	2.640	
	11	37.70	247.8	0.2478	18.20	2.460	1.00	2.460	
	12	38.30	251.5	0.2515	18.61	2.520	1.083	2.323	
II	1	3.31	21.50	0.0215	1.580	0.214	0.0833	2.57	
	2	8.60	56.7	0.0567	4.170	0.565	0.167	3.38	
	3	18.30	123.5	0.1235	9.080	1.230	0.250	4.93	
	4	25.10	172.0	0.1720	12.62	1.710	0.333	5.14	
	5	32.42	224.0	0.2240	16.45	2.220	0.416	5.44	1.5
	6	37.10	258.0	0.2580	18.95	2.560	0.500	5.12	
	7	43.00	302.0	0.3020	22.20	3.00	0.583	5.14	
	8	44.30	312.0	0.3120	22.90	3.10	0.667	4.64	
	9	45.30	322.0	0.3220	23.60	3.20	0.750	4.27	
	10	45.30	325.0	0.3250	24.85	3.34	0.833	4.01	
III	1	4.7	30.6	0.0306	2.25	0.304	0.0833	3.67	
	2	13.3	87.0	0.087	6.40	0.865	0.167	5.16	
	3	25.1	166.5	0.1665	12.25	1.655	0.250	6.63	
	4	31.4	210.0	0.2100	15.45	2.085	0.333	6.27	
	5	39.4	266.0	0.266	19.60	2.645	0.416	6.36	
	6	44.6	304.0	0.304	22.40	3.030	0.500	6.06	2.0
	7	47.4	324.0	0.324	23.90	3.220	0.583	5.54	
	8	50.0	346.0	0.346	25.4	3.44	0.667	5.16	
	9	54.2	378.0	0.378	27.80	3.76	0.750	5.00	
	10	56.2	391.0	0.392	28.80	3.90	0.833	4.69	

Note: T = 160°C; P = 5 atm; catalyst, 0.02% by weight; M = 650 g; volume = 7.39×10^{-4} m³.

The values of ω_{I}, ω_{II} and, ω_{III} obtained correspond to the following air consumption in the experiments: $Q_I = 0.2\ m^3/h$, $Q_{II} = 0.3\ m^3/h$, $Q_{III} = 0.4\ m^3/h$, and hence

$$\frac{W'_I}{Q_I} = \frac{3.4}{0.2} = 17$$

$$\frac{W'_{II}}{Q_{II}} = \frac{5.03}{0.3} = 16.77$$

$$\frac{W'_{III}}{Q_{III}} = \frac{6.25}{0.4} = 15.625$$

As we can see from Figure 9.5, the points corresponding to ω_I, ω_{II}, and ω_{III}, lie on a straight line in $(\omega - W'/Q)$ coordinates as required by equation (9.47), and consequently $C_2 = 0.15$ was chosen correctly. From this plot we find that $C_2/C_1 \simeq 10.5$ and it follows that $C_1 = \dfrac{0.15}{10.5} = 1.43 \times 10^{-2}$, and

$$\omega_{max} = \frac{C_2 W'}{C_1 Q} + \omega = 10.5 \times 15.625 + 16.1 \simeq 180$$

Also,

$$\frac{\omega_{max}}{P_{air}} = \frac{180}{5} = 36$$

Figure 9.4 Oxidation rate of *o*-xylene W' versus concentration *o*-toluyl acid C at $P = 5$ atm for various air supplies (reduced liters per hour): I, $Q = 1000$; II, $Q = 1500$; III, $Q = 2000$.

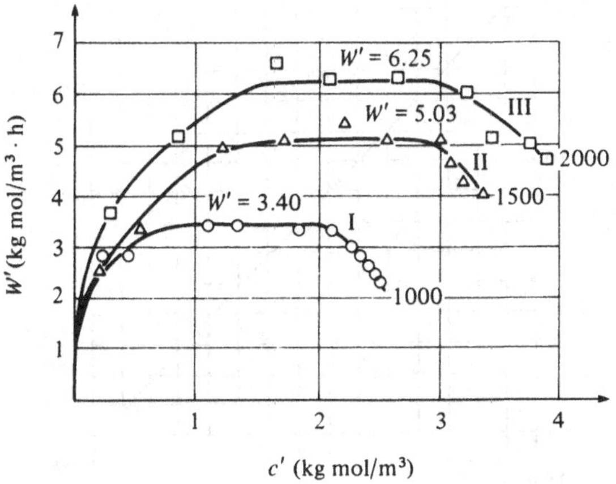

Based on equation (9.46), the volumetric air flow rate for the reaction in the kinetic zone is then $Q \simeq 0.37\ m^3/h$. So $0.37 \times 5 = 1.85$ reduced m^3/h of air corresponds to $7.39 \times 10^{-4}\ m^3$ of *o*-xylene, and for $1.0\ m^3$ *o*-xylene we need 2510 reduced m^3/h of air.

Solving equation (9.46) with respect to Q for a given air pressure, we can find the air supply necessary for reaction in the kinetic zone. Thus at an air pressure $p = 10$ atm, $Q \simeq 0.182\ m^3/h$ or 1.82 reduced m^3/h; at air pressure $p \simeq 7.5$ atm, $Q \simeq 0.245\ m^3/h$ or 1.83 reduced m^3/h; and at air pressure $p = 15$ atm, $Q \simeq 0.12\ m^3/h$ or 1.8 reduced m^3/h.

In Table 9.2 we present experimental data for various pressures and air supplies, and the kinetic curves for these experiments are presented in Figure 9.6. Experiments (i) and (ii) were conducted under conditions just sufficient for a process in the kinetic zone, and in experiment (i) the air supply exceeded by far the necessary amount; no increase in product output is observed with this increase. Experiment (iii) was conducted under stimulated kinetic conditions, which yielded the concentration–time curve located in the vicinity of this zone. Experiment (iv) was conducted with the air supply inadequate for a process in the kinetic zone, and the concentration–time curve in this experiment is well below the rest. The accumulation of reaction products is in this case dependent on the air supply.

Figure 9.5 Dimensionless oxygen concentration ω versus W'/Q parameter.

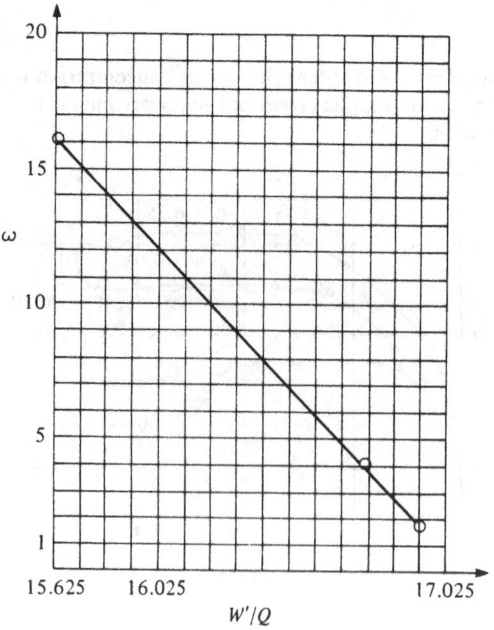

9.6 Ideal displacement, and ideally mixed, reactors for liquid-phase oxidation of hydrocarbons

At the point of entrainment of a continuous mixing process, the transfer of reagents along the flow significantly affects the kinetics of oxidation, and this effect may be considered theoretically in two limiting cases, the first being plug flow (ideal displacement) and the second ideal mixing of the reacting materials.[1,2,17–19]

Let us first consider the oxidation process that occurs in an ideally mixed reactor in which liquid is, essentially, instantaneously and completely mixed with the reaction products. The change in concentration of the reacting matter, at any section in the reactor where turbulent diffusion along the flow takes place, is due to three factors: diffusion, liquid flow, and chemical reaction. Now consider a sparged reactor with volume V, which is fed a hydrocarbon mixture with concentration c_i at a volumetric flow rate Q. During time dt, the reactor thus receives a hydrocarbon amount $Qc_i\ dt$, and the

Figure 9.6 Kinetic curves for the accumulation *o*-toluyl acid for various pressures and air supplies.

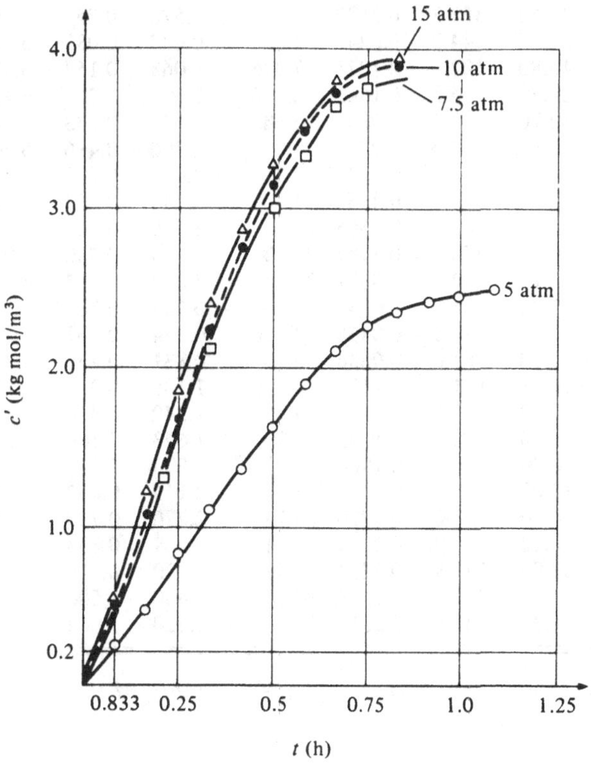

Table 9.2. Experimental data for various pressures and air supplies

(1)	(2)	(3)	(4)	(5)	(6)	(7)	(8)	(9)	(10)
		Concentration of *o*-toluyl acid							
Run	Point	wt % of acid	grams of acid	kg of acid	kg mol $\times 10^{-4}$	kg mol $\cdot$ m^3	Time (h)	W' (kg mol $\cdot$ m^3/h)	Q (m^3/h)
i	1	8.33	54.2	0.0542	3.99	0.541	0.0833	6.50	
$P = 15$	2	18.75	123.0	0.123	9.04	1.221	0.167	7.32	
	3	27.9	186.2	0.1862	13.70	1.855	0.250	7.32	
	4	35.7	242.2	0.2422	17.81	2.410	0.333	7.23	
	5	42.50	290.0	0.290	21.30	2.88	0.416	6.93	3.0
	6	46.30	329.0	0.329	24.20	3.28	0.500	6.56	
	7	55.0	374.0	0.374	27.55	3.734	0.583	6.41	
	8	56.6	382.0	0.362	28.00	3.80	0.667	5.69	
ii	1	7.70	50.0	0.050	3.680	0.498	0.0833	5.97	
$P = 10$	2	16.00	106.5	0.1065	7.840	1.062	0.167	6.372	
	3	24.67	168.5	0.1685	12.40	1.68	0.25	6.720	
	4	32.17	234.0	0.224	16.50	2.285	0.333	6.715	
	5	38.50	276.0	0.276	20.30	2.50	0.416	6.610	2.0
	6	44.00	316.0	0.316	23.20	3.140	0.500	6.280	
	7	48.00	350.0	0.350	25.80	3.490	0.583	5.990	
	8	49.30	358.0	0.358	26.33	3.565	0.667	5.350	
	9	49.60	358.2	0.3582	26.40	3.570	0.75	4.760	
iii	1	6.77	44.0	0.044	3.235	0.437	0.0833	5.246	
$P = 7.5$	2	15.80	107.1	0.1071	7.875	1.063	0.167	6.377	
	3	24.20	164.2	0.1642	12.075	1.603	0.26	6.412	
	4	30.70	213.3	0.2133	15.68	2.12	0.333	6.367	
	5	38.50	272.0	0.2720	20.00	2.706	0.416	6.505	
	6	41.80	303.0	0.3030	22.28	3.000	0.500	6.000	1.5
	7	45.40	334.6	0.3346	24.60	3.330	0.583	5.712	
	8	49.1	366.6	0.3666	26.95	3.650	0.667	5.472	
	9	50.5	378.7	0.3787	27.80	3.760	0.750	5.00	
	10	51.1	382.7	0.3827	28.14	3.808	0.833	4.571	
iv	1	3.70	24.0	0.024	1.765	0.239	0.0833		
$P = 5$	2	7.00	45.8	0.0458	3.360	0.454	0.167		
	3	12.70	83.6	0.0836	6.140	0.831	0.250	3.330	
	4	16.83	111.1	0.1111	8.170	1.108	0.333	3.324	
	5	20.50	135.5	0.1355	9.950	1.349	0.416	3.240	
	6	24.70	163.5	0.1335	12.00	1.623	0.500	3.246	
	7	28.70	190.0	0.1900	13.95	1.890	0.583	3.240	1.0
	8	32.00	212.1	0.2121	15.61	2.115	0.667	3.170	
	9	34.70	228.0	0.2280	16.75	2.270	0.750	3.030	
	10	36.20	235.9	0.2359	17.34	2.349	0.833	2.820	
	11	37.00	242.3	0.2423	17.68	2.420	0.916	2.640	
	12	37.70	247.8	0.2478	18.20	2.460	1.000	2.460	
	13	38.30	251.5	0.2515	18.61	2.520	1.083	2.323	

Note: T = 160°C; catalyst, 0.02% by weight; v_s = 0.044 m/s; M = 650 g; volume = 7.39 $\times 10^{-4}$ m^3.

reaction mass that leaves the reactor is $Qc'\,dt$. Also, for this period, the accumulation of products (in moles) as a result of chemical reaction will be $VW'\,dt$. The concentration of the final product will then change by the amount dc', and a material balance of the process in this case may be written in the form[4]

$$Qc_i\,dt - Qc'\,dt + VW'\,dt = V\,dc'$$

or

$$V\frac{dc'}{dt} = Qc_i - Qc' + VW' \tag{9.49}$$

Because the reactor feed consists of hydrocarbons (and no recycle products), according to equation (9.49), for a steady-state regime ($dc'/dt = 0$), we have

$$\frac{c'}{W'} = \frac{V}{Q} \tag{9.50}$$

The concentration c' of the reaction products, in the continuous reactor, can be determined using data that are available on the concentration kinetics, and a typical kinetic rate curve for a batch process is shown in Figure 9.7. Here we see a family of points satisfying equation (9.50). The numbers

Figure 9.7 Kinetic curve for the accumulation *o*-toluyl acid in batch oxidation process of *o*-xylene at a temperature of 160°C, a pressure of 10 atm, an air flow rate of 2000 reduced liters/h, and an *o*-xylene charge of M = 0.9 kg.

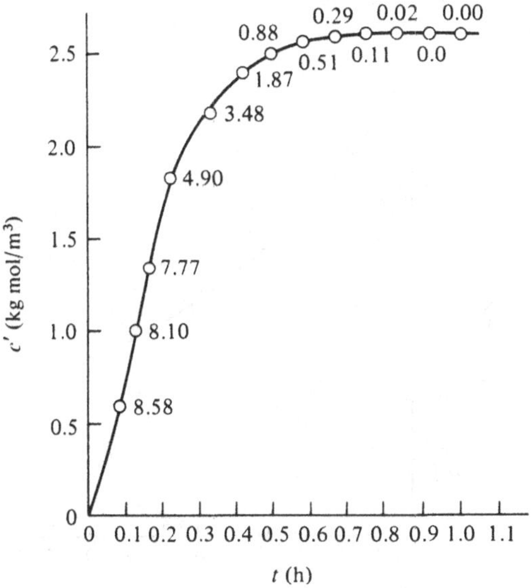

on the curve correspond to reaction rates computed according to the following numerical difference formula:[20]

$$Y'_n = \frac{2}{3h}(Y_{n+1} - Y_{n-1}) - \frac{1}{12h}(Y_{n+2} - Y_{n-2}) \qquad (9.51)$$

Here Y'_n is the formation rate of the reaction products at a point n (kg mol/m$^3\cdot$h), h is the time interval between adjacent points (hours), and Y_n is the concentration of reaction products (kg mol/m^3).

For the case considered ($h = \frac{1}{12}$), formula (9.51) becomes

$$Y'_n = 8(Y_{n+1} - Y_{n-1}) - (Y_{n+2} - Y_{n-2}) \qquad (9.52)$$

From the first point, $Y'_n = \tan\alpha = W'$, using Figure 9.7 and formula (9.52), the concentration of *o*-toluyl acid at the exit of the reactor as a function of the space-averaged velocity Q/V of the hydrocarbon can be determined, as seen in Figure 9.8.

Now, the reactor output[4] is

$$W' = c'Q = f\left(\frac{V}{Q}\right)Q \qquad (9.53)$$

and dividing equation (9.53) by V, we get the *relative output:*

$$\theta = \frac{W'}{V} = \frac{Q}{V}f\left(\frac{V}{Q}\right) = \frac{Q}{V}c'$$

Figure 9.8 Concentration of *o*-toluyl acid at the reactor exit versus space velocity of hydrocarbon.

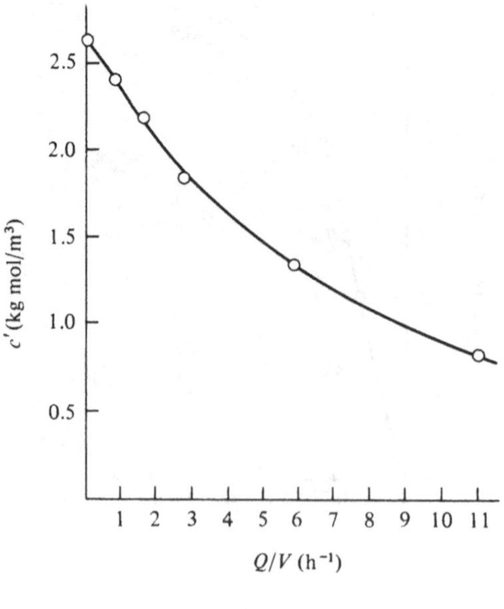

Physically, θ is the reactor output per unit reactor volume, and Figure 9.9 shows a curve of the reactor relative output as a function of the space velocity Q/V of the hydrocarbon derived from the plot in Figure 9.8, and we see that this curve has a maximum that corresponds to the concentration of reaction products at the reactor exit, which is equal to 45 percent of maximum value achieved in the reaction.

It is known[4] that the overall reactor efficiency is determined not only by its output but also by the concentration of reaction products at the reactor exit, and therefore we can, in the first approximation, evaluate the overall reactor efficiency by analyzing the maxima of the function[4]

$$\delta = \theta f\left(\frac{V}{Q}\right) = \theta f(t) = \frac{Q}{V} c'^2$$

In Figure 9.10 we show the relationship of the overall efficiency of the reactor of ideal mixing δ as a function of the space velocity of hydrocarbon Q/V, again computed with the use of Figure 9.8. Comparing the curves in Figures 9.9 and 9.10, we can see that the value of Q/V corresponding to the extremum value is moved in the direction of smaller values. The acid concentration at the reactor exit is about 50 to 70 percent of the optimal value achieved in the reaction.

Let us now consider the oxidation process of *o*-xylene in a plug-type reactor. In this case the relative output of the reactor[4] is

$$\theta = \frac{W'}{V} = \frac{Q}{V} f\left(\frac{V}{Q}\right) = \left(\frac{Q}{V}\right) f(t)$$

Figure 9.9 Relative throughput of the reactor versus space velocity of hydrocarbon for ideal mixing.

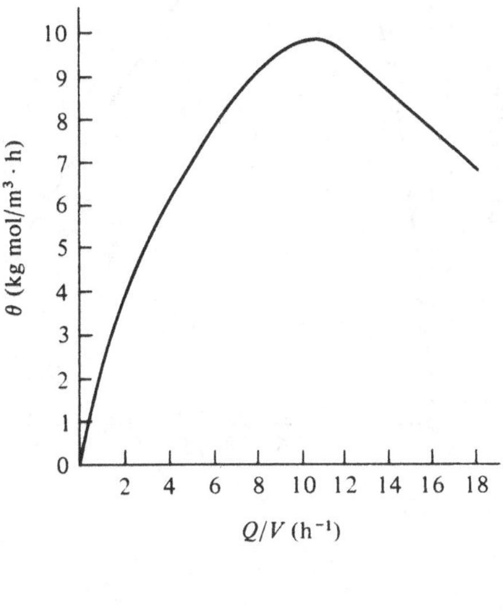

and the relative output of the plug-type reactor is obtained by dividing the concentration curve of Figure 9.7 by a corresponding value of time. In Figure 9.11 is shown the relationship of the plug-type reactor relative output θ to the space velocity Q/V of the hydrocarbon. The efficiency of the equipment, as for the ideal mixing reactor case, is determined by analyzing the maximum of the following function:

$$\delta = \theta f(V/Q) = \theta f(t)$$

and Figure 9.12 presents the relationship of the overall efficiency δ of the plug-type reactor as a function of the space velocity of the hydrocarbon, using Figure 9.8. Comparison of Figures 9.10 and 9.12 shows that the maximum for the plug type occurs in the region of smaller values of Q/V. The maximum concentration of products at the reactor exit equals 80 percent of optimal value achieved in the reaction. In the case of the overall output of the two types of reactors, Figures 9.9 and 9.11 show that θ_{max} in the ideally mixed reactor is 20 percent higher than for the plug-type reactor.

In the first case, $\theta_{max} \simeq 9.80$, and in the second case, $\theta_{max} \simeq 8.20$ (kg mol/m$^3 \cdot$h)

The concentration for the plug-type reactor exit is $c'_{ex} \simeq 0.80$ (kg mol/m^3) for $Q/V = 10.5\ h^{-1}$.

Figure 9.10 Overall efficiency of the reactor versus space velocity of hydrocarbon for ideal mixing.

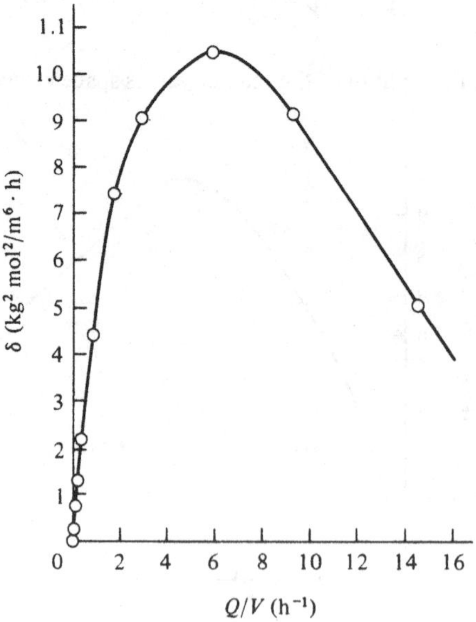

The higher output of the ideally mixed-type reactor, however, does not eliminate the inherent defect of this equipment. This defect is the low concentration of reaction products at the reactor exit, the influence of which on the efficiency is taken into account by the δ function, and when we consider this concentration, the plug-type reactor is more effective than the complete mixing reactor, as indicated in Figures 9.10 and 9.12. Comparison of the efficiency δ of these two types of reactors shows that δ_{max} in the equipment of complete displacement is about 13.7 percent higher than for the case of complete mixing.

In the first case, $\delta_{max} \simeq 14.2$ (kg^2 mol^2/m$^6 \cdot$h), and in the second case, $\delta_{max} \simeq 10.4$ (kg^2 mol^2/m$^6 \cdot$h).

The concentration at the plug-type reactor exit is

$$c'_{ex} \simeq 1.80 \text{ kg mol/m}^3 \text{ at } Q/V = 3.0 \text{ h}^{-1}$$

and the concentration at the exit of the complete mixing-type reactor is

$$c'_{ex} \simeq 1.32 \text{ kg mol/m}^3 \text{ at } Q/V = 6.0 \text{ h}^{-1}$$

In analyzing Figures 9.7 through 9.12 it appears that the best oxidation conditions can be achieved in the *maximum efficiency regime*, and in a continuous process operating in this regime the reaction product concentration

Figure 9.11 Relative throughput of the reactor versus space velocity of hydrocarbon for ideal displacement.

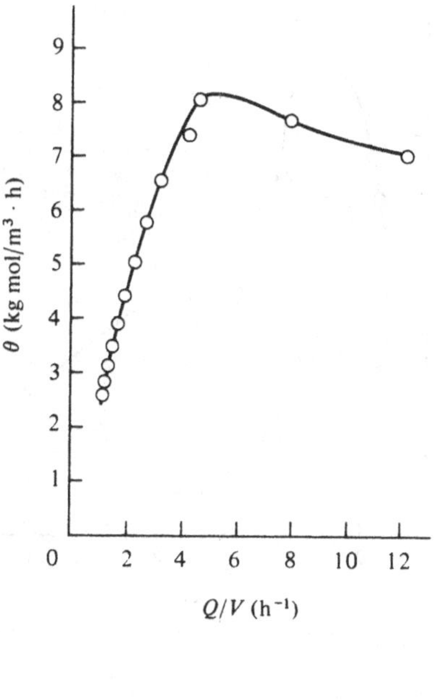

is higher in the plug-type reactor than in the complete mixing-type equipment, because the maximum concentration is achieved there in 10 minutes instead of 20 minutes in the plug-type equipment. Consequently, in the complete mixing-type apparatus we get 8 (kg mol/$m^3 \cdot$h) of product, and in plug-type apparatus only 5.4 (kg mol/$m^3 \cdot$h).

9.7 Calculation of optimal gas and liquid supply to a continuous sparged reactor for the liquid-phase oxidation of hydrocarbons

From Section 9.5 we can choose the following oxidation regime: pressure p = 10 atm, temperature T = 160°C, charge of *o*-xylene M = 0.650 kg, air supply Q = 2000 reduced liters/h, and catalyst–acetate cobalt = 0.02 percent by weight.

On the basis of the kinetic curve (shown in Figure 9.13) and formula (9.52), we can calculate the relationship of the concentration of *o*-toluyl acid at the reactor exit to the space velocity Q/V of the hydrocarbon, and this is shown in Figure 9.14. Also in Figures 9.15 and 9.16 we give the results of analyzing the relative output θ and overall efficiency δ, and we can use these for studying the occurrence of a maximum in the complete mixing-type reactor. The data for computing these curves are given in Table 9.3.

Figure 9.12 Overall efficiency of the reactor versus space velocity of hydrocarbon for ideal displacement.

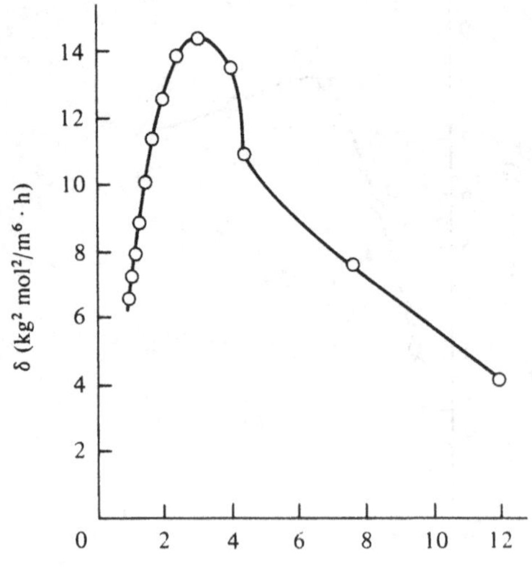

Table 9.3. *Data on catalytic oxidation of o-xylene*

Run	wt %	M (kg mol $\times 10^{-4}$)	c' (kg mol·m^3)	Time (h)	$\theta = W'$ (kg mol ·m^3/h)	c'/W' (h)	Q/V (h^{-1})	δ $\left(\frac{\text{kg}^2\ \text{mol}^2}{\text{m}^6 \cdot \text{mol}}\right)$
1	4.7	2.25	0.304	0.0833	5.24	0.058	17.24	1.593
2	13.2	6.32	0.856	0.167	7.46	0.1147	8.72	6.386
3	22.7	11.10	1.50	0.250	7.69	0.195	5.15	11.54
4	31.7	15.58	2.11	0.333	7.05	0.299	3.34	14.88
5	39.2	19.57	2.65	0.416	5.39	0.492	2.03	14.28
6	44.5	22.27	3.01	0.500	3.86	0.780	1.282	11.62
7	48.5	24.47	3.31	0.583	3.06	1.082	0.924	10.13
8	51.2	26.04	3.53	0.667	2.63	1.342	0.7451	9.284
9	54.0	27.66	3.75	0.750	2.24	1.674	0.5974	8.400
10	56.2	28.83	3.90	0.833	1.65	2.364	0.4230	6.435
11	57.4	29.60	4.03	0.917	1.21	3.330	0.300	4.88
12	59.2	30.46	4.12	1.00	1.00	4.120	0.243	4.12
13	60.8	31.04	4.20	1.0833	0.85	4.941	0.202	3.57
14	61.8	31.48	4.26	1.167	0.600	7.100	0.140	2.556
15	62.5	31.78	4.30	1.250	0.32	13.44	0.0744	1.376

Analysis of these curves (Figures 9.13 through 9.16) shows that, when we operate in the regime of maximum output, we obtain 1.28 kg mol/m^3, or 19.4 percent by weight *o*-toluyl acid, and operating in the regime of maximum efficiency, we obtain 2.4 (kg mol/m^3), or 35.5 percent by weight in 22.5 minutes. Operation in the regime of maximum efficiency is economically more advantageous than the regime of maximum output and in the first case, because the concentration is higher, the *o*-xylene recycled would be 25 percent less. Consequently, the volume of the equipment for extraction of *o*-xylene can also be smaller.

It follows from Section 9.5 that for oxidation of 0.65 kg of *o*-xylene ($p = 10.0$ atm), the air supply must be $Q = 1820$ reduced liters/h. In the regime of maximum efficiency the conversion is 35.5 percent by weight and

Figure 9.13 Kinetic curve for accumulation of *o*-toluyl acid in a batch process of oxidation *o*-xylene at a temperature of 160°C, a pressure of 10 atm, an air flow rate of 2000 reduced liters/h, and an *o*-xylene charge of $M = 0.65$ kg.

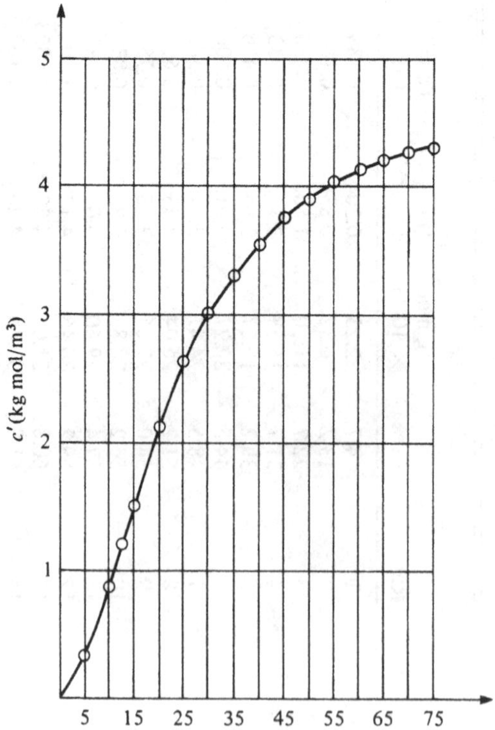

is achieved within 22.5 minutes, so, for achieving the same conversion, we need to have an air supply of about 0.7 reduced m^3 ($1.820 \times 22.5/60 =$ 0.683 reduced m^3). For oxidation of 0.65 kg of *o*-xylene, with conversion of 35.5 percent by weight, the theoretical consumption of oxygen will be

$$\frac{0.65 \times 48 \times 0.355}{106} = 0.104 \text{ kg}$$

or of air,

$$\frac{0.104 \times 100}{23.1 \times 29} \simeq 0.35 \text{ reduced m}^3$$

Thus for the oxidation process in the regime of maximum efficiency, doubling of the air supply, in comparison with the theoretical amount, is required.

During a run of 8 h in the optimal regime, as shown in Figure 9.17 ($p = 10$ atm, $T = 160\,^\circ$C, air supply $Q = 200$ reduced liters/h, volume of *o*-xylene $V = 0.75$ l) in the continuous process, the conversion of reaction product *o*-toluyl acid at the reactor exit was maintained at the stable level of 35.5 percent by weight.

Figure 9.14 Concentration of *o*-toluyl acid at the reactor exit versus space velocity of hydrocarbon for ideal mixing.

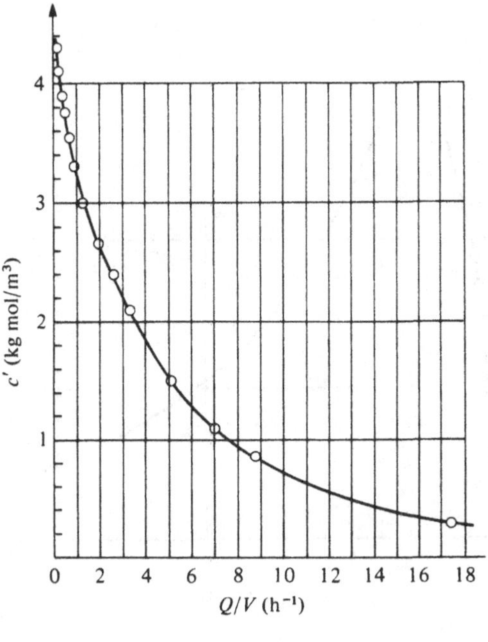

Figure 9.15 Relative throughput of the reactor versus space velocity of hydrocarbon for ideal mixing.

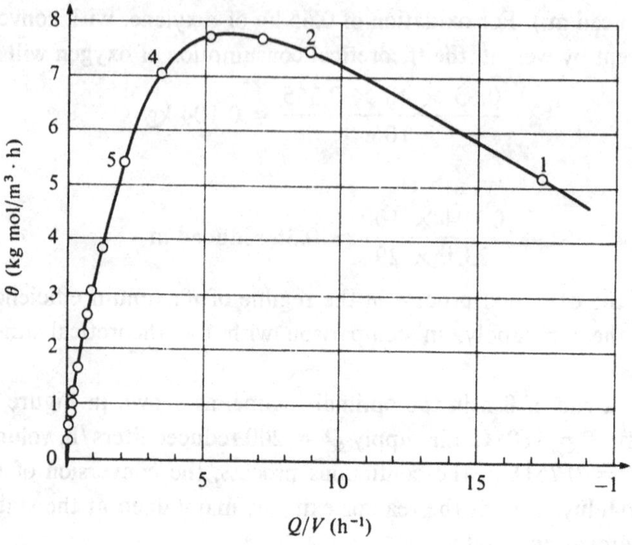

Figure 9.16 Overall efficiency of the reactor versus space velocity of hydrocarbon for ideal mixing.

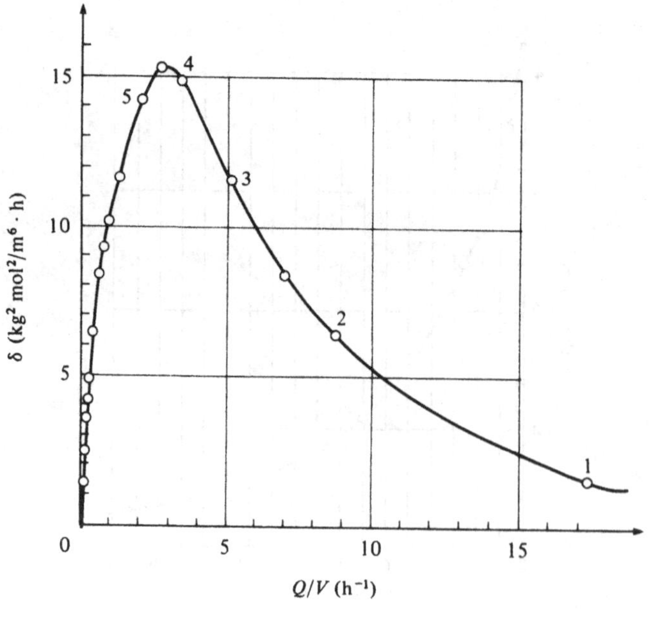

9.8 Calculation for a sparged-type reactor for the liquid-phase oxidation of hydrocarbons

Laboratory investigations on small batch reactors have preceded their commercial installation for the continuous process of the liquid-phase oxidation of hydrocarbons with molecular oxygen. As a result of these investigations, optimal conditions of temperature, pressure, concentrations of reactants and catalyst, and so on, have been determined, ensuring the occurrence of the reaction in the kinetic regime. In the simulation of such processes it is necessary to consider the complicated effects of chemical diffusion and hydrodynamic factors as we have previously noted. The kinetic principles for catalytic liquid-phase oxidation of hydrocarbons, via a chain mechanism, are dependent on the nature of the complicated interaction of the initial reacting materials with the products of the catalyzed reaction.

Realization of a continuous process based on the assumption of plug flow (ideal displacement) does not ensure self-propagation of the chain reactions, inasmuch as the initial materials here are not mixed with the components of the reaction mass, and so continuous processes of liquid-phase oxidation are most effectively carried out by instantaneous and complete mixing of the reacting materials with the catalyzing reaction products, in other words, in ideally mixed sparged reactors.

Figure 9.17 Kinetic curve of the accumulation *o*-toluyl acid in the reactor in the regime of maximum efficiency for ideal mixing.

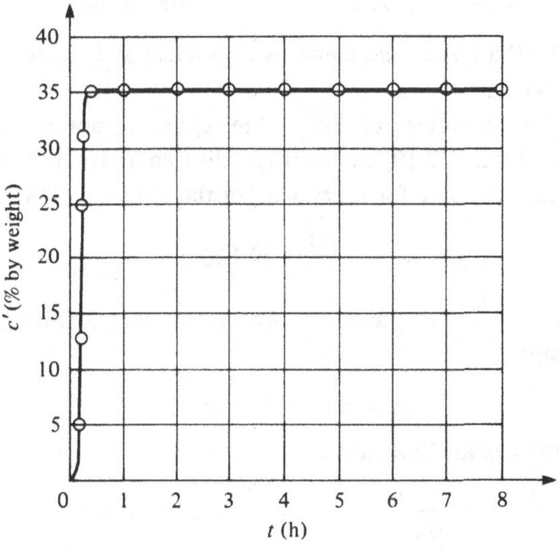

As an example, let us consider a reactor for the oxidation of *o*-xylene to *o*-toluyl acid, the unit having output of phthalic anhydride on the order of 1000 metric tons/yr. We assume the on-stream time of the reactor to be equivalent to 8000 h/yr, so that the phthalic anhydride output in an hour is 125 kg/h and the final product output in an hour is 125.4 kg/h because the phthalic anhydride content is 99.7 percent by weight in the final product.

The theoretical rate of *o*-xylene for manufacturing 125 kg of phthalic anhydride is then

$$\frac{106 \times 125}{148 \times 0.847} \simeq 105.65 \text{ kg/h}$$

where the molecular weight of *o*-xylene is 106, the molecular weight of phthalic anhydride is 148, and 0.847 is a coefficient covering losses of *o*-xylene occurring in three stages. The required amount of 98 percent *o*-xylene free from admixtures is then

$$\frac{105.65}{0.98} = 107.8 \text{ kg/h}$$

or 1.01 kg mol/h, and this amount corresponds to the *o*-xylene feed consumed in the formation of *o*-toluyl acid. Thus the reactor output of *o*-toluyl acid is

$$1.01 \times 8000 \times 136 = 1100 \text{ tons (metric)/yr}$$

where the molecular weight of *o*-toluyl acid is 136.

The oxygen feed rate for oxidation of *o*-xylene is

$$Q_{O_2} = 1.01 \times 1.5 \times 2.0 = 3.03 \text{ kg mol/h}$$

where 1.5 and 2.0 are the stoichiometric coefficient and oxygen excess coefficient, respectively.

In the optimal oxidation regime, or the regime of optimal efficiency as seen in Figures 9.9 and 9.10, and as was indicated above, 6.95 kg mol/h of *o*-toluyl acid can be made for every 1 m³ of the reactor, so that

$$V = \frac{1.01}{6.95} = 0.146 \text{ m}^3$$

Now for $Q/V = 6 \text{ h}^{-1}$ (see Figure 9.4), the volumetric flow rate of recirculating *o*-xylene is

$$Q = 6V = 0.88 \text{ m}^3/\text{h}$$

Also, the volumetric air flow rate is

$$Q_{O_2} = \frac{3.03 \times 32}{0.23 \times 1.29} = 326 \text{ reduced m}^3/\text{h}$$

or, at a pressure of 10 atm and $T = 160\,^\circ\mathrm{C}$,

$$Q_{O_2} = \frac{326(273 + 160)}{273.10} = 51.6\ \mathrm{m^3/h}$$

For calculating the reactor diameter d, we compute the superficial gas velocity at the reactor exit $v_{s_{ex}}$ using the following equation:

$$3600AA_1v_{s_{ex}}(\omega_{max} - \omega) = \frac{W'}{W'_{max}}$$

or

$$v_{s_{ex}} = \frac{W'}{3600AA_1(\omega_{max} - \omega)W'_{max}}$$

Here $v_{s_{ex}}$ (m/s) is the superficial gas velocity at the reactor exit A_1 ($= 12.56 \times 10^{-4}\ \mathrm{m^2}$) is the cross-sectional area of the laboratory reactor, $A = 1.43 \times 10^{-2}\ \mathrm{m^2}$, and $\omega_{max} = \alpha P_{air} = 36.10$. Hence

$$v_{s_{ex}} \simeq 0.0462\ \mathrm{m/s}$$

After computing $v_{s_{ex}}$, we can calculate the cross-sectional area of the reactor:

$$A = \frac{Q_{O_2}}{v_{s_{ex}}} \simeq 0.310\ \mathrm{m^2}$$

and the diameter is then 0.628 m.

The height of liquid in the apparatus is

$$h = \frac{V}{A} = \frac{0.146}{0.310} = 0.427\ \mathrm{m}$$

and for calculating the height of the reactor we define the height of the bubble layer x_1, which in turn can be computed knowing the average gas content ϕ_{av} of the layer, using equation (4.53):

$$\phi_{av} = ab - \frac{(ab)^{1/2}}{2}(1 + ab)\left[\sin^{-1}\left(\frac{1 - ab}{1 + ab}\right) + \frac{\pi}{2}\right] \qquad (9.54)$$

where $a = 3C_D/8$ and $b = v_s^2/gr_b$, and symbols are defined in Chapter 4. Hence, in order to find ϕ_{av}, we need to be able to calculate the bubble radius r_b beforehand. Because the rising velocity of a single bubble is[21]

$$v_r = \left(\frac{4\sigma^2 g}{\alpha\mu\rho_f}\right)^{1/5} \simeq 0.242$$

(where $\sigma = 2.83 \times 10^{-3}$ kg/m is the surface tension, $g = 9.81$ m/s is the acceleration due to gravity, $\alpha \simeq 12\pi$, $\mu = 1.225 \times 10^{-4}$ is the dynamic liquid viscosity, and $\rho_f = 90$ is the density of the liquid), the bubble radius[15] will be

$$r_b = C_D v_r^2/4g \simeq 0.0017 \text{ m}$$

Here C_D is the drag coefficient.

Substituting into equation (9.54) the values $v_s = 0.0462$ m/s, $r_b = 0.0017$ m, and $C_D = 0.2$, we find that $\phi_{av} \simeq 0.1315$; and hence

$$x_1 = h/(1 - \phi) = 0.545 \text{ m}$$

The overall height of the reactor is then

$$H = \frac{x_1}{K} = \frac{0.545}{0.5} = 1.09 \text{ m}$$

where K is the fraction of the reactor filled.

Now let us check to see whether the reactor designed here satisfies the requirements set forth for an apparatus with perfect (ideal) mixing. We calculate the order of the reaction rate with respect to oxygen:

$$\eta = \frac{[O_2]}{W'}\frac{\partial W'}{\partial [O_2]} \tag{9.55}$$

Using equations (9.13), (9.22), and (9.40), we find that

$$W' = \frac{\alpha^2[O_2]^2}{\beta(\alpha^2[O_2]^2 + \alpha[O_2] + 1)} \tag{9.56}$$

(α and β are constants independent of oxygen concentration and oxidation rate).

Substituting W' from equation (9.56) into equation (9.55) and carrying out the differentiation, we find that the reaction order is

$$\eta = \frac{\alpha[O_2] + 2}{\alpha^2[O_2]^2 + \alpha[O_2] + 1} = \frac{\omega + 2}{\omega^2 + \omega + 1}$$

At an assumed value of $\omega = 10$ the reaction order with respect to oxygen is $\eta = 0.108$, which is quite small. The calculated reaction order is confirmed by experimental data.

It is seen from Figure 9.13 that the accumulation curve for *o*-toluyl acid in the vicinity of the optimal regime is represented by a straight line, this corresponding to a zero-order reaction. In this case the variation in the concentration of the reaction product along the height of the apparatus, taking place as a result of diffusion and chemical reaction under continuous-flow conditions, is described by the equation

$$c_i/c_o = z + \chi_1(1 - e^{-(1-z)/\chi_1})$$

Here c_i and c_o are the concentrations of the reaction product at the inlet and outlet from the reactor, respectively, $z = x/h$ the dimensionless distance from the bottom of the reactor, x the distance along the height of the reactor, $\chi_l = A\overline{D}/Qh$ a dimensionless parameter characterizing the effect of convective diffusion on the distribution of the reaction product along the height of the reactor, and $\overline{D}$ (m^2/s) a longitudinal mixing coefficient.

From Figure 9.18, in which the dependence of c_i/c_o on z for various values of χ_l is presented, it is seen that for values of χ_l close to 10, the reactor is essentially operating in the ideal mixing regime. To determine χ_l we use the longitudinal turbulent diffusion coefficient value obtained [22] in a continuous sparged reactor with a diameter of 77 mm and a height of 8 m. Because the turbulent diffusion coefficient $\overline{D}$ is proportional to the macroscopic scale of the turbulent motion, which is in turn proportional to the diameter of the apparatus, we can, with a sufficient degree of accuracy, say that

$$\overline{D} \approx 0.008\left(\frac{0.628}{0.077}\right) = 0.816 \text{ m}^2/\text{s}$$

and thus $\chi_l \simeq 2424$; that is, a perfect mixing regime is achieved in our reactor.

Figure 9.18 Variation of the concentration of the reaction product along the height of the apparatus for various values of convective diffusion parameter χ_l.

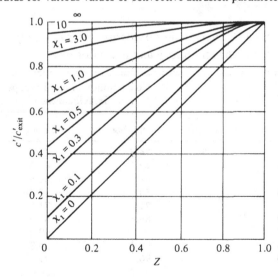

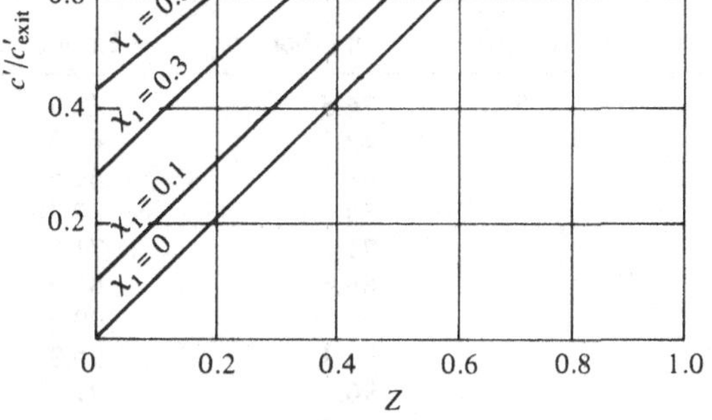

9.9 Influence of scaling-up calculation of a sparger reactor for liquid-phase oxidation of hydrocarbons

Investigations of the process of liquid-phase oxidation of hydrocarbons, as noted in Sections 9.1 through 9.8, allow us to find optimal operating conditions and to make an approximate design for the sparged reactor. We supposed earlier that the oxidation rate W' of hydrocarbon must be equal to oxygen diffusion flux, per unit reaction volume. In other words, we considered the problem of a diffusion limitation imposed by the molecular oxygen transfer mechanism on the process of liquid-phase oxidation of hydrocarbons. We made an assumption of uniform air distribution on each cross-sectional area of apparatus, implying an absence of channeling and stagnant zones. However, in equipment with large diameters, transversal irregularity has been observed,[23–26] which may result in a significant lowering of the mass-transfer efficiency, and therefore we need to experimentally evaluate the role of scale-up factors when we design a sparged reactor, in order to be sure of the appropriateness of the chosen model for describing a real process of liquid-phase oxidation in a commercial sparged reactor.

As a model for checking the effect of scaling up, the catalytic oxidation of diethylbenzene by air, in the laboratory environment, and in a pilot unit[27,28] is now considered. The oxidation process in the laboratory was carried out in a glass sparged reactor with a diameter of 20.0 mm and height of 250 mm, the reactor being filled with 400 ml of diethylbenzene with a supply of 72.0 liters/h. Oxidation in the pilot unit was carried out in a

Table 9.4. *Comparison of oxidation rates of diethylbenzene in the laboratory and in a pilot unit*

	Consumption of ethyl groups of diethylbenzene (%)	
Time from the run (h)	In laboratory	In pilot unit
1.0	26.5	27.7
2.0	43.1	43.0
3.0	49.2	49.5
4.0	59.8	60.2
5.0	71.0	71.2
6.0	84.6	85.0
7.0	90.1	90.2
8.0	95.4	95.8
9.0	96.8	97.5
10.0	99.2	100.0

reactor with d = 313 mm, h = 2700 mm, filled with 100.0 liters of diethylbenzene, and an air supply of 18.0 m^3/h. Thus both in laboratory and pilot systems, for each liter of liquid 180 liters of air was provided; also, the superficial air velocity in both systems was v_s = 0.064 m/s. Oxidation was performed for 8 h at a temperature T = 130 °C, and the experimental data obtained from these two systems are presented in Table 9.4. It is seen that the oxidation rate of diethylbenzene achieved in the pilot reactor is not essentially different from data obtained in the laboratory reactor, even though the diameter ratio is about 15:1.

Oxidation of isopropylbenzene[29] (similar to the oxidation of diethylbenzene) is well modeled when transferring from d = 16 mm to a commercial reactor with d = 2200 mm. Experimental data on continuous oxidation of diethylbenzene in a commercial reactor for 50 hours have confirmed the obtained regularity in a pilot unit.

Thus the design of commercial sparged reactors for liquid-phase oxidation may be made on the basis of experimental data obtained in the laboratory, without taking into account the effect of scaling up.

References

1. Levenspiel, O., *Chemical Reaction Engineering*, Wiley, New York, 1972.
2. Denbigh, K., *Chemical Reactor Theory*, Cambridge University Press, Cambridge, 1965.
3. Semenov, N. N., *Some Problems of Chemical Kinetics and Reactivity*, Izdatelstvo AN SSSR, Moscow.
4. Beresin, I. V., Denisov, E. T., and Emanuel, N. M., *Oxidation of Cyclohexane*, Izdatelstvo MGU, Moscow.
5. Emanuel, N. M., Denisov, E. T., and Maysus, Z. K., *Chain Reaction of Oxidation of Hydrocarbons in Liquid Phase*, Izdatelstvo Nauka, Moscow, 1965.
6. Denisov, E. T., Doctoral dissertation, Moscow University, 1964.
7. Knorre, D. G., Maysus, Z. K., Obukhova, L. K., and Emanuel, N. M., *Usp. Khim.*, *26*(4), 416 (1957).
8. Trayball, P. I., and Bartlett, P. D., *Tetrahedron Lett.*, *30*, 24 (1960).
9. Muzychenko, L. A., Candidate's dissertation, D. I. Mendeleev Moscow Institute of Chemical Technology, 1963. (In Russian.)
10. Azbel, D. S., and Muzychenko, L. A., *Khim. Prom. (Moscow)*, no. 12, 881 (1964).
11. Azbel, D. S., and Muzychenko, L. A., *Vsesoyusnaya konferentzia po khimicheskim reactoram*, vol. 1. Izdatelstvo AN SSSR, Novosibirsk, 1965, p. 190.
12. Kamneva, A. I., Doctoral dissertation, D. I. Mendeleev Moscow Institute of Chemical Technology, 1960. (In Russian.)
13. Denisov, E. T., and Emanuel, N. M., *Usp. Khim.* *29*, 1409 (1960).
14. Emanuel, N. M., *Voprosy khimicheskoy kinetiki, katalisa i reactzionnoy sposobnosti*, Izdatelstvo AN SSSR, Moscow, 1955, p. 112.
15. Azbel, D. S., *Khim. Prom. (Moscow)*, no. 7, 523 (1965).
16. Azbel, D. S., and Muzychenko, L. A., "Design of a bubble-type reactor for hydrocarbon oxidation in the liquid phase," *20th IUPAC, Moscow, 1965*.
17. Planovsky, A. N., and Gurevich, D. A., *Apparatura promyshlennosty poluproduktov i krasiteley*, Goskhimizdat, Moscow, 1961.

18. Kirilov, N. M., *Zh. Fiz. Khim.*, *13*, 1978 (1940).
19. Philipov, Uy. V., *Vliganie perenosa vetchestv vdol potoka na kinetiku reaktali v potoke, kinetika i katalis*, Izdatelstvo AN SSSR, 1960.
20. Polozhy, G. N., *Matematichesky praktikum*, Fizmatgis, Moscow, 1960.
21. Levich, V. G., *Physicochemical Hydrodynamics*, Prentice-Hall, Englewood Cliffs, N.J., 1962.
22. Dilman, V. V., and Aysenbud, M. B., *Khim. Prom. (Moscow)*, no. 8, 607 (1962).
23. Rozen, A. M., Lapovok, L., and Elatomtzev, V., *Khim. Neft. Mashinostr.*, *4*, 14 (1964).
24. Rozen, A. M., *Khim. Prom. (Moscow)*, no. 2, 85 (1965).
25. Rozen, A. M., *Dokl. Vyssh. Shk., Ser. Energ.*, *3*, 173 (1958).
26. Rozen, A. M., *Teoria rasdelenia isotopov v kolonakh* (Theory of Isotope Separation in Columns), Atomizdat, Moscow, 1962.
27. Khcheyan, Kh. E., Azbel, D. S., Ioffe, A. E., and Mak, N. E., Autorskoe svidetelstvo (patent), no. 141149, Moscow, 1961.
28. Azbel, D. S., Ioffe, A. E., and Mak, N. E., *Sintez i svoystvo monomerov*, Izdatelstvo ANSSSR, Moscow, 1964, p. 184.
29. NIISS, *Annotatziony otchet*, NIISS, Moscow, 1964.

10

Reactor design for microbiological processes

10.1 Introduction

In Chapter 6 we made a theoretical investigation of the hydrodynamic and mass-transfer modes for the solid particles–liquid system. This chapter is a logical continuation of the theory and includes some important practical applications of it. We will theoretically and experimentally investigate the effects of hydrodynamic and mass-transfer factors on the process of cultivation of microorganisms, with the goal of obtaining relationships useful for the engineering design of such processes occurring in the microbiological industry.

As a model of the process of cultivation of microorganisms, the manufacture of enzymic product will be investigated, and we will consider the effect of mixing and aeration on the accumulation of enzymic activity, the experimental determination of the mass transfer in microheterogeneous systems, and the influence of mass transfer on the biochemical kinetics. Hence we will gain insights into the optimum design of biochemical reactors.

10.2 The limiting stage in biochemical processes

In constructing a model of a biochemical process, the essential functional relations must be taken into account, but such a model cannot be universal because the field of its application is determined by the properties of the system. If the process studied is complicated, an effective model must be "collapsible," permitting a separate study of different aspects of the process while allowing consideration of their interaction.[1] In a microbiological synthesis process involving a widespread heterogeneous process, manufacture of reaction products accompanies the cultivation of microorganisms, and intensive heat and mass transfer usually occur. It is known that the transfer of mass in this case includes the effects of many reactions tied together as chain links.

For instance, biosynthesis of basic cell polymers and the growth of a cell are determined by three fundamental factors:[2]

1. The availability of sufficient amounts of structural units in a cell, which precedes the synthesis process.
2. The supply of construction materials, in the form of amino acids and mononucleotides.
3. A sufficient influx of energy.

Conversely, the growth rate of microorganisms may be limited by each of these factors, the first determining the intercellular properties, and the second and the third determining the heat and mass transfer. We need not be concerned about the presence, in normally metabolizing cells, of either a lack, or a significant excess, of structural units, because in such conditions self-regulation of the process rate is realized without the induction synthesis of enzymes, this rate being determined by the rate of substrate supply (the reagent).

For the case of exogenous substrate with a larger reaction rate, the induction mechanism of enzymic synthesis does take place.[3,4] For example, the bacterial cells continue growing at the same rate after being transferred into richer nutrient medium. However, the synthesis of RNA and its content in the cells is immediately intensified, and subsequently the growth of the cell is quickened. Similarly, DNA synthesis is intensified.[5]

Presently, there has been a sufficient amount of theoretical and experimental work[3] which points to the possibility of regulatory capabilities, permitting synthesis of missing structural elements using the methods of straight and reversal liaison. Limitation of the process rate by the presence of structural units takes place only in the transient period and in real steady-state systems, to all appearances, incomplete enzyme saturation will be observed.

The second and the third factors may be considered as mass-transfer processes, because the carbohydrates, which the cell utilizes for synthesis of cell structure, become an energy source at the same time.[6] Entering into a cell, nutrient substances play a double role: first, they are inductors of enzyme biosynthesis, and second, they feed substrate for reactions. Clearly, then, entering of nutrient substances into the cell is the determining stage of the whole process, and the feed rate of required substances into the cell is determined by the enzymes, which regulate the intensity of transfer of ingredients from the outer medium. The equilibrium concentrations in the medium and in the cell are different; for example, microelements may be accumulated by some microorganism in concentrations two orders of magnitude greater than their content in the nutrient medium,[7,8] and this experimental fact corroborates the conclusion that the transport of microelements is bound either by changeable adsorption or by the effects of special carriers.[9–11]

There are two approaches to solving the problem of cell permeability. In the first one, we can explain the transfer mechanism from the point of view of membrane theory, which essentially neglects all the properties of individual cell structures except those of the membrane. The second approach is founded on the predominant role of protoplasm and the cell albumens. To a considerable extent, these points of view have drawn closer in recent times; however, the enzyme contribution in the transfer process remains an indispensable factor.

The effects of various transport phenomena (for the nutrient substances and metabolism products) around the cell zone have been mentioned in several studies[12–14] but with no apparent success in providing a good estimation of the actual process. For example, it has been mentioned[15] that the application of a mass-transfer model allowing for molecular diffusion around the cell zone, for intensive microbiological processes, is difficult because the diffusion coefficient around the cell zone must depend on the concentration in this zone and, consequently, on cell spacing. Concentration values for nutrient substances and metabolism products in the medium, without consideration of their distribution in the thin-cell zone (the diffusional layer), are used to evaluate the process only on the basis of kinetics of enzyme reaction. According to a study[14] based on the cell model, assuming that mass transfer is carried out only by molecular diffusion,

$$\dot{m} = DS\frac{dc}{dr}$$

(where $\dot{m}$ is mass transfer per unit time, D the diffusivity, S the cell surface area, c the local concentration, and r the radial coordinate), an equation can be derived for the average concentration near the surface of a cell suspended in a stagnant medium:

$$c_{av} = c_0 + \frac{\dot{m}}{4\pi D}\left[\frac{1}{r_c} - \frac{3(R^2 - r_c^2)}{2(R^3 - r_c^3)}\right] \tag{10.1}$$

where R is the radius of an imaginary sphere equivalent to the cell spacing, r_c the cell radius, and c_0 the concentration in the bulk (or substrate). This equation was derived assuming continuous mass feed from the outside, but if the mass taking part in the reaction is not being fed from the outside, equation (10.1) may be rewritten as

$$c = c_0 - \frac{\dot{m}}{8\pi D(R^3 - r_c^3)}\left(\frac{2R^3}{r} - r^2 + 3R^2\right)$$

It has been found[13,14] that for cells with a diameter of 10 mm (0.39 in.) and intercell distances of 20 mm (0.79 in.) [which corresponds to a cell concentration $\simeq 10^{-6}$ unit/ml (2×10^{-3} unit/in.^{3})], the zones of lowered concentration will close in 0.1 s, and it should be noted that, in order of magnitude, the calculations made will not only be correct for this case but will also be correct in the case of intensive mixing, because the rate of convective transfer to microparticles is three to four times higher than for molecular diffusion.

Theoretical attempts at solving the problem of unsteady diffusion to a biological particle suspended in liquid have been made,[16] but it was usually assumed that there was a homogeneous reaction of first order inside the particle and that there was constant oxygen diffusion into the particle, both claims being dubious.

According to Ierusalimsky,[17] "none of the enzyme processes in the cell is developed at full power, and they cannot be because an extremely high concentration of substrate would be needed. The maximum rate is a merely mathematical, never achieved limit. In actual fact, the rate of biochemical reactions is always limited by lack of substrate, or by a retroinhibited action of the metabolism products."

Thus only in the transient period may one observe the limiting of the process rate by the presence of structural units. Limitation of the process by its intercell reactions and active transfer of reagent is unlikely because the substrate feeding rate or evacuation of metabolites is significantly less. Consequently, the process, as a whole, is limited by the feed rate of nutrient substances to the reaction surface, or in other words it is limited by the mass-transfer process.

10.3 Effect of mixing intensity and air feed on the rate of oxygen dissolution

To determine the effect of the rate of oxygen dissolution on the specific growth rate of microorganisms, it is necessary to find the mass-transfer characteristics for oxygen absorption from air of the equipment. Absorption of oxygen from air by a culture medium is determined by mixing and aeration, and also by equipment design. The rate of oxygen dissolution is significantly affected by physicochemical factors such as the viscosity, surface tension, solubility diffusion, the presence of surface-active agents, and so on; and this may be written in the form

$$W' = k_v \Delta c$$

Here W' is the diffusion flux of oxygen per unit volume of the apparatus, Δc the concentration gradient between the bulk flow and the interface, and k_v the volumetric mass-transfer coefficient.

We can write k_v in the functional form

$$k_v = f(D, \nu, \sigma, \rho_f, \rho_g)g(\Gamma, \phi)$$

where D is the diffusivity; ν the kinematic viscosity; σ the surface tension; ρ_f and ρ_g are the density of liquid and gas, respectively; Γ represents system geometry factors; and ϕ is the void fraction. Here $f(D, \nu, \sigma, \rho_f, \rho_g)$ is a function characterizing the effect of the properties of the gas and liquid, and $g(\Gamma, \phi)$ is a function characterizing the effect of hydrodynamic and design factors. Now, for each medium, either in water or in culture liquid, the function $f(D, \nu, \sigma, \rho_f, \rho_g)$ has a fixed value: $f = m_j$, where m_j is the numeral value of the function for medium j, and so the relative value of k_v (or W') can be represented in the form

$$\frac{k_{v_1}}{k_{v_2}} = \frac{m_1 g(\Gamma_1, \phi_1)}{m_2 g(\Gamma_2, \phi_2)}$$

Hence characteristics obtained in experiments with water, for different regimes of mixing, will also apply, under the appropriate design conditions, to a culture medium.

A system comprised of a sodium sulfite–water solution and oxygen has been utilized for studying mass transfer from gas to liquid. The process in this case is the oxidation of sodium sulfite to sulfate in the presence of a copper catalyst, and when the concentration of sodium sulfite in water solution is higher than 0.03 kg mol/m^3 and ions of Cu^{2+} higher than 10^{-4} kg mol/m^3, the absorbing oxygen is bound in the liquid and its concentration in the solution is equal to zero.[18] This allows us to ignore the effect of mixing in the liquid phase while calculating the driving force of the process, and to obtain a mass-transfer coefficient in the liquid phase that is not distorted by liquid back mixing. The concentration of sodium sulfite in the experiments[18] varied from 20 to 30 g/liter (0.13 to 0.19 lb/ft^3) to 4 g/liter (0.25 lb/ft^3), and under these conditions the oxidation rate turns out to be a linear function of time.

The rate of oxygen dissolution was determined according to the equation[19]

$$Q' = \frac{1.88 \times 10^{-3} \Delta V}{\tau V} \qquad \text{liters of } O_2/\text{liter} \cdot \text{s}$$

where ΔV is the volume difference of a 0.02 N solution of hyposulfite between a given point and the zero point, τ the time interval between a given point and the zero point, and V the volume of the sample (usually $V = 2.0$

ml). The computed values of Q' are in good accordance with experimental data[19] and with the equation

$$k_v = \text{const.} \left(\frac{v_s d}{\nu} \right)^{1.95}$$

where v_s is the superficial velocity of the flow, d the diameter of the apparatus, and k_v the volumetric mass-transfer coefficient from gas to liquid.

Experimental correlations for a sparged apparatus with a mixer and for a Hemap apparatus, respectively, are shown in Figures 10.1 and 10.2, which indicate the effects of the numbers of revolution of the mixer and air supply on the dissolution rate of oxygen. It is clear from these figures that with an increase in the air supply the rate of dissolution increases approximately linearly, then approaches a maximum, after which point an increase of air supply leads to a decrease of the process rate. By varying the number of revolutions of the mixer, the form of curves is not changed; however, the dissolution rate is increased or decreased in proportion to increase or decrease of number of revolutions. From Figures 10.1 and 10.2, the optimal air supplies for high dissolution rates are: for a sparger with a mixer 20.0 liters/min (0.7 ft³/min), for the Hemap apparatus 45 liters/min (1.6 ft³/min), and for the tubular apparatus 7 liters/min (2.47 ft³/min).

Let us now compare these three types of apparatus as to their absorption

Figure 10.1 Effect of revolution number of mixer and air flow rate on the oxygen dissolution rate in sparger with a mixer.

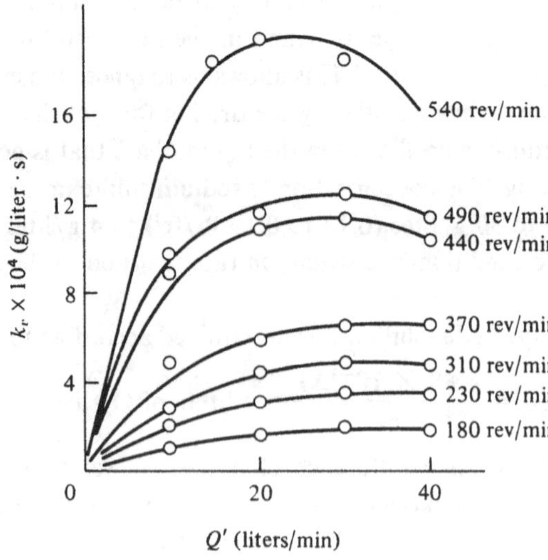

as a function of the power consumed in mixing. Estimation of the power consumption for sparged equipment, with agitators for mixing the liquids (which are assumed to have properties close to those of water), was made according to[18]

$$P = \frac{0.9 n^3 d_m^5}{d^{0.4} v_g^{0.2}}$$

and for tubular apparatus with a spiral mixer,

$$P = 1.29 \left(\frac{\sigma}{\rho_f g d^2} \right)^{0.19} \rho_f n^3 d^5 \, c^{1.45}$$

Here n is the number of revolutions, d_m the mixer diameter, c the solute concentration, v_g the air velocity, and σ the surface tension.

For the Hemap apparatus, power consumption was determined experimentally by measuring the power consumption during mixing of 15 liters of the liquid and during idle running. The difference of these power consumptions (minus 10 percent for electrical losses) determines the power consumption in question. Decrease of power consumption when air was injected is indicated by the term $1/v_g^{0.2}$. In Figure 10.3 the effect of power consumption during liquid mixing on the rate of oxygen dissolution (at the optimal air supply) is shown and we see that, for equal power consumption, the high-

Figure 10.2 Effect of revolution number of mixer and air flow rate on the oxygen dissolution rate in Hemap apparatus.

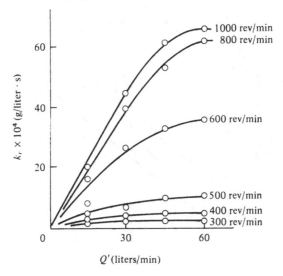

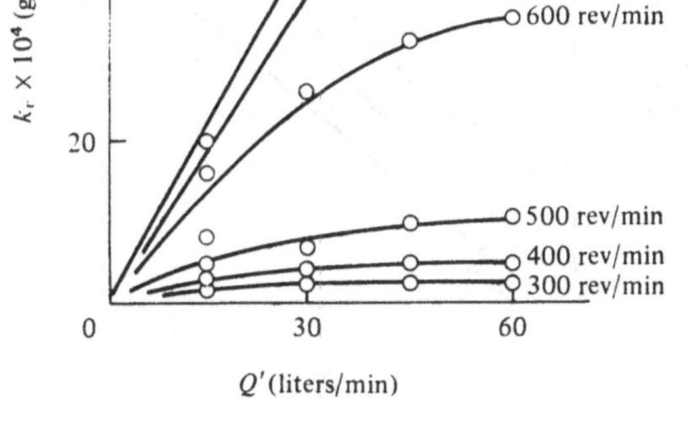

est oxygen dissolution rate is achieved in a sparger apparatus with a mixer, and the lowest rate is obtained from the tubular apparatus.

10.4 Effect of various parameters on accumulation of enzymic activity

Choice of an optimal design for a biochemical reactor (fermenter) necessarily involves study of the deep cultivation of microorganisms. Many studies have indicated that the growth of biomass and accumulation of enzymic activity are significantly affected by mixing aeration, and the influence of number of revolutions and air supply on accumulation of proteolytic activity in a sparger apparatus with a mixer and in a Hemap apparatus are presented in Figures 10.4 and 10.5. These plots have an extremum character. Also, we see that increase of the air supply results in decreasing of enzymic activity for a sparger apparatus with a mixer at the optimal number of revolutions, but in the case of microorganism cultivation, this relationship is reversed.

The growing culture of microorganisms realizes energy and mass exchange (metabolism) with the surrounding medium as nutrients are supplied to the growing cell and metabolism products are withdrawn from it.

Figure 10.3 Effect of mixing power consumption of liquid on the oxygen dissolution rate at the optimal air flow rate.

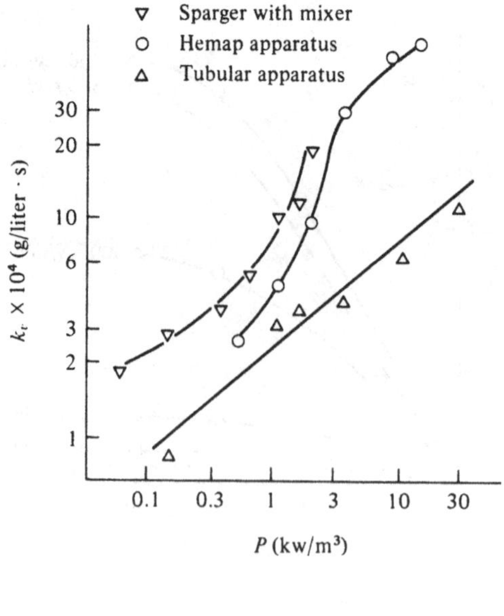

By maintaining the initial substrate, parameters determining the accumulation of enzymic activity may be found in such factors as the supply of mechanical energy and oxygen, and withdrawal of volatile metabolic products with waste gases.

Figure 10.4 Accumulation of enzymic activity versus number of revolutions of mixer and specific air rate for Hemap.

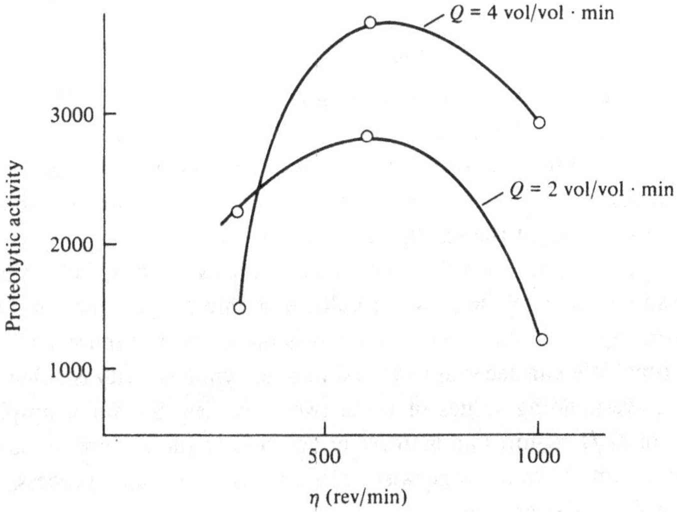

Figure 10.5 Accumulation of enzymic activity versus number of revolutions of mixer and specific air rate for sparger with mixer.

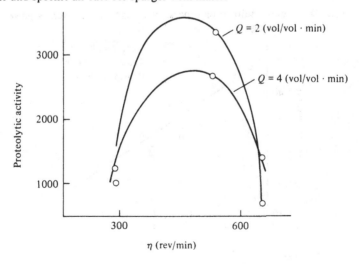

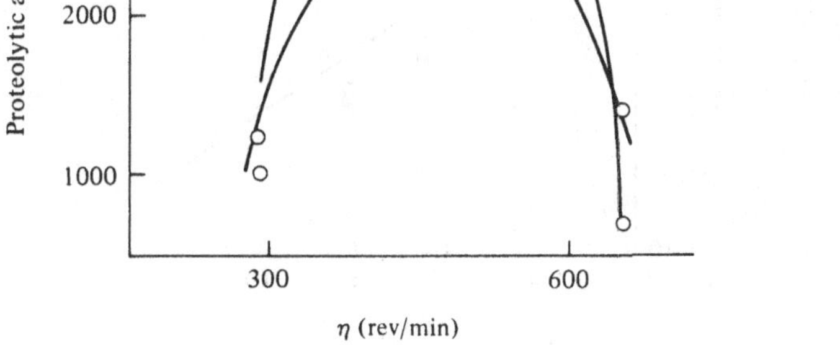

A graph of *Bacillus subtilis* 103 cultivation time versus power consumption for liquid mixing is plotted on a logarithmic scale in Figure 10.6, taking into account not only the phases of active growth, but also lag periods and stationary phases of growth. The time interval was defined as that from sowing to maximum enzymic activity, and Figure 10.6 shows experimental data obtained in three different kinds of apparatus. The experimental points are approximated by a straight line, and this relationship may be written in the form

$$t_{\max} \simeq 33P^{-0.26}$$

We see that an increase of power consumption from 0.1 to 10 kW/m^3 (2.68 $\times 10^{-3}$ to 2.68 $\times 10^{-1}$ $Btu/ft^3 \cdot s$) decreases cultivation process time from 60 to 18 h. Because a change of hydrodynamic regime gives rise to a change in the process rate of biosynthesis, we must relate all parameters to a unit rate of biosynthesis of the enzymes.

Proteolytic activity as a function of two factors – Q'/t, indicating the oxygen supply rate to the growing culture of microorganisms, and W'/t, characterizing the intensity of aeration – is presented in Figure 10.7. Here t is the time. We can see that the maximum enzymic activity is achieved at certain corresponding values of these two variables. So, for example, an increase of Q'/t requires an increase of aeration of the culture to maintain this maximum, because otherwise, after maximizing the process, rates would be sharply decreased.

It is a well-established fact that the permanent displacement of aeration

Figure 10.6 Duration of cultivation process of *B. subtilis* 103 versus power consumption for liquid mixing.

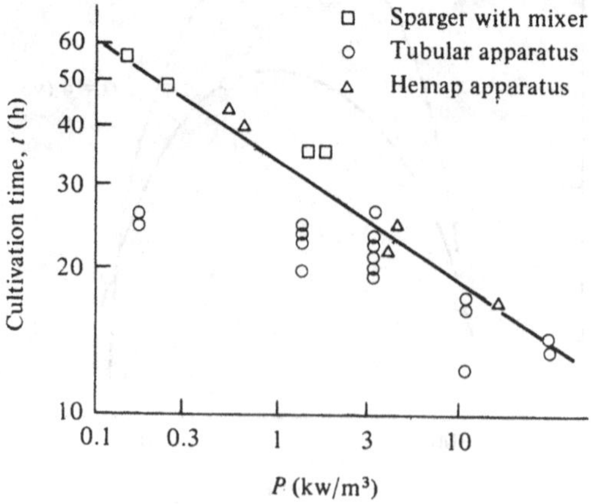

of the culture with increase of the Q'/t parameter is determined by the structural and energy exchanges between the substances, and the ratio of these effects may be indirectly determined by use of a respiration coefficient:

$$K = \frac{m_{CO_2}}{m_{O_2}}$$

where m_{CO_2} is the amount of CO_2 respirated and m_{O_2} is the amount of consumed oxygen.

It is known[17] that this respiratory coefficient is dependent on the chemical content of the oxidizing substances and on the physiological state of the microorganisms, which in turn is determined by the cultivation conditions.

The respiratory coefficient versus time is given in Figure 10.8 and clearly with acceleration of the process the respiratory coefficient continuously increases, testifying to the increase of mass transfer in the direction of energy exchange. As the energy exchange increases, so does the amount of volatile acid evacuated by a cell into the nutrient medium, in its turn making necessary an increased aeration of the growing culture of microorganisms in order to maintain the cultivation condition on the optimal level. Based

Figure 10.7 Accumulation of proteolytic activity against intensity of oxygen supply and intensity of aeration.

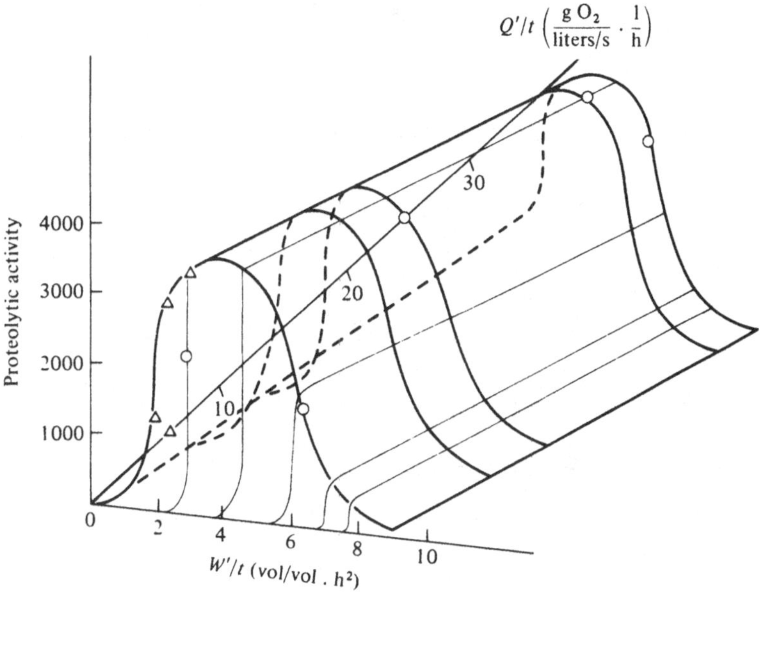

upon Figure 10.7, we must wait so that the increase of the solution rate of oxygen, for other parameters fixed, also brings an increase in the respiratory coefficient and, subsequently, displacement in the direction of energy exchange.

10.5 Mass-transfer effects in the kinetics of biochemical processes

The special technology for microbiological processes, which are multiphase indeterminate systems with polyvalent stochastic relations between the process parameters, calls for new approaches and solutions in designing the equipment and developing the technology. The majority of these processes depend on the action of enzymes, biological catalysts that frequently display activity only for one single reaction, whereas the overall functioning of the cell depends on the integration of a large number of reactions whose sequence and direction is determined by genetic information embodied in the nucleic acids. Growth of the living cell is essentially different from the growth of a crystal or the growth of a gas bubble, and although factors that control these two processes are obviously also encountered in the process of cell growth, in this case they are subject to growth laws, characteristic of living matter, whose mathematical formulation for a totality of cells (a cell population) may be represented as[20]

$$N = N_0 e^{\mu t}$$

Figure 10.8 Respiratory coefficient of microorganisms versus time of cultivation.

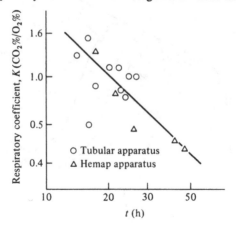

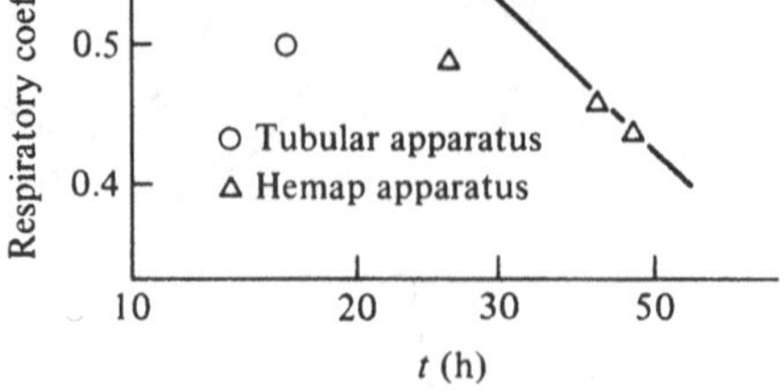

where N_0 and N are the original and current number of cells, and μ is a measure of the growth rate. The rate of biochemical processes is determined by the biochemical kinetics and by the rate of supply and evacuation of reagents in the reaction zone. Being heterogeneous, biochemical processes always occur with intensive motion of the reagents taking place, and thus molecular diffusion in the bulk of the flow is insignificant. The nature of the diffusional boundary layer near the active surface depends on the concentration of reagents in the bulk of the flow, the mass-transfer coefficient, the kind of reaction surface, and the rate of biochemical reaction. The concentration distribution of reagents and reaction products in the diffusional boundary layer significantly affects the rate of biochemical reactions. The concentration of a substance supplied to the reaction zone, at a given order of reaction, is determined by hydrodynamic conditions and by physical properties of the medium, and in its turn the change of reaction rate on the surface has an effect on the diffusional flux into the reaction zone. Therefore, only the combined consideration of transfer to the reaction surface and biochemical kinetics will determine the true rate of biochemical process.

All the principal biochemical processes in the cell, which are associated with energy exchange and structural rebuilding, are catalyzed by ferments. The living cell is a highly organized system of chemically active substances, and this means that, in contrast to lifeless systems, all the chemical reactions taking place in the cell take place in a precise order; that is, every reaction occurs at a particular time and place. Biosynthesis of the internal structures of the cell, and its growth, are determined by the presence in the cell of structural units in which the synthetic process takes place, and also by the supply of materials and energy.

The general concepts of the kinetics of enzyme reactions have been formulated[20] and they suggest that a molecule participating in the reaction (substrate concentration c_o) is reversibly adsorbed by an enzyme to form the stable complex of enzyme–substrate. Solving the system of kinetic equations for reactions of the first order, a relation can be obtained[20] between the concentration of the substrate and the rate of the enzyme reaction:

$$W_e = \frac{W_{max} c_0}{C_{w_1} + c_0} \tag{10.2}$$

where W_e is the enzyme reaction rate, W_{max} the maximum rate of fermentation, c_o the substrate concentration, and C_{w_1} a constant numerically equal to the maximum reaction rate.

It can be shown that there is a relation similar in form to equation (10.2) between the concentration of an individual nutrient and the rate of growth of the microorganisms. Because the process of growth of the microorganisms is accompanied by an increase in the amount of enzymes, the absolute rate is replaced by the relative rate:[21]

$$\frac{1}{N}\frac{dN}{dt} = \mu = \frac{\mu_{max} c_0}{C_{w_2} + c_0} \tag{10.3}$$

An equation has been proposed[5] and experimentally verified which characterizes the influence of the metabolites on the rate of growth of microorganisms:

$$\frac{1}{N}\frac{dN}{dt} = \mu = \frac{\mu_0 C_{w_3}}{C_{w_3} + c_p} \tag{10.4}$$

and this resembles the equation for nonconcurrent retardation of enzyme reactions.[22] Here μ_{max} and μ_0 are the respective maximum rates and c_p is the concentration of the products of exchange. An important deficiency of equations (10.3) and (10.4) is the absence of any terms that would take into account transfer phenomena, whereas actual biochemical processes in cell populations occurring in multiphase heterogeneous systems are comprised of several stages: the first seems to be the stage in which substrate is transported to the surface of the cell; the second is active transfer of substrate within the cell; the third is adsorption, occurring simultaneously with the biological reaction at the active center of the enzyme; the fourth is desorption of the reaction products; the fifth is active transfer of metabolic products to the surface of the cell; and the sixth is transport of the metabolic products into the body of the solution. Since the biochemical process takes place, to a first approximation, in six successive stages, its overall rate will be determined by the slowest stage.[22–24] The growing bacterial cell has very great potential resources. In 1 s there are synthesized as many as 40,000 amino acids of a single type and up to 150,000 peptide bonds are formed, and if growth proceeds under favorable conditions, the cell may reject a mass of material 50 times as great as the weight of the cell at the beginning of the cell cycle.[25] Thus the rate of the biochemical process is limited by the velocity of reactants and discharge of reaction products; in other words, the overall biochemical process is limited by transfer phenomena. This influence of transfer phenomena has been noted elsewhere.[2,15,26] Quantitative estimates of this effect have not been made except for studies in which molecular-kinetics theory was used to locate zones of increased and decreased concentration of materials, in the neighborhood of cells suspended in an immobile medium. In Chapter 8 the influence of hydrodynamic parameters and physical properties of the medium on the rate of mass transfer to a cell suspended in a turbulent flow was determined, and the results given there permit us to calculate the mass flux to a cell surface.

It has been[20] suggested that in various enzyme oxidation–reduction processes there is participation by free radicals, whose formation may be due to the presence of free valencies in the cells. A freshly formed radical inside the cell may collide with one of the macromolecular chains, leading to removal of an atom of hydrogen or some similar atom, with resultant production of a new free valency. It follows, if this model is accepted, that a flickering picture of movement of free radicals may be retained for a considerable time inside the cell, appearing and disappearing now at one point, now at another.[3,27] If we postulate the presence of a limited number of identical mutually independent *active centers* on the molecular chains inside the cell, we may introduce the concept of a "density" of active centers, θ_k, as used in Langmuir's kinetics. The kinetic equation for reaction of the enzyme with substance i, for a multicomponent mixture, has the form

$$\frac{d\theta_{c_i}}{dt} = m_i\left(1 - \sum_k \theta_{c_k}\right) - k_i\theta_{c_i}$$

where k_i is the rate constant of desorption, θ_{c_k} the density of the active center, and m_i the diffusional flow of material to the particle. If we write this set of equations for all components of the mixture and equate the right-hand sides to zero (for steady state), solving for the relative quantities θ_c, we find

$$(m_1 + k_1)\theta_{c_1} + m_1\theta_{c_2} + \cdots = m_1$$

$$\cdot \quad \cdot \quad \cdot$$

$$m_2\theta_{c_1} + (m_2 + k_2)\theta_{c_2} + \cdots = m_2$$

The solution of this system of equations may be written as

$$\theta_{c_i} = \frac{1}{\Delta}\sum_k m_k A_{ki} \qquad (10.5)$$

where Δ is the determinant of the system of equations and A_{ki} is the cofactor of the elements with subscripts k and i in the determinant Δ. Since the intracellular reactions consist of a series of successive and parallel stages, it is hard to suppose that they will all proceed at the same rate, and it is therefore reasonable to introduce the concept of a limiting stage for the different substances. In this case equation (10.5) takes the form

$$\theta_c = \frac{m_1}{m_1 + k_1}$$

For the intracellular processes the reaction rates may be taken as proportional to θ_c, and then, when recalculated on the basis of a unit of growing biomass, the equation characterizing the rate of growth of the microorganisms takes the form

$$\frac{1}{N}\frac{dN}{dt} = \mu = \frac{km_1}{m_1 + k_1} \tag{10.6}$$

where k is the surface mass-transfer coefficient. Solving simultaneously equations (8.35) and (10.6), we obtain

$$\frac{1}{N}\frac{dN}{dt} = \mu = \frac{kA(U^{1/4}/l^{1/12})c_0}{A(U^{1/4}/l^{1/12})c_0 + k_1} \tag{10.7}$$

where $A = 8D^{2/3}\nu^{1/12}a^{4/3}(\Delta\rho/\rho)^{1/6}$ is a quantity characteristic of the particular medium and microorganisms.

The special property of equation (10.7) is that in it the limiting factor is the rate of transfer of material to the cell surface, and it indicates that if the intensity of mixing is raised by increasing the flow velocity U or by decreasing the characteristic dimension l, the diffusional flow of material to the cell surface is increased, leading to an increase in the specific rate of growth μ and, as a final result, to a shortening of the culture time for the microorganisms.

10.6 Experimental verification of theoretical equations

Experimental examination of the theoretical framework we have developed has been carried out by analyzing kinetic curves obtained during cultivation of microorganism–enzyme processes, under various mixing conditions.

Specificity of the technology of microbiological processes and the presence of stochastic liaisons among the process parameters together determine a slight drift of population, by cell sowing, toward their maximum accumulation.

It is convenient here to define a dimensionless cell count

$$\zeta = \frac{N - N_0}{N_{max} - N_0}$$

where N is the instantaneous cell count, N_0 the initial cell count, and N_{max} the maximum cell count.

The kinetic curves for microorganism growth at various flow rates are represented in Figure 10.9, where data are plotted as dimensionless cell counts per unit volume against the time t from the beginning of the cultivation. It appears from Figure 10.9 that the shape of the curve does not change from one flow rate to the next.

The lag period, in other words the period when there is no gemmation reproduction of cells, decreases, with increase of the flow rate, from 10 to 5 h. In this period there are no respiratory processes, and the concentration of CO_2 and O_2 in the gaseous phase and dissolved oxygen in the liquid are not changed. The direction of change of the pH and Eh values depends on the type of culture and composition of the substrate, but the level of pH and Eh is regulated by the culture itself. Decrease of lag period with mixing rate gives support to the notion that an intensive mass-transfer process take place in this period, and this indirectly corroborates the hypotheses suggested in various studies.[2,28]. An increasing flow rate causes the slope of the curves to be greater, and after accumulation of a certain number of cells (per unit volume), the rate of growth decreases and becomes stationary. This stationary state is usually continued no more than 1 hour, and then the process rate is sharply decreased.

The specific growth rate, μ, as we mentioned earlier, can be expressed in the form

$$\mu = \frac{1}{N}\frac{dN}{dt}$$

or in the integral form

$$\mu = \frac{\ln N_2 - \ln N_1}{t_2 - t_1}$$

Figure 10.9 Kinetic curves of growth of microorganisms in apparatuses with capacity of 25, 100, and 1000 liters.

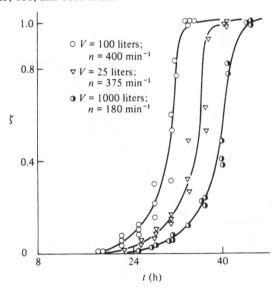

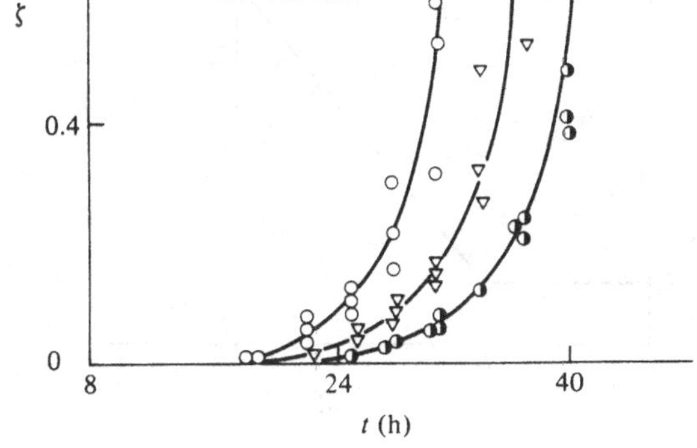

and so for an estimation of this specific rate it is convenient to plot the curve in the semilogarithmical coordinates shown in Figure 10.10, where the curve slope will then give the specific rate of microorganism growth. Also, the experimental relationship $\mu = \mu(U)$ given in Figure 10.11 corresponds to the equation

$$\mu = 0.21U^{1/4} \tag{10.8}$$

One of the most important parameters that must be obtained in process estimation is the time necessary for accumulation of the maximum number of cells. If we plot the process time as logarithmic ordinate against liquid flow rate as logarithmic abscissa, the $t_{max} = t_{max}(U)$ curve in Figure 10.12 is obtained, and we see that the relationship between the data in these axes is good, approximated by a straight line and expressed by the equation

$$t_{max} = 20.1/U^{1/4} \tag{10.9}$$

Some processes of growing yeast are limited by the rate of oxygen supply into the reaction zone, which in turn is determined by its dissolution. However, for the culture *B. subtilis* 103, oxygen does not limit the reaction rate, because the concentration of dissolved oxygen is never decreased to the critical level. In addition, the dissolution rate of oxygen is proportional to the liquid flow rate to the power of 1.95, and the specific rate of microorganism growth changes according to the "law of $\frac{1}{4}$"; in other words, the powers differ by a factor of seven. Analysis of equations (10.8) and (10.9) shows that the increased intensity of mixing with increase of the superficial velocity U of

Figure 10.10 Growth curve of microorganisms in semilogarithmic coordinates.

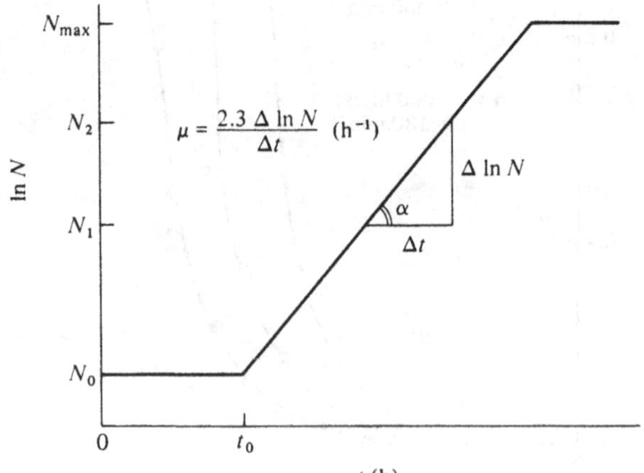

the flow increases the diffusional mass flux to the cell surface, resulting in an increase of specific rate of growth μ and, in the final analysis, to a decrease in the time required for cultivation of microorganisms. The specific rate of microorganism growth increases with average flow rate U by the law of $\frac{1}{4}$.

For verification of the theoretical equations, and for estimating the effect of scaling up on the mass-transfer rate in microheterogeneous systems, experimental investigations for various regimes in a baffled apparatus with

Figure 10.11 Specific growth rate versus superficial velocity of liquid flow.

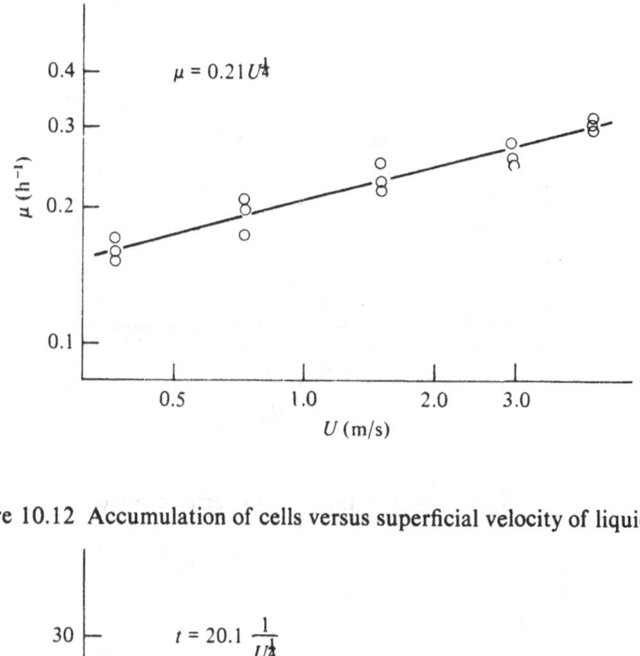

Figure 10.12 Accumulation of cells versus superficial velocity of liquid flow.

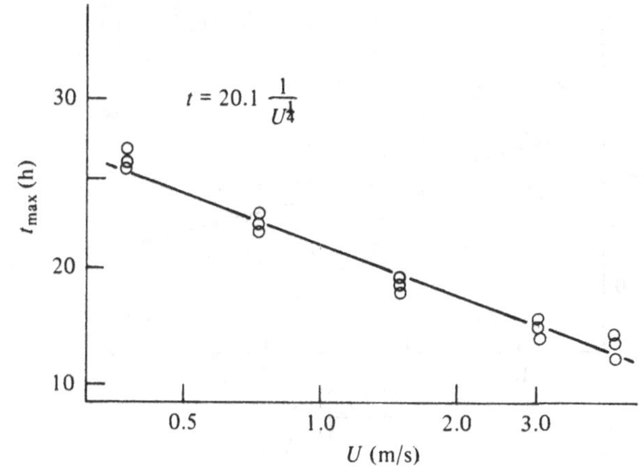

an agitator of the same design but different capacities have been conducted. The results are summarized in Table 10.1 and the experimental data presented in Figures 10.9, 10.13, and 10.14.

It follows from Figure 10.9 that the character of the kinetic curves remains the same throughout, but the duration of the process significantly increases for intensive mixing as compared to the process in a tubular apparatus. As can be seen in Figures 10.13 and 10.14, the experimental points

Figure 10.13 Specific growth rate of microorganisms versus complex:

$$\mu = 0.135 \left[\frac{nD_m^2}{(T^2 x_1)^{\frac{1}{3}}} \right]^{\frac{1}{4}} \frac{1}{D_m^{\frac{1}{12}}}$$

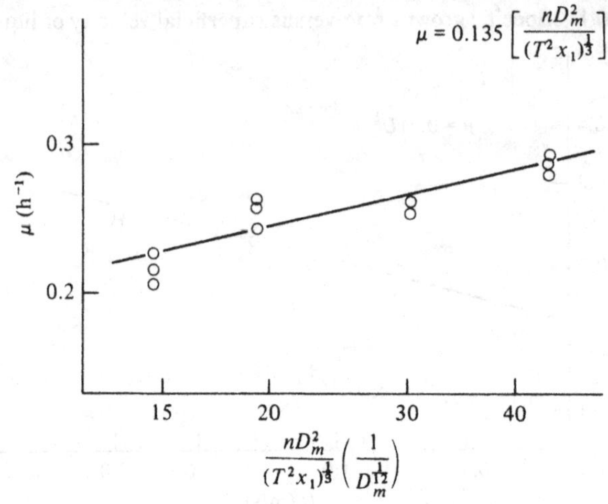

Figure 10.14 Accumulation of cells versus complex:

$$t = 73.4 \left[\frac{(T^2 x_1)^{\frac{1}{3}}}{nD_m^2} \right]^{\frac{1}{4}} \frac{1}{D_m^{\frac{1}{12}}}$$

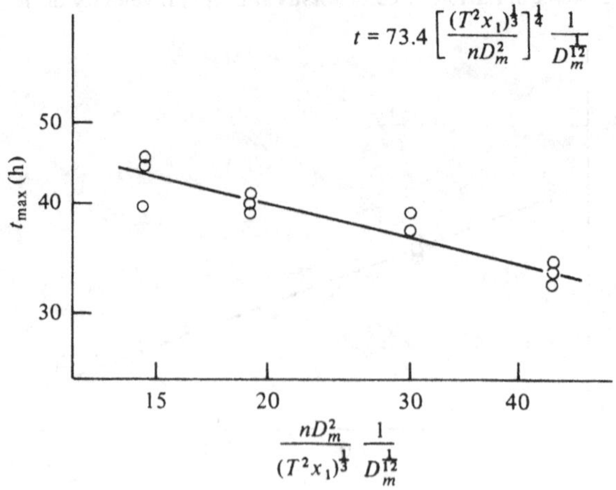

lie very close to a straight line, and this indicates the validity of the theoretical equations for use in scaling up to a geometrical similar apparatus of larger capacity.

The experimental data for baffled equipment with an agitator are fairly well described by the following equations:

$$\mu = 0.135 \left(\frac{nD_m^2}{T^2 x_1} \right)^{1/4} \frac{1}{D_m^{1/12}}$$

(where n is the revolution rate, D_m the mixer diameter, T the tank diameter, and x_1 the height of the mixture), and

$$t_{\max} = 73.4 \left[\frac{(T^2 x_1)^{1/3}}{nD_m^2} \right]^{1/4} D_m^{1/12}$$

which corroborate the theoretical analysis of the process.

10.7 Method of design of a biochemical reactor

In practice we often have insufficient data for a theoretical study of the whole process, and in these cases it is advisable to make a compromise between the purely theoretical approach and the assumption of several simplifications, as indicated by the particular experimental data available.

For example, let us design a reactor (Figure 10.15) with a capacity of 40.0 m^3 (1400 ft^3) for the cultivation of *B. subtilis* 103, the producer of proteolytic enzymes. The reactor has a lower drive and a three-blade agitator, and a mechanical foam breaker is built into the apparatus. For improving heat transfer, the reactor also has a built-in heat exchanger, which at the same time serves as an inner tube diffuser. Control of the mixing rate is achieved with changeable propellers and pulleys. Equation (10.7),

Table 10.1. *Experimental data from baffled apparatus of various capacities*

Characteristic	Capacity of equipment (liters)		
	25	100	1000
Coefficient of filling up	0.6	0.6	0.6
Number of revolutions	280; 375	400	180; 360
Diameter of equipment (mm)	300	500	1000
Liquid height (mm)	210	450	8000
Ratio of blade height to diameter of mixer	$\frac{1}{4}$	$\frac{1}{4}$	$\frac{1}{4}$
Diameter of mixer (mm)	100	200	300

which we give below, was obtained for estimation of the specific rate of microorganism growth as a function of the physicochemical and hydrodynamic factors of the process:

$$\mu = \frac{kA(U^{1/4}/l^{1/12})c_0}{A(U^{1/4}/l^{1/12})c_0 + k_1} \qquad (10.10)$$

where

$$A = \text{const.} \times D^{2/3}\nu_f^{1/12}a^{4/3}(\Delta\rho/\rho)^{1/6}$$

For the case when the reaction rate is limited by diffusion, equation (10.10) takes the form

$$\mu = \frac{kAU^{1/4}c_0}{k_1 l^{1/12}}$$

From these equations it is evident that, for an exact solution, we must know the constants k and k_1 characterizing the reaction, the diffusivity D

Figure 10.15 Commercial biochemical reactor with capacity $V = 40\ m^3$.

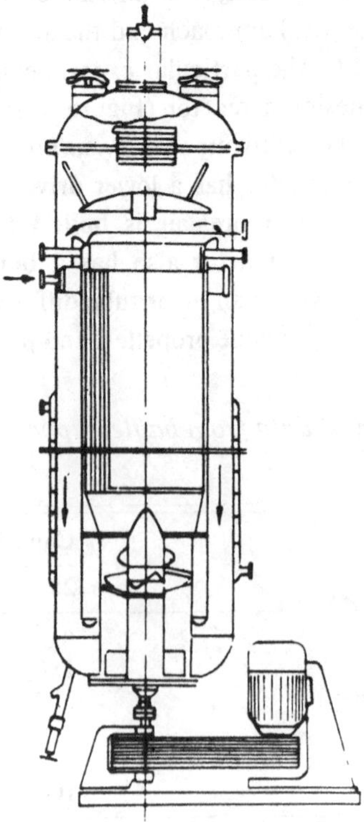

of the limiting substance, and its concentration c_0. As these values are usually unknown, it is convenient to determine the value of factor (kAc_0/k_1) experimentally, and on the basis of experimental data obtained in an apparatus with a capacity of 10.0 liters (0.35 ft^3), we find that

$$\mu = 0.17\left(\frac{U^{1/4}}{d_{pr}}\right)\ h^{-1}$$

(where $l = d_{pr}$, the diameter of the propeller) or, for the process time,

$$t_{max} = 24.9\left(\frac{d_{pr}^{1/12}}{U_{pr}^{1/4}}\right)\ h \qquad (10.11)$$

Thus from experiments we find that

$$\frac{kAc_0}{k_1} = 0.17$$

Because geometric similarity in scaling up to the apparatus with capacity of 40.0 m^3 (1400 ft^3) is retained, the specific rate of microorganism growth and the time of cultivation for maximum accumulation of bacterial cells and enzymic activity can be estimated.

Substituting the propeller cross section and diameter in equations (10.10) and (10.11), as well as a velocity value equal, for example, to 4.5 m/s, we obtain

$$\mu = 0.17\,\frac{U_{pr}^{1/4}}{d_{pr}^{1/12}} = 0.243\ h^{-1}$$

and

$$t_{max} = 24.9\,\frac{d_{pr}^{1/12}}{U_{pr}^{1/4}} = 17.3\ h$$

From these values we can see that, for scaling up to an apparatus of larger capacity but with the same flow rate, the specific growth rate decreases and the duration of the process increases, respectively.

References

1. Levich, V. G., *Doklady 5 mezhvusovskoy konferenzii po physich. e matemat. modelirovaneyu*, Izdatelstvo Vysshaya Shkola, Moscow, 1968, pp. 35–48.
2. Ierusalimsky, N. D., *Osnovy fisiologii mikrobov* (Physiology of Microbes), Izdatelstvo AN SSSR, Moscow, 1963.
3. Sytnik, K. M., Kordum, V. A., and Kok, I. P., *Regulatortye mekhanismy* (Regulator Mechanisms of a Cell), Izdatelstvo AN SSSR, Kiev, 1969.
4. Monod, J. Y. J., *Mol. Biol., 3*, 318 (1961).
5. Ierusalimsky, N. D., *Upravlyaemy biosintez* (Biosynthesis under Control), Izdatelstvo AN SSSR, Moscow, 1966.

6. Bezborodov, A. M., *Biosintez biologicheski activnykh vetchestv microorganismami* (Biosynthesis of Biologically Active Substances with Microorganisms), Izdatelstvo AN SSSR, Moscow, 1969.
7. Webb, L., *Ingibitory fermentov metabolisma* (Inhibitors of Metabolism Enzymes), Izdatelstvo Miz, Moscow, 1966.
8. Tosic, I., and Mitchel, R. L., *Nature, 162,* 552 (1948).
9. Imada, *Doklad na konferenzii agrofisicheskogo obtckestva, Tokyo, 1963,* vol. 1, p. 4.
10. Metechell, P., *Giba Found. Stud. Group, 5,* 53 (1969).
11. Epstein, A., *Giba Found. Stud. Group, 5,* 53 (1960).
12. Aiba, S., and Hara, M., *Proc. 5th Symp. Inst. Appl. Microbiol., Univ. of Tokyo,* 1963, p. 298.
13. Bayer, *Biophysika* (Biophysics), Izdatelstvo IL, Moscow, 1962.
14. Kovrov, B. G., *Sbornik upravlyaemy biosintez* (Biosynthesis under Control), Izdatelstvo Nauka, Moscow, 1966.
15. Sokolov, D. P., Candidate's dissertation, D. I. Mendeleev Moscow Institute of Chemical Technology, 1969. (In Russian.)
16. Mueller, J., Boyle, W., and Leghftoot, E., *Appl. Microbiol., 15,* 672 (1967).
17. Ierusalimsky, N. D., *Biokhimia mikrobov* (Biochemistry of Microbes), Gorki, USSR Gosizdat, 1964.
18. Eremin, V. A., Candidate's dissertation, Moscow Institute of Chemical Engineers, 1968. (In Russian.)
19. Osipov, A. V., Candidate's dissertation, Leningrad Institute of Chemical Technology, Leningrad, 1970. (In Russian.)
20. Michaelis, L., and Menton, M., *Biochem. Z., 49,* 333 (1913).
21. Monod, J., *La Croissance des cultures bacteriennes,* Paris, 1942.
22. Dikson, M., and Webb, L., *Fermenty* (Enzymes), Izdatelstvo IL, Moscow, 1961.
23. Denbigh, K. G., *Chemical Reactor Theory,* Cambridge University Press, Cambridge, 1965.
24. Levenspiel, O., *Chemical Reaction Engineering,* Wiley, New York, 1972.
25. Setlow, R., and Pollard, E., *Molecularnia biofisika* (Molecular Biophysics), Izdatelstvo Miz, Moscow, 1964.
26. Kaler, V. L., Sergeev, A. A., *Dokl. Acad. Nauk SSSR (Minsk), 11*(6), 551 (1967).
27. Khinshelwood, S. N., *Khimicheskaya kinetika i tzepnye reaktzii* (Chemical Kinetics and Chain Reactions), Izdatelstvo, AN SSSR, Moscow, 1966.
28. Rabotnova, I. L., *Rol fisiko-khimicheskikh uslovi v zhisnedeyatelnosty microorganismov* (The Influence of Physicochemical Conditions on the Activity of Microorganisms), Izdatelstvo AN SSSR, Moscow, 1957.

INDEX